Information Systems Engineering and Management

Volume 55

The book series "Information Systems Engineering and Management" (ISEM) publishes innovative and original works in the various areas of planning, development, implementation, and management of information systems and technologies by enterprises, citizens, and society for the improvement of the socio-economic environment.

The series is multidisciplinary, focusing on technological, organizational, and social domains of information systems engineering and management. Manuscripts published in this book series focus on relevant problems and research in the planning, analysis, design, implementation, exploration, and management of all types of information systems and technologies. The series contains monographs, lecture notes, edited volumes, pedagogical and technical books as well as proceedings volumes.

Some topics/keywords to be considered in the ISEM book series are, but not limited to: Information Systems Planning; Information Systems Development; Exploration of Information Systems; Management of Information Systems; Blockchain Technology; Cloud Computing; Artificial Intelligence (AI) and Machine Learning; Big Data Analytics; Multimedia Systems; Computer Networks, Mobility and Pervasive Systems; IT Security, Ethics and Privacy; Cybersecurity; Digital Platforms and Services; Requirements Engineering; Software Engineering; Process and Knowledge Engineering; Security and Privacy Engineering, Autonomous Robotics; Human-Computer Interaction; Marketing and Information; Tourism and Information; Finance and Value; Decisions and Risk; Innovation and Projects; Strategy and People.

Indexed by Google Scholar. All books published in the series are submitted for consideration in the Web of Science.

For book or proceedings proposals please contact Alvaro Rocha (amrrocha@gmail.com).

Lalitha Krishnasamy · Rajesh Kumar Dhanaraj ·
Dragan Pamucar · Mariya Ouaissa
Editors

Distributed Deep Learning and Explainable AI (XAI) in Industry 4.0

Springer

Editors
Lalitha Krishnasamy
Department of Artificial Intelligence and Data Science
Nandha Engineering College
Erode, Tamil Nadu, India

Dragan Pamucar
Department of Operations Research and Statistics
University of Belgrade
Belgrade, Serbia

Rajesh Kumar Dhanaraj
Symbiosis International (Deemed University)
Pune, Maharashtra, India

Mariya Ouaissa
Cadi Ayyad University
Marrakech, Morocco

ISSN 3004-958X ISSN 3004-9598 (electronic)
Information Systems Engineering and Management
ISBN 978-3-031-94636-3 ISBN 978-3-031-94637-0 (eBook)
https://doi.org/10.1007/978-3-031-94637-0

This Springer imprint is published by the registered company Springer Nature Switzerland AG
The registered company address is: Gewerbestrasse 11, 6330 Cham, Switzerland

Contents

Introduction to Industry 4.0: Practical Issues and Challenges

S. Mohana Saranya, Dinesh Komarasamy, S. Mohanapriya, and M. R. Prasanndh Raaju

Abstract Industrial 4.0 made an industrial revolution by integrating digital technologies in various industrial sectors to increase the production of industries. This integration of digital technologies includes Cloud Computing (CC), Cyber-Physical Systems (CPS), Internet of Things (IoT), and Artificial Intelligence (AI) and so on. Industry 4.0 provides significant advancements and benefits, such as increased efficiency, flexibility and customization, it also has wide range of practical issues and challenges. This chapter gives an overview to Industry 4.0, focusing on the practical issues and challenges faced by organizations adopting these technologies. It explores the complexity of integrating new technologies into existing infrastructures. Industrial 4.0 also need for skilled labours who are skilled to integrate the advanced technologies with the existing one, and the challenges of security and privacy in an ever-changing interconnected environment. Moreover, the chapter discusses about the overcoming these challenges to realize the importance of Industry 4.0 and outlines approaches to overcome them.

Keywords Industry 4.0 · Digital technologies · Manufacturing · Cyber-physical systems (CPS) · Internet of things (IoT) · Cloud computing · Artificial intelligence (AI) · Data security

S. M. Saranya (✉) · D. Komarasamy · S. Mohanapriya · M. R. P. Raaju
Department of CSE, Kongu Engineering College, Erode, Tamilnadu, India
e-mail: mohanasaranyaa@gmail.com

D. Komarasamy
e-mail: dinesh.cse@kongu.ac.in

S. Mohanapriya
e-mail: mohanapriyas.cse@kongu.edu

L. Krishnasamy et al. (eds.), *Distributed Deep Learning and Explainable AI (XAI) in Industry 4.0*, Information Systems Engineering and Management 55,
https://doi.org/10.1007/978-3-031-94637-0_1

1 Introduction to Industry 4.0

The Fourth Industrial Revolution is the combination of progressed computerized innovations with manufacturing and industrial process. Coined in 2011 at the Hannover Fair by the German government, the Fourth Industrial Revolution speaks to a move towards smart factories where cyber-physical frameworks, Internet of Things (IoT), Artificial Intelligence (AI), and big data analytics. These advances empower independent decision-making in machines, real-time optimization of production process, and upgraded adaptability in manufacturing environments [1]. The Fourth Industrial Revolution includes workforce upskilling and strong change management strategies. It presents numerous benefits like increased efficiency, innovation, and cost savings. Though it has numerous benefits, it also has challenges such as include technological integration, data security and workforce adaptation [2, 3].

1.1 Definition and Overview of Industry 4.0 and Its Key Technologies

The 4th Industrial Revolution signifies a significant shift in manufacturing powered by advanced computer technologies with the help of its key features (discussed in the upcoming part). The advancement of the Fourth Industrial Revolution has progressed through distinct stages of conceptualization, technological integration, and optimization. At first it was conceptualized within the early 2010s, where the Fourth Industrial Revolution emerged as a vision for digitizing manufacturing process through interconnected frameworks and data-driven insights. As IoT innovations developed, manufacturers started integrating smart sensors and devices to accumulate real-time data, laying the basis for AI-driven analytics and automation.

In the mid-2010s, the Fourth Industrial Revolution picked up energy with advancements in cloud computing, empowering adaptable information storing capacity and preparing capabilities across global manufacturing networks. AI and Machine Learning (ML) developed as essential advances for prescient support, quality confirmation, and flexible manufacturing techniques, improving overall efficiency and cost reduction. Nowadays, the Fourth Industrial Revolution continues to proceeds in advancements of edge computing, 5G network, and computerized twins, promising future optimization of manufacturing process and improved customization capabilities [4]. The integration of sustainability principles, such as circular economy concepts and eco-friendly manufacturing processes underscores the Fourth Industrial Revolution's part in fostering economic development, natural stewardship, and societal well-being. Figure 1 represents major technologies used in the Fourth Industrial Revolution. Cyber-physical systems, cloud computing, the Internet of Things, big data analytics, artificial intelligence, advanced robotics, augmented reality, and virtual reality are the main technologies of the Fourth Industrial Revolution. The key technologies will be discussed later in this chapter.

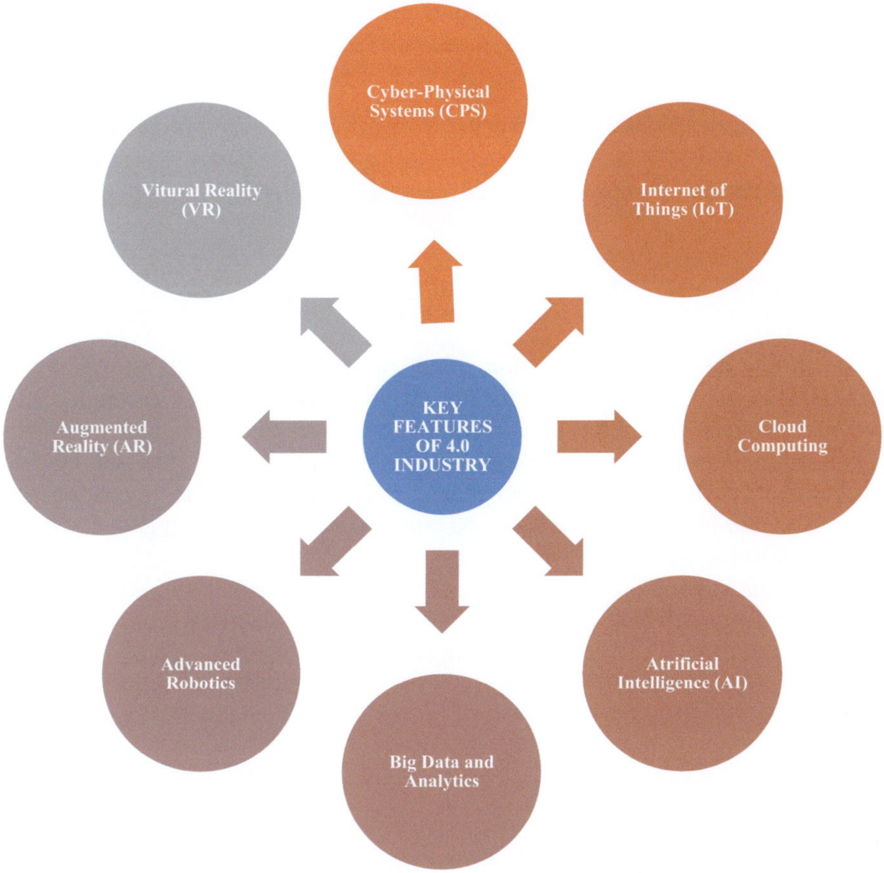

Fig. 1 Key technology of industry 4.0

1.2 Historical Context Leading to the Fourth Industrial Revolution

The first industrial transformation, progressed in the late eighteenth century, stamped the move from agrarian economies to mechanized generation encouraged by steam motors and mechanization. This period saw the development of industrial facilities and mass generation, on a very basic level changing financial and social structures. The moment mechanical transformation started, the second industrial transformation, worked during the late 19th and early twentieth centuries, where electric control and assembly lines, advance quickening generation capabilities and empowering mass utilization are integrated in the industrial 4.0. Advancements such as the telegram, telephone, and early forms of industrial automation laid the foundation for more proficient and interconnected manufacturing process.

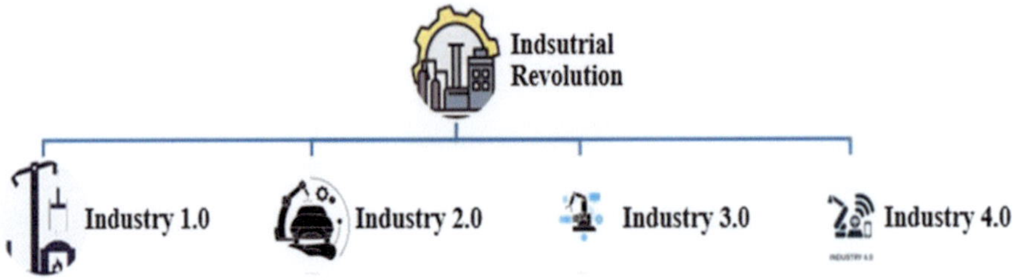

Fig. 2 Emerging of fourth industrial revolution

The third industrial transformation, often known as the digital transformation or digital transformation as shown in the Fig. 2, which was witnessed during the mid-twentieth century. During this transformation, there was advancement in the advent of computers, robotics, and telecommunications, fundamentally changing how we process and transfer data. The improvement of integrated circuits, microchip, and the web revolutionized data handling and communication, fostering global network and started businesses towards digital transformation.

By the late twentieth century, progressions in computer control, organizational systems, and program integration paved the way for the collaboration of physical and computerized domains. This merging, combined with the expansion of data-driven bits of knowledge and mechanization, set the arrange for the development of the Fourth Industrial Revolution which was formally introduced in 2011 as discussed earlier.

1.3 Importance of Industry 4.0 in the Context of Modern Manufacturing and Industrial Processes

In the setting of advanced manufacturing, the Fourth Industrial Revolution (i4.0) tells us about the shift from the traditional methods by integrating cutting-edge computer technologies with industrial operations. This transformative approach not only improves operational efficiencies but too redefines the whole manufacturing landscape by cultivating agility, customization, and sustainability. At its core, i4.0 leverages interconnected cyber-physical frameworks (CPS), cloud computing, artificial intelligence (AI), and the Internet of Things (IoT) to make keen production lines and interconnected supply chains. These advances empower on-time data collection, analyse, and making decisions, subsequently optimizing production process and asset utilization. By fostering consistent communication between machines, products, and people, i4.0 empowers versatile manufacturing systems capable of reacting dynamically to showcase market demands and operational challenges.

One of the key focal points of i4.0 lies in its capacity to boost the move from mass production to richly customized, on-demand manufacturing. Through advanced data analytics and AI-driven knowledge, manufacturers can tailor products and administrate more precisely to customers' needs, subsequently improving customers' fulfilment and advertise competitiveness. This customization is further encouraged by agile manufacturing systems that can rapidly reconfigure operations in response to changing market trends.

2 Key Technologies in Industry 4.0

As we explored earlier in the introduction, the industrial landscape is ongoing a tremendous change by the fourth industrial revolution. With the help of modern informatica, the fourth industrial revolution has been characterized with the following key features,

- Cyber-Physical Systems (CPS)
- Internet of Things (IoT)
- Cloud Computing
- Artificial Intelligence (AI)
- Big Data and Analytics
- Advanced Robotics
- Augmented Reality (AR)
- Virtual Reality (VR)

2.1 Cyber-Physical Systems (CPS)

Cyber-Physical Frameworks (CPS), representing the merging of physical and digital world are the basics for the fourth industrial revolution [5–7]. These frameworks integrate comping, networking, and physical processes, empowering on-time interaction between the physical and digital realms.

Figure 3 represents the components of CPS. CPS are characterized by their capacity to monitor, control, and facilitate physical processes with precision and versatility, cultivating the advancement of smart industries and progressed manufacturing environments. Basically, CPS comprise of three essential components: Embedded Systems, Communication Systems, and Physical Processes. A characterizing feature of CPS is their capability for real-time checking and control. Sensors continuously collect information from the physical environment, which is handled and analysed in-real-time by the embedded systems. This information gives quick feedback on the status of equipment and the processes, empowering a convenient intercession to anticipate the breakdowns, optimize performance, and ensure the quality. Consider an situation where CPS can detect the inconsistencies in a manufacturing system and naturally alter the parameters to preserve ideal working conditions. In overall the benefits of implementing CPS are as follows,

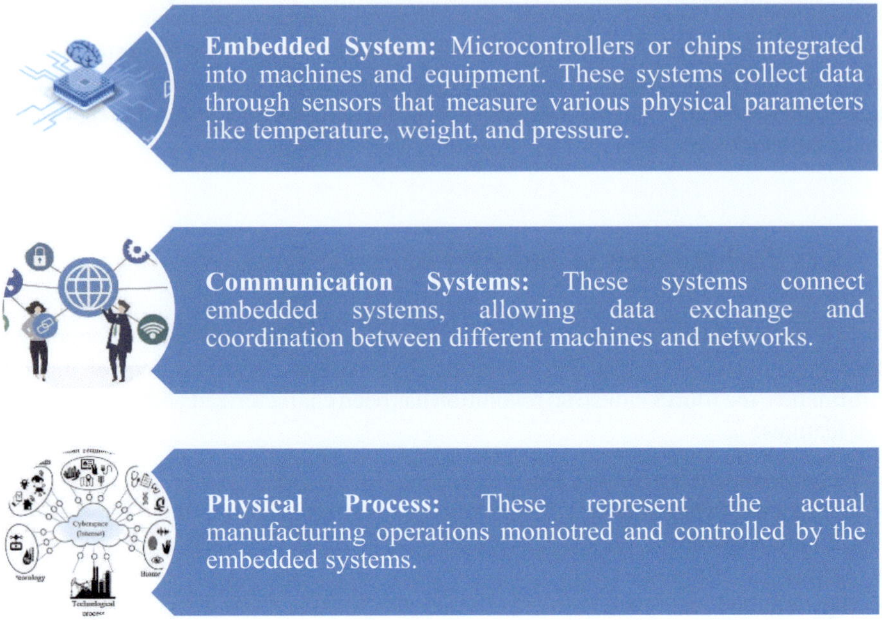

Fig. 3 Components of CPS

- Streamline Operations
- Agile Manufacturing
- Enhanced Quality Control
- Predictive Maintenance
- Sustainable Manufacturing

Figure 4 represents challenges in implementing CPS. In spite of all these advantages and benefits in using CPS in a manufacturing process, at the same time it has few challenges in implementing those to the manufacturing process and making it to run smoothly without any discrepancies.

2.2 Internet of Things (IoT)

An essential component of the fourth industrial revolution is the IoT. It describes setting up a system of connected devices to gather, exchange, and evaluate data in order to streamline production procedures [8]. IoT makes an interface between physical world to the computerized world, permitting them to communicate and collaborate in real-time. This interconnectivity transforms conventional manufacturing facilities into smart industries, where machines, products, and peoples are associated consistently to upgrade efficiency, effectiveness, and adaptability.

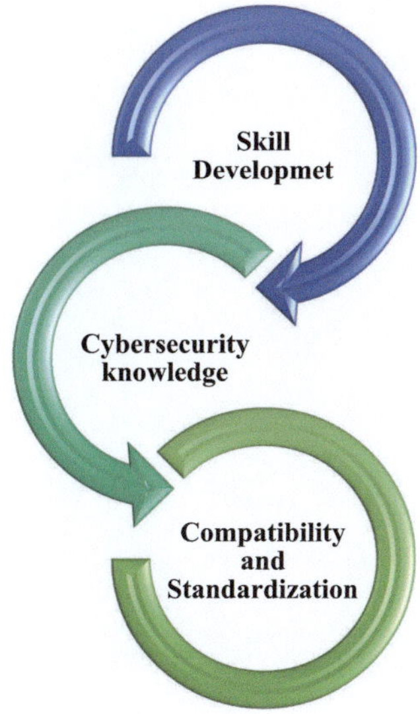

Fig. 4 Challenges in implementing CPS

IoT in the fourth industrial revolution action includes sensor powers and Data-Driven insights in the manufacturing processes [9, 10]. Sensors assemble information on key factors such as temperature, humidity, vibration, and operational status. This information is at that point transmitted over communication systems to centralized platforms for analysis and interpretation. The insights inferred from this information empower manufactures to screen operations persistently, recognize area of waste, and make informed decisions. For example, IoT sensors on a production line can distinguish variations in machine execution, prompting immediate alterations to preserve ideal output.

Figure 5 represents the benefits of IoT in industrial 4.0. As discussed for CPS, IoT also has some challenges and considerations in manufacturing processes while implementing it in spite of all the benefits and its advantages. The benefits of IoT in manufacturing process are predictive maintenance, connected supply chain management, smart manufacturing and efficiency of the system.

Figure 6 represents the challenges and considerations on implanting IoT. The main challenges are security concerns, interoperability and Infrastructure & training.

2.3 Cloud Computing (CC)

CC is also one of the significant components in the fourth industrial revolution, giving the foundational framework that supports the integration and optimization of different progressed technologies [11, 12]. By maximizing the capability of CC, businesses can achieve significant enhancements in efficiency, collaboration, and operational effectiveness. CC empowers the improvement and deployment of modern applications

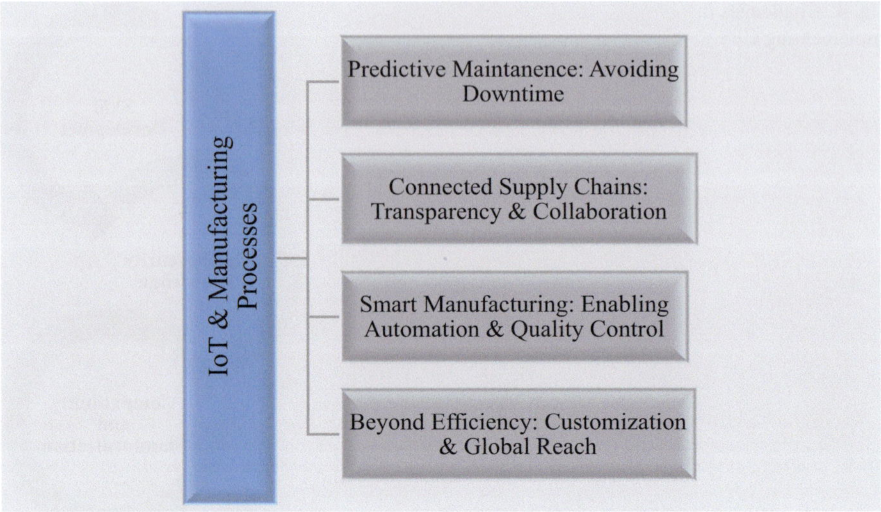

Fig. 5 Benefits of IoT in industry 4.0

Fig. 6 Challenges & considerations on implanting IoT

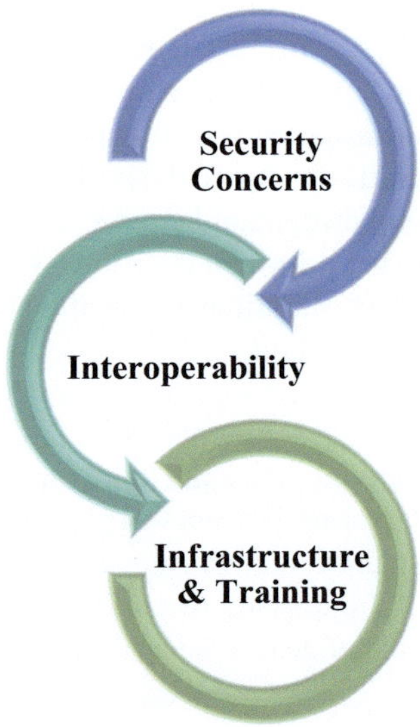

and administrations with exceptional speed and adaptability, supporting different programming languages and offer integrated development tools. This capability is especially profitable within the fast-paced, innovation-driven environment of modern manufacturing. As discussed for the key components above CC also contains its own benefits, superiority and the CC overcome all the earlier discussed challenges while

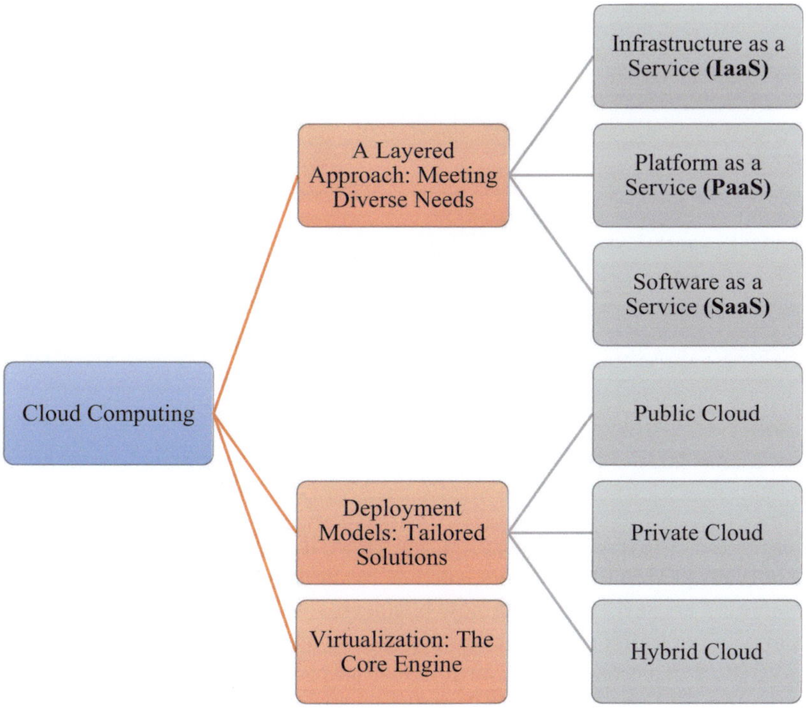

Fig. 7 Models & layers of cloud computing

implementing the key factors in the manufacturing processes. The benefits of CC are,

- Enhancement Collaboration
- Easy Accessibility
- Unlimited Storage
- Cheap Maintenance
- Robust Safety Features

Figure 7 represents the models and layers of CC. It also describes the various cloud such as private cloud, public cloud and hybrid cloud. It also gives everything as a service most preferably Infrastructure as a Service (IaaS), Platform as a Service (PaaS) and Software as a Service (SaaS).

2.4 Artificial Intelligence (AI)

Artificial Intelligence (AI) is a very important key in foundation of the fourth industrial revolution, driving to the change of constructing processes through its ability to

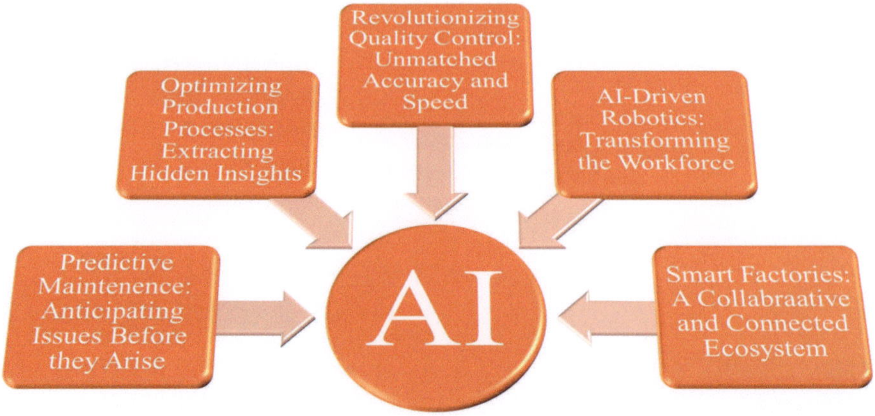

Fig. 8 AI in industry 4.0

analyse tremendous amount of information, learn from it, and make independent decisions like humans [13]. AI key factors, such as machine learning, neural networks, and deep learning are revolutionizing how businesses work by empowering predictive maintenance, optimizing production plans, and upgrading quality control. AI is a transformative constrain in the fourth industrial revolution, driving exceptional levels of effectiveness, adaptability, and intelligence in manufacturing processes. Its applications in prescient maintenance, production optimization, quality control, and workforce collaboration are reshaping the industrial landscape. As AI proceeds to advance, its integration with other the fourth industrial revolution keys will further upgrade the capabilities of smart industrial facilities, empowering producers to meet the requests of future with dexterity and advancement.

The Fig. 8 shows the part of AI in the fourth industrial revolution. Grasping that AI is fundamental for businesses aiming to flourish within the digital age, because it offers a pathway to feasible development and competitive advantage.

2.5 Big Data and Analytics

Figure 9 represents, BDA and the fourth industrial revolution. The first light of the fourth industrial revolution ushers in a modern period where digital change and intelligent technologies are profoundly coordinated into manufacturing and production processes. At the heart of this transformation lies Big Data Analytics (BDA), an effective tool that tackles the tremendous amount of information produced by interconnected devices and systems to infer significant bits of knowledge [14]. The combination of BDA with the fourth industrial revolution key factors—such as the IoT, AI, and ML—enables companies to upgrade their operational effectiveness, progress product quality, and convey personalized client experience. Big Data Analytics is one

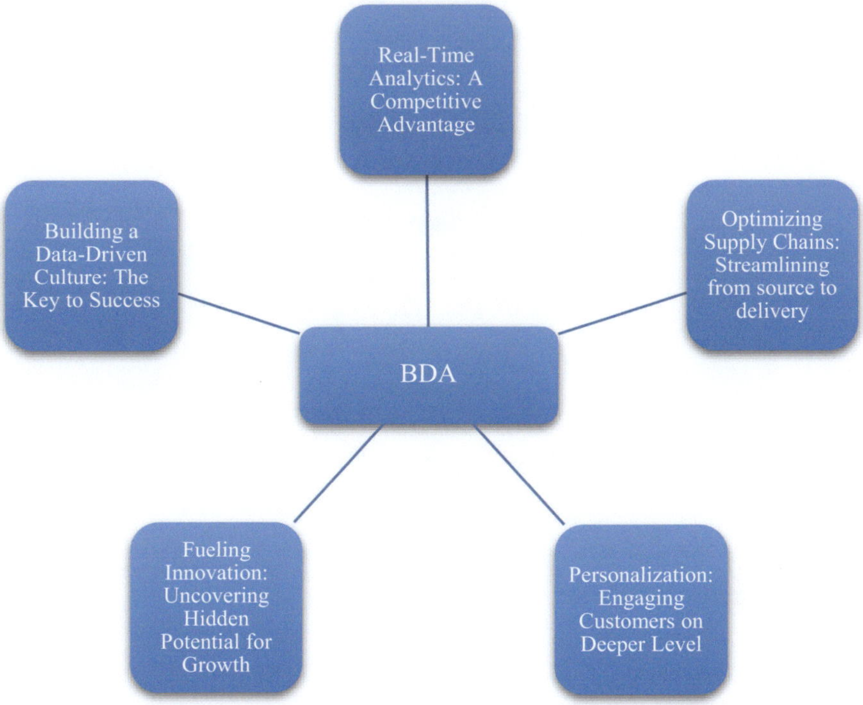

Fig. 9 BDA & industry 4.0

of foundation key factor of the fourth industrial revolution, empowering companies to convert their operations, optimize their supply chains, engage more effectively with customers, and drive development. As firms proceed to explore the difficulties of the digital era, the key application of BDA will be vital in achieving operational excellence, competitive advantage, and long-term success. The consistent integration of BDA into the fourth industrial revolution hones not as it improves the efficiency and effectiveness of businesses processes but moreover sets the stages for a future where data-driven insights of knowledge are at the centre of each key decision.

2.6 Advanced Robotics

The appearance of the fourth industrial revolution has introduced in an unused day of processing and mechanical operations, with progressed robotic automation playing a central part in this change. These cutting-edge automated frameworks are characterized by their upgraded capabilities, including improved dexterity, accuracy, and autonomy [15]. By integrating progressed robotic technology with other the fourth industrial revolution advances such as the IoT, AI, and ML, companies can

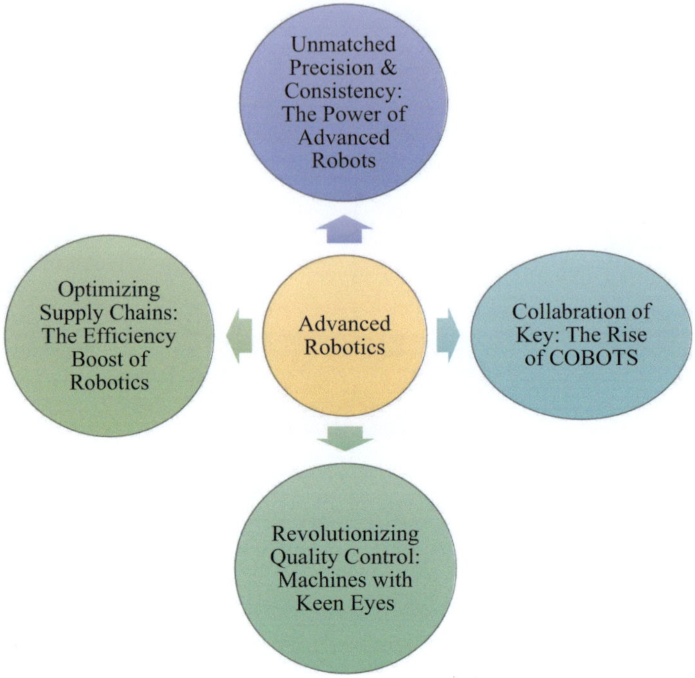

Fig. 10 Advanced robotics & industry 4.0

accomplish phenomenal levels of robotization, productivity, and adaptability in their production processes and also with the benefits of Advanced Robotics as shown in the Fig. 10.

Advanced robotic inventions are a foundation of the fourth industrial revolution, driving unique advancements in automation, quality control, supply chain administration, and human–robot collaboration. The vital arrangement of these advances empowers companies to upgrade their operational effectiveness, diminish costs, and keep up a competitive advantage. As the field of advanced robotic technologies proceeds to advance, addressing the related challenges and capitalizing on ongoing progressions that will be basic for opening their full potential and forming the future scene of manufacturing and mechanical operations.

2.7 Augmented Reality (AR) and Virtual Reality (VR)

Augmented Reality (AR) and Virtual Reality (VR) innovations is transforming the fourth industrial revolution by upgrading different angles of mechanical operations. Similar to the all above key factors, here both these strong pillars are binding the hole between the digital and physical worlds, giving better approaches to visualize,

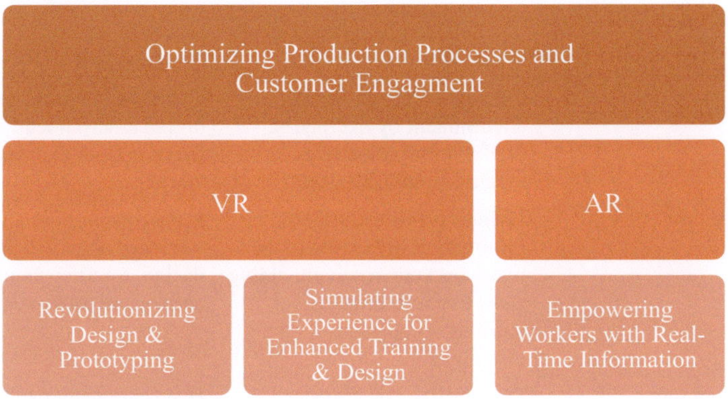

Fig. 11 AR—VR in industry 4.0

interact with, and manage complex mechanical and manufacturing processes. AR and VR are not only the changing how tasks are performed but moreover how labourers are prepared, and products are outlined, and maintenance is conducted, driving to critical enhancements in efficiency, accuracy, and collaboration. The key features of both the AR and VR are clearly displayed in the Fig. 11.

In spite of the various benefits, the collaboration of AR and VR into the fourth industrial revolution does come with objections. One of the essential obstacles is the more implementation cost, including the equipment cost, software, and the improvement of customized applications. Companies must also address the need related to the information safety and security measures, as these innovations often involve int the collection and preparing of sensitive information. Additionally, the adoption of AR and VR requires critical changes in organizational culture and workforce training, as workers need to be comfortable while utilizing these unused apparatuses and innovations.

AR and VR are critical technologies within the advancement of the fourth industrial revolution, advertising transformative benefits over different mechanical segments. By upgrading training, maintenance, design, and production processes, AR and VR are driving productivity, diminishing costs, and cultivating development.

As these innovations proceed to advance and become more available and affordable, their integration will be essential for companies looking to retain the competition within the instant changing digital world. Table 1 shows the comparison of AR and VR's role on Industry4.0.

Table 1 AR versus VR

Aspect	Augmented reality (AR)	Virtual reality (VR)
Definition	Enhances the real-world environment by overlaying digital information (e.g., images, sounds, or data)	Creates an entirely simulated digital environment, isolating the user from the physical world
Primary use case	Used to provide additional information or context to the real-world environment, aiding tasks like maintenance, training, and design visualization	Immerses users in a virtual space for activities like product design, training simulations, and virtual prototyping
Equipment required	Devices like AR glasses, smartphones, or tablets with cameras and AR-enabled software	Head-mounted displays (HMDs) like Oculus Rift, HTC Vive, or PlayStation VR, often paired with motion controllers and sensors
Interaction with environment	Integrates and overlays digital content into the physical environment, allowing users to interact with both	Replaces the physical environment with a fully digital one, where all interactions occur within the virtual space
Cost	Generally more affordable due to the use of existing devices like smartphones, but specialized AR equipment can increase costs	Higher initial investment due to the need for sophisticated hardware and software
Information security	Collects and processes data from the real environment, raising concerns about privacy and security of sensitive information	Focused on internal simulation data but may involve secure data exchange for realistic training or design
Challenges	Integration cost, compatibility with existing infrastructure, and workforce training for efficient usage	Requires cultural and organizational shifts, high-cost hardware/software, and addressing potential health concerns like motion sickness

3 Challenges in Adopting Industry 4.0

Adopting the fourth industrial revolution, characterized by binding of Artificial Intelligence, cyber-physical systems, cloud computing, Internet of Things (IoT), and cognitive computing, presents a variety of practical challenges for organizations. These challenges span technological, organizational, infrastructure and socio-economic dimensions [16–18]. Figure 12. Shows the major concerns in adoption of the fourth industrial revolution.

Fig. 12 Key challenges in adoption of industry 4.0

Figure 12 represents the major challenges in implementation of the fourth industrial revolution. The major challenges are technological challenges, organizational challenges, socio economic challenges, economic barriers, infrastructure & connectivity and standards & protocols.

3.1 Technological Challenges

There are several technological challenges to adopt the changes with the existing technologies. In the recent world, it is very difficult to directly replace the existing technique. Therefore, there are several challenges to adapt the proposed techniques with the existing technology. Technological challenges are integration and interoperability, data management and skill gaps.

Integration: Integration of the fourth industrial revolution technologies with existing applications could be difficult and expensive. **Interoperability**: Standardization must be established for different devices and systems to interact and use of technologies which will be compatible. Data management has very concern about the data quality, data volume and data security and privacy. **Data Quality**: Data may contain missing values, noise and also scarcity of data leads to wrong prediction. **Data Volume**: Managing the vast data produced by connected applications and sensors needs the robust information storage and processing solutions. **Data Security and Privacy**: Due to increased networking and sharing of data, it is necessary to safeguard data from cyber threats in the fourth industrial revolution.

Adaptation of new technologies need a skilled employ for operating the industrial 4.0. **Skilled Labor Shortage**: There are often minimum workers available with the necessary skills to implement and cope up with state-of-the-art technologies such as IoT, robotics and AI. **Continuous Training**: Providing ongoing training and development to keep the workers updated with changing technological advancements. **Collaboration of various professionals:** Professionals from different fields has to work collaboratively to carry out the work.

3.2 Organizational Challenges

Similar to technological challenges, the organization also face several huddles to integrate the advanced technologies with the existing one. There are several challenges faced by organization for integrating the new technologies with the existing technologies. **Refuse to Change**: Employees may avoid taking in new technologies and fear of job loss or inexperience in new domains. **Leadership Commitment**: To promote change and guarantee alignment with the strategic goals of the organisation, strong leadership is necessary. Organisations also struggle to make the necessary investments and at the same time to replace emerging sectors. **High Initial Costs**: The cost for the fourth industrial revolution technologies can be a considerable investment. **Uncertain ROI**: Calculating the return on investment can be challenging due to the long-term nature of some benefits and the evolving nature of the technology. The new industrial 4.0 supports for scalable resources. **Pilot to Production**: Scaling up from pilot projects to full-scale implementation can be difficult, requiring careful planning and resource allocation. **Customization**: Tailoring the fourth industrial revolution solutions to specific business needs and contexts without compromising scalability.

3.3 Socio-economic Challenges

There are several socio-economic challenges faced by the environment such as workforce impact, ethical and regulatory issues and sustainability. **Job Displacement**: Computerized and AI can take-up to job displacement, particularly in routine and manual roles. **Job Transformation**: Making workers to learn new skills and adapt to new roles by which existing jobs will change according to technologies. There are **Ethical and Regulatory Issues in the environment. Ethical Considerations**: Ensuring the moral application of AI and data, taking into mind concerns about accountability, transparency, and prejudice. **Regulatory Compliance**: Navigating the complicated web of laws pertaining to environmental effect, safety requirements, and data protection. **Environmental Impact**: Ensuring that the adoption of new technologies aligns with sustainability goals. **Circular Economy**: Implementing practices that support a circular economy, such as recycling and reusing materials.

3.4 Economic Barriers

There are some economical barriers for adopting the industrial 4.0 such as Economic Uncertainty and global competitions. Market volatility and economic downturns can impact the ability of companies to invest in new technologies. Global Competition: Staying competitive in a global market where other companies may already be leveraging the fourth industrial revolution technologies.

3.5 Infrastructure and Connectivity

Implementation of new technologies need a huge infrastructure and connectivity issues. Also, the new fourth industrial revolution requires consistent and fast-speed internet connectivity which is essential for the real-time data exchange required in the fourth industrial revolution. **Infrastructure Investment**: Significant investments in upgrading infrastructure, such as 5G networks, are often required. Moreover, the exiting techniques also lack in **Fragmented Standards and protocol harmonization.** The absence of universal standards for the fourth industrial revolution technologies can hinder interoperability and integration efforts. **Protocol Harmonization**: Developing and adopting common protocols to ensure seamless communication between different systems and devices.

4 Case Studies Highlighting Challenges Faced by Organizations Implementing Industry 4.0

Artificial Intelligence has achieved great success in many areas like natural language processing, medical analysis and computer vision. In [19] author identifies and classifies the main challenges faced by industries in adopting AI. Various challenges like system integration, data, workforce and guarantee of trustworthy AI have been explored in this paper. It also provides various existing solutions to these challenges and gives direction to the future research for integration of AI in industries. In the paper [20] author proposed the incorporation of Internet of Things (IoT) and data analytics in Industry for the proactive maintenance. In the fourth industrial revolution predictive maintenance is necessary for the operation and maintenance of equipment in a cost-effective way. The paper provides a review of the current practices for the proactive maintenance within the fourth industrial revolution. It also reviewed how data analytics techniques from traditional to advanced methods help in processing and analysing the data. Various challenges in adoption of IoT and Data analytics for the fourth industrial revolution have also addressed in this paper.

In [21] author proposed the approaches used by manufacturing industries to create a learning principle for shifting to the fourth industrial revolution. These approaches are mainly related to training and education. In this paper a case study of a manufacturing ecosystem in Queensland, Australia was considered. In [17] author proposed the barriers in adoption of the fourth industrial revolution in manufacturing organizations. The fourth industrial revolution in manufacturing helps to automate and manage manufacturing systems effectively. However, adopting the fourth industrial revolution makes to face numerous hurdles and, in this paper, author proposed those major hurdles. 16 barriers were proposed in various dimensions like technology, economy, regulatory and organizational perspectives. Among these barriers, author concluded that economic dimension plays a strong role and affecting other dimensions. Final the author come up with good number of strategies to mitigate these challenges.

4.1 Importance of Addressing Challenges for Successful Industry 4.0 Adoption

There are several main challenges for adopting Industrial 4.0 with the existing techniques. Some of the challenges are integrating cutting edge digital technologies, build intelligent system. Thus, the major issues are need to be addressed to promote the acceptance of the fourth industrial revolution. In order to implement the fourth industrial revolution, maximize the cutting-edge technologies, efficiency and competitiveness in the manufacturing, the issue need be addressed.

The following are some of the main justifications for why it's critical to solve these issues:

1. Resolve the integration problem to maximize the production by ensuring smooth functionality among adoption of new technologies.
2. In new revolution, vast data are collected through a digital devices and sensors, thereby maintaining and safeguarding the data plays a vital role.
3. After storing vast data, it is crucial to manipulate with the data to make real time decision by analysing the data. Moreover, the data collected through sensors which are very difficult to maintain in a same format and so difficult to perform advanced analytics.
4. Advanced technologies also made a huge impact in education sector. Due the industrial 4.0, the education system has to be reoriented in order to cope with the industrial requirements.
5. Adoption of new technologies with the existing technologies must be compatibility and seamless integration to satisfy the needs of industry wide standards.
6. Technological connection needs to be established smoothly in Phased implementation, clear communication, and stakeholder participation. So that can minimize resistance and provide a soft change in excellent change management strategies.

7. Securing continued funding and gaining the support of stakeholders requires careful financial planning and evidence of the investments' long-term return on investment
8. In order to stay out of legal trouble, preserve operational legitimacy, and keep customers' trust, compliance with pertinent legislation is crucial.
9. Focusing on eco-friendly practices and minimizing environmental impact can improve a business's standing and competitiveness while also supporting global sustainability objectives.
10. Ensuring that solutions are scalable helps achieve consistent advantages across the organization and encourages long-term growth and flexibility.

5 Future Outlook and Trends

Future outlook and trends in the industrial 4.0 are

a. **Emerging technologies and their potential impact on industrial 4.0**

In addition to the previously discussed technology, there are few upcoming technologies and their key impact on industrial 4.0.

i. **Horizontal and vertical integration:**

The integration of horizontal and vertical is crucial for industrial 4.0. When there is horizontal integration, all of the supply chain management's procedures are closely coordinated at the production level across several manufacturing locations. All organizational layers are connected in vertical integration to enable unrestricted information process flow from the top floor to the bottom floor. As a result, there are fewer data and information silos and operations are streamlined since production is closely linked with corporate activities [22].

ii. **Industrial Internet of Things (IIoT)**

While both terms can be reciprocally used, Industrial Internet of Things (IIoT) is preferred over Industrial 4.0 [23, 24]. In Industrial 4.0, devices, robotics, machinery, equipment, and products are the most common objects. Industrial 4.0 provides real-time performance and condition data via RFID (Radio Frequency Identification) tags and sensors. As a result, businesses can operate efficiently to this technology, which also speeds up product conception and modification and reduces equipment downtime.

iii. **Additive Manufacturing or 3D printing**

Initially developed as a rapid model, layered manufacturing, also known as 3D printing, has many uses, ranging from large-scale production to dispersed manufacturing. By storing parts and products as model files in a digital inventory and printing them as wanted, 3D printing reduces the cost and remove the need for both on-site and off-site international manufacture. There has been an increase in the variety of uses for 3D printing in the last few decades. Originally intended as a tool for rapid prototyping, additive manufacturing (3D printing) has many uses today, ranging from mass production to distributed manufacturing. Some of the major areas in 3D printing are.

- Digital twins
- Cybersecurity

A digital twin is a digital reproduction of an actual machine, product, process, or system, based on IoT sensor information's. This core component of the fourth industrial revolution allows businesses to assess, understand, and improve the operation and maintenance of industrial products and systems. An asset operator, for example, can utilise a digital twin to identify a particular malfunctioning part, identify future issues, and boost uptime. Furthermore, considering the growing connectivity and usage of big data in the fourth industrial revolution, cybersecurity is essential. By using cutting-edge technologies like blockchain and machine intelligence to automate threat detection, prevention, and response, businesses can lower the risk of data breaches and production delays across their networks.

iv. **Blockchain**

Blockchain is a distributed ledger system that safely and openly records numerous machines transactions. The impact of blockchain plays an important role in transparency in the supply chain, data integrity, smart contracts and enhanced security. Among these, Blockchain technology offers a permanent and lucid transaction data's, hence improving the supply chain's traceability and accountability transparency. The storage of data in the blockchain rise a concern in data integrity. The storage in blockchain is decentralized, safe and impervious to manipulation, which lowers the risk of fraud and mistakes. Initially, the data stored in blockchain made a smart contract with different vendors to fix the number of redundant data. Smart contracts simplify procedures and cut down on administrative expenses by automatically executing and enforcing its conditions. Blockchain makes this possible. Industrial 4.o needs to make the data secure in the distributed environment. Industrial transactions and data exchanges are highly secure because to blockchain's cryptographic features [25, 26].

v. **Industry 4.0 Digital Transformation Outcomes**

The manner that the fourth industrial revolution is changing our global economies has an ongoing impact on other countries. According to Microsoft COO Judson Althoff, the ability to create "better products more effectively, more efficiently, with lower carbon footprint, lower water utilization, more sustainably than ever before" is the main advantage of the fourth industrial revolution and the industrial metaverse. According to a McKinsey study, businesses who implemented the fourth industrial revolution digital transformation experienced increases in Key Performance Indicators (KPIs) in five key growth areas: sustainability, productivity, agility, speed to market, and personalization.

The most worth able benefits expect from the manufacturing are.

- Optimized processors
- Greater assest utilization
- Higher labor and productivity
- Supply Chain Visibility
- Sustainability

vi. **Barriers of Industry 4.0**

The integration of these new technology and processes presents a number of possible problems for enterprises seeking to finish their digital transformation toward the fourth industrial revolution. These can be classified as operational, cultural, or technical. The following are some of the typical obstacle's organizations encounter when implementing the fourth industrial revolution technologies:

- Combining information technology (IT) with operational technology (OT).
- Inadequate Funding for Infrastructure and Technology Implementation.
- Managing the Risks of Cybersecurity.
- Inadequate Knowledge and Experience in Using and Maintaining Technology.
- Overcoming Change-Resistant Cultures Not every use case can be applied to every industry, and each business has different obstacles in implementing the fourth industrial revolution successfully.

b. **Predictions for the future of industry 4.0 and its evolution**

The fourth industrial revolution is about to undergo a major metamorphosis, propelled by ongoing technological progress and changing consumer needs. The following are some significant forecasts and developing patterns for the fourth industrial revolution:

- Enhanced AI and machine learning capabilities
- Increased connectivity and advanced IoT integration
- Widespread adoption of digital twins
- Increased emphasis on cybersecurity
- Growth of smart factories
- Sustainable and green manufacturing
- Increased customization and personalization
- Data-driven decision making
- Human–machine collaboration
- Global collaboration and innovation networks

c. **Recommendations for organizations looking to embrace Industry 4.0**

Adopting a thorough and strategic approach is essential for firms aiming to embrace the fourth industrial revolution in order to successfully navigate this technological change. The following are some essential suggestions:

- Establish a clear vision and plan
- Invest in the development of talent and skills
- Use data and analytics
- Adopt a flexible and scalable technology infrastructure
- Implement pilot projects and scale them gradually
- Strengthen cybersecurity measures
- Foster an innovative culture
- Work with outside partners
- Prioritize customer-centric innovations
- Monitor and adjust to technological trends.

6 Strategies to Overcome Challenges

Figure 13 shows the strategies to overcome challenges in adoption of Industry 4.0.

- Strategic Planning: Draft a comprehensive strategy defining the goals, phases, and standards for the fourth industrial revolution
- Small-scale Programs: Before expanding, begin with small-scale initiatives to test and improve technologies.
- Cooperation and Partnerships: Work together with educational institutions, business consortiums, and technology suppliers to obtain information and resources.
- Workforce Development: It is possible to implement upskilling and training programs to prepare employees for new roles and technology improvements.
- Sturdy Cybersecurity Protocols: Implement meticulous cybersecurity practices, like regular audits and upgrades.
- Flexible and Modular Solutions: Use technologies that are adaptable, modular, and easy to grow and adjust to meet changing needs.

By tackling these real-world concerns via strategic planning, capital expenditure on human resources to effectively overcome the challenges for integrating the industrial 4.0 with the existing technologies.

7 Conclusion

The fourth industrial revolution denotes a change shift in industrial practices by incorporating the key factors such as cyber-physical systems, IoT, cloud computing, and AI. While it offers numerous benefits, including enhanced efficiency, flexibility, and customization, the transition to the fourth industrial revolution is not without its challenges. These include the complexity of integrating new technologies into existing infrastructures, the demand for a completely ready workforce proficient in advanced technologies, and the critical issues of data security and privacy. Adapting these obstacles is crucial for organizations to fully leverage the potential of the fourth

Fig. 13 Strategies to overcome challenges in adoption of industry 4.0

industrial revolution. This chapter emphasises on the importance of understanding and overcoming these hurdles to ensure a successful and sustainable implementation of the fourth industrial revolution initiative. This chapter also discussed the various challenges faced to integrate the industrial 4.0 with the existing technologies. Also, this chapter discuss about the various strategies followed to overcome the difficult to integrate the existing techniques with the new technology.

References

1. Babu, C.V.S., Sriram, E., Matthai, P.A., Sudharshan, S.: Shaping sustainable manufacturing: IoT and industry 4.0 integration for transformation, in Futuristic Technology for Sustainable Manufacturing, IGI Global, pp. 216–247. (2024)
2. Sharma, M., Paliwal, T., Baniwal, P.: Challenges in digital transformation and automation for industry 4.0, in AI-Driven IoT systems for industry 4.0, CRC Press, pp. 143–163 (2024)
3. Ghobakhloo, M.: Industry 4.0, digitization, and opportunities for sustainability. J. Clean. Prod. **252**, 119869 (2020)

4. Javaid, M., Khan, S., Haleem, A., Rab, S.: Adoption of modern technologies for implementing industry 4.0: an integrated MCDM approach. Benchmarking An Int. J. **30**(10), 3753–3790 (2023)
5. Piardi, L., Leitão, P., Queiroz, J., Pontes, J.: Role of digital technologies to enhance the human integration in industrial cyber–physical systems. Annu. Rev. Control. **57**, 100934 (2024)
6. Kantaros, A., Ganetsos, T.: Integration of cyber-physical systems, digital twins, and 3D printing in advanced manufacturing: a synergistic approach, Kantaros, A. Ganetsos, pp. 1–22, (2024)
7. Zhou, J., Zhou, Y., Wang, B., Zang, J.: Human–cyber–physical systems (HCPSs) in the context of new-generation intelligent manufacturing. Engineering **5**(4), 624–636 (2019)
8. Parashar, B., Sharma, R., Rana, G., Balaji, R.D.: Foundation concepts for industry 4.0, in New Horizons for Industry 4.0 in Modern Business, Springer, pp. 51–68. (2023)
9. Agrawal, K., Nargund, N.: Deep learning in industry 4.0: transforming manufacturing through data-driven innovation, In: International Conference on Distributed Computing and Intelligent Technology, Springer, pp. 222–236. (2024)
10. Khang, A., Rath, K.C., Satapathy, S.K., Kumar, A., Das, S.R., Panda, M.R.: Enabling the future of manufacturing: integration of robotics and IoT to smart factory infrastructure in industry 4.0, In: Handbook of Research on AI-Based Technologies and Applications in the Era of the Metaverse, IGI Global, pp. 25–50. (2023)
11. Azadi, M., Moghaddas, Z., Cheng, T.C.E., Farzipoor Saen, R.: Assessing the sustainability of cloud computing service providers for industry 4.0: a state-of-the-art analytical approach. Int. J. Prod. Res. **61**(12), 4196–4213 (2023)
12. Kolasani, S.: Revolutionizing manufacturing, making it more efficient, flexible, and intelligent with Industry 4.0 innovations. Int. J. Sustain. Dev. Through AI. ML IoT. **3**(1), 1–17 (2024)
13. Abulibdeh, A., Zaidan, E., Abulibdeh, R.: Navigating the confluence of artificial intelligence and education for sustainable development in the era of industry 4.0: challenges, opportunities and ethical dimensions, J. Clean. Prod., 140527, (2024)
14. Dhanaraj, R.K., Jhaveri, R.H., Krishnasamy, L., Srivastava, G., Reddy, P.K., Maddikunta, P.K.: Black-hole attack mitigation in medical sensor networks using the enhanced gravitational search algorithm, Int. J. Uncertainty Fuzziness Knowl. Based Syst. World Sci. 29 (2021)
15. Ahmed, I., Jeon, G., Piccialli, F.: From artificial intelligence to explainable artificial intelligence in industry 4.0: a survey on what, how, and where. IEEE Trans. Ind. Inform. **18**(8), 5031–5042 (2022)
16. Sayem, A., Biswas, P.K., Khan, M.M.A., Romoli, L., Dalle Mura, M.: Critical barriers to industry 4.0 adoption in manufacturing organizations and their mitigation strategies. J. Manuf. Mater. Process. **6**(6), 136 (2022)
17. Elnadi, M., Abdallah, Y.O.: Industry 4.0: critical investigations and synthesis of key findings. Manag. Rev. Q. **74**(2), 711–744 (2024)
18. Dieste Gracia, M., Sauer, P., Guido, O.: Organizational tensions in Industry 4.0 implementation: a paradox theory approach, In: Book of abstracts of the 2021 Annual Conference Decision Sciences Institute, (2021)
19. Krishnasamy, L., Somasundaram, K., Quadir, M., Dhanaraj, R.K., Roopa, C., K.P, A deep learning model for intelligent energy management, In: 2022 4th International Conference on Advances in Computing, Communication Control and Networking (ICAC3N), Greater Noida, India, pp. 1307–1311 (2022)
20. Soori, M., Jough, F.K.G., Dastres, R., Arezoo, B.: Internet of things and data analytics for predictive maintenance in industry 4.0, a review
21. Thoben, K.-D., Wiesner, S., Wuest, T.: 'Industrie 4.0' and smart manufacturing-a review of research issues and application examples. Int. J. Autom. Technol. **11**(1), 4–16 (2017)
22. Kumar, R., Rani, S., Khangura, S.S.: Machine learning for sustainable manufacturing in industry 4.0: Concept, concerns and applications. CRC Press, (2023)
23. Peter, O., Pradhan, A., Mbohwa, C.: Industrial internet of things (IIoT): opportunities, challenges, and requirements in manufacturing businesses in emerging economies. Proced. Comput. Sci. **217**, 856–865 (2023)

24. Tan, S.F., Samsudin, A.: Recent technologies, security countermeasure and ongoing challenges of Industrial Internet of Things (IIoT): a survey. Sensors **21**(19), 6647 (2021)
25. Alladi, T., Chamola, V., Parizi, R.M., Choo, K.-K.R.: Blockchain applications for industry 4.0 and industrial IoT: a review. IEEE Access **7**, 176935–176951 (2019)
26. Balaji, S., Jeevanandham, S., Choudhry, M.D., Sundarrajan, M., Dhanaraj, R.K.: Data aggregation through hybrid optimal probability in wireless sensor networks. In ICST Transactions on Scalable Information Systems. European Alliance for Innovation N.O. (2024). https://doi.org/10.4108/eetsis.4996

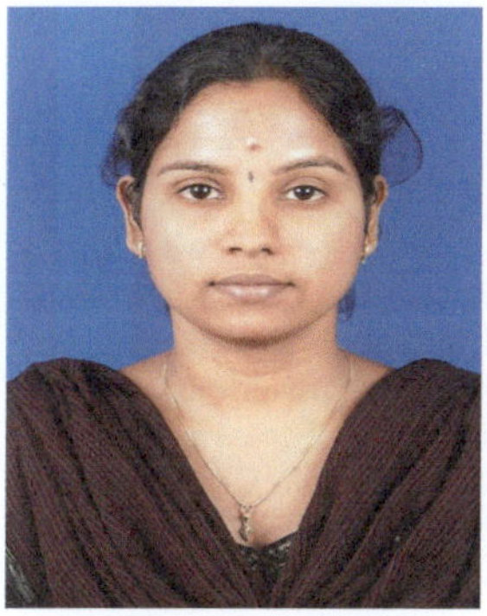

S. Mohana Saranya earned her master's degree from the College of Engineering Guindy (CEG), Anna University, Chennai. She is currently an Assistant Professor at Kongu Engineering College in Erode, with a focus on research in deep learning and computer vision. She has secured a project grant from the Ministry of Education (MoE) and has successfully completed a student project funded by AICTE. Additionally, she has organized two national-level seminars sponsored by CSIR and NBHM. Her academic contributions include publishing nine research papers in international journals and presenting 21 articles at international conferences. She is also a lifetime member of the Computer Society of India.

Dinesh Komarasamy presently working as a Assistant Professor (SRG) in the department of computer science and engineering, Kongu Engineering College, perundurai. He received the Bachelor of Engineering and Master of Engineering in Computer Science and Engineering under Anna University, Chennai, India in 2010 and 2012 respectively. He has received his Doctorate of Philosophy (Ph.D) under the title of "A Framework For Scheduling Of Heterogeneous Workloads In Cloud Management System" in the year 2019. He has around 3.6yrs. of research experience in College of Engineering, Anna University, Guindy. He has completed his research in optimal scheduling of jobs in cloud environment. He has published many Papers which are indexed by SCI and Scopus. His area of research includes cloud computing and machine learning. Presently, he is working as a Assistant Professor in the department of computer science and engineering in Kongu Engineering College, perundurai for the past 8 years. He has currently doing research projects funded by Indian Council of Social Science Research (ICSSR). He also presented papers in several national and international conferences.

S. Mohanapriya received Bachelor's degree and Master's degree in Computer Science and Engineering during the year 2012 and 2014 respectively from Anna University, Chennai. Currently she is working towards the Ph.D. degree in Computer Science and Engineering at Anna University, Chennai in the area of Few-shot Object Detection using Deep Learning. She has been currently working as Assistant Professor in the Department of Computer Science and Engineering at Kongu Engineering College (Autonomous), Perundurai, Erode. She has conducted various workshops and published several research papers in well reputed journals and conferences.

M. R. Prasanndh Raaju is currently pursuing a Bachelor's degree in Computer Science and Engineering (3rd year) at Kongu Engineering College (Autonomous), Perundurai, Erode. His areas of interest include Data Science, IoT, Machine Learning, and he has actively participated in various academic projects, technical workshops and doing a government sanctioned project Tomatix (for Tomato Grand Challenge). As an enthusiastic learner, he aspires to contribute to innovative solutions in the field of Computer Science.

Explainable AI Principles of Building Industry 4.0

N. Sanjana, R. Immanual, K. M. Kirthika, and S. Sangeetha

Abstract The full potential of Industry 4.0's deep learning models cannot be realized without explainable AI (XAI). Such models operate as black boxes, although they are known for their excellent predictive abilities, making it hard to build trust, ensure accountability, or align them with human values. This chapter looks at some XAI principles that can help address the challenges presented by these black-box models in industrial settings, including but not limited to potential biases, difficulties in debugging, production delays, and regulatory compliance problems. The various XAI techniques are discussed here, which range from model-agnostic methods like feature importance analysis, LIME, SHAP, etc., to model-specific methods such as saliency maps for CNNs and layer-wise decomposition for RNNs, showing that they can be used to demystify deep learning models across different industrial domains. It also considers the needs to be considered when implementing XAI successfully, like the trade-off between explanation and accuracy computation cost, among others, thus encouraging collaboration between AI engineers and subject matter experts who can provide valuable domain knowledge integration. This chapter helps reader to know the model to make trustworthy, transparent, and accountable AI systems that promote innovation while still upholding ethical values in Industry 4.0 and beyond by accepting XAI principles.

Keywords Explainable artificial intelligence · Model agnostic methods · Model specific methods · Deep learning · Industry 4.0

N. Sanjana (✉) · K. M. Kirthika
Department of Computer Science and Engineering, Sri Ramakrishna Institute of Technology, Coimbatore, Tamilnadu, India
e-mail: Sanjananithykumar06@gmail.com

R. Immanual
Department of Mechanical Engineering, Sri Ramakrishna Institute of Technology, Coimbatore, Tamilnadu, India

S. Sangeetha
Department of Electrical and Electronics Engineering, Sri Ramakrishna Institute of Technology, Coimbatore, Tamilnadu, India

L. Krishnasamy et al. (eds.), *Distributed Deep Learning and Explainable AI (XAI) in Industry 4.0*, Information Systems Engineering and Management 55,
https://doi.org/10.1007/978-3-031-94637-0_2

1 Introduction

When it comes to Industry 4.0, where data-driven technologies are transforming manufacturing, deep learning models have become essential. By using these systems, which are highly intelligent machines created by experts, a lot of automation can be done at unimaginable levels, leading to real-time decision-making. But as such models increase in complexity, they frequently become 'black boxes' leaving many stakeholders baffled about the logic and reasoning behind the decisions. This lack of transparency is a major issue for manufacturers. Picture pouring substantial resources into rolling out an advanced deep-learning system for predictive maintenance whose recommendations defy human explanation and trustworthiness. Or think about a situation when there is a highly accurate model for spotting defects in products that are discriminatory, thereby resulting in expensive recalls and regulatory inquiries. The good news is that Explainable AI (XAI), an upcoming field, offers some way out of this maze. XAI techniques can reduce the divide between such super-predictive mechanisms and the inherent human need for intelligibility and faith by clarifying how deep learning models make decisions. To unlock the full potential of deep learning in Industry 4.0, this chapter will take on a journey that embraces XAI principles. It addresses the issues black-box models face in industrial settings and explains why explainability is necessary to reduce risk, foster human-AI collaboration, and ensure responsible AI development.

This chapter will delve into different XAI techniques, including feature importance analysis and LIME for model-agnostic methods, as well as saliency maps that are used in computer vision and layer-wise decomposition, which is applied to natural language processing. This section will expose practical examples that show how these methods can be employed in demystifying deep learning models across various industrial applications. Finally, this chapter examines some of the important factors for successfully implementing XAI, such as balancing between accuracy and interpretability, computational costs associated with XAI approaches, as well as integration of domain knowledge and collaboration between subject matter experts and AI engineers. It will present the example of the usage of such methods in the demystification of deep learning models within numerous industries. Finally, this chapter presents some of the major considerations of effective XAI implementation including accuracy-interpretabity trade-off, computational costs associated with XAI methods, and incorporation of prior knowledge and teamwork between the domain specialists and data scientists.

By the end of this chapter, you will understand the importance of XAI in Industry 4.0 and how to apply its principles to create reliable, transparent, and accountable AI systems that will drive innovation while supporting ethical values. So, together, embark on this exciting journey and unlock the true potential of deep learning in technology.

1.1 *Overview of Industry 4.0 and the Role of Deep Learning Models*

The symbiosis of digital and physical systems has led to a drastic revolution of industrial processes, known as Industry 4.0 or The Fourth Industrial Revolution [1]. Industry 4.0 is the process by which cloud computing, artificial intelligence, cyber-physical systems, and the Internet of Things are incorporated into industries. Industry 4.0 consists of several components is illustrated in the Fig. 1. One of the main concepts of the Industry 4.0 is "smart factories"—completely digitally built structures containing advanced systems which assure complete automatization and self-maintenance. These systems stand up with help of machine learning technologies.

Deep learning is a general type of machine learning that uses artificial neural networks of many layers and can be applied as a primary intelligent component of smart manufacturing [2]. Moreover, deep learning techniques can extract valuable patterns and insights from vast and complicated datasets. As a result, the tool can be utilized in various industrial applications. For instance, in predictive maintenance, deep learning can analyze sensor data to forecast when equipment will malfunction long before it does. This method reduces labor costs since maintenance is done when an asset is unavailable. With deep learning-accelerated computer vision, automated quality control, defect detection, and real-time process tracking are enhancing manufacturing. Additionally, deep learning is beneficial in the intelligent optimization of other processes, such as scheduling, resource allocation, and logistics [3].

In addition to the above, deep learning also plays a vital role in improving the production processes by providing an intelligent platform to schedule, allocate, and manage resources and the supply chain. Natural language processing combined with deep learning opens the possibility to create models that are capable of filtering

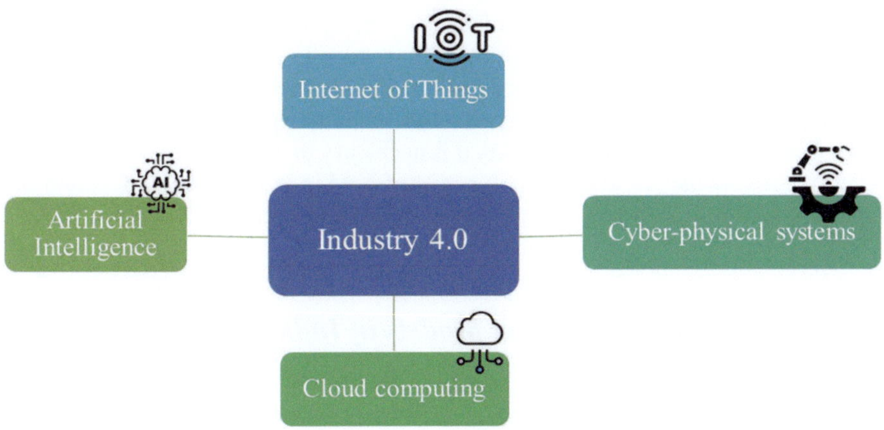

Fig. 1 Industry 4.0

Fig. 2 Objectives of XAI

through technical documents, customer feedback, and maintenance diaries and make it much easier to sort knowledge and design knowledge-based decision support systems. However, the application of deep learning in an industrial environment comes with a few challenges. The utilization of deep learning models on industrial data results in unintelligible and incomprehensible models, making it hard to build trust and application by domain experts and stakeholders. Therefore, to tap deep learning's full potential in Industry 4.0, it is necessary to address these with the principles of explainable AI. The main objective of XAI in Industry 4.0 is Trustworthy, transparency, Informativeness, and Confidence, as shown in Fig. 2.

1.2 Challenges of Black-Box Models in Industrial Settings

Advanced computer models like deep learning do amazing things. But they work like black boxes, making it tricky at factories and plants. First off, nobody knows why these models say what they say. That makes folks unsure if they can trust the model. Think of relying on new AI to say when equipment needs to be fixed or check quality. But you don't understand why it recommends things. This "opacity" means it's hard to know if the outputs makesense, especially for important jobs [4]. Another challenge arises when these black-box models exhibit unexpected or counterintuitive behaviors, such as making biased or discriminatory decisions. Thus, without understanding of the internal decision making process it turns into a real challenge to identify the root causes of such poor results, not to mention improving them [5].

This can lead to costly errors, production delays, and regulatory penalties or legal liabilities [6]. Furthermore, deploying black-box models in industrial settings often requires close collaboration between AI experts and domain specialists. However, the lack of interpretability poses a significant barrier to effective communication and knowledge transfer between these two groups [7]. Domain experts may struggle to provide meaningful feedback or insights to improve the model's performance, while AI engineers may find it difficult to incorporate domain knowledge into the model's architecture or training process [8]. Lastly, the opaque nature of black-box models can hinder their scalability and long-term maintenance in dynamic industrial environments. As production processes, equipment, or data distributions evolve, diagnosing and addressing any performance degradation or concept drift becomes increasingly challenging without a clear understanding of the model's underlying mechanisms [6].

Thus, although deep learning and other modern machine learning schemes undeniably possess strong prognosticative power, their computational uncertainty hampers their responsible and reliable use in industrial application. Many of these advanced technologies have the strength in the real-world manifestation in applying and utilizing innovations which ensures transparency, accountability and ethical compliance in such a breakthrough conceptually using the ideas of Explainable AI (XAI) [9]. Manufacturing and industrial processes are being revolutionized through Industry 4.0 that is characterized by the integration of cyber-physical systems, Internet of Things, artificial intelligence among other things. This change has seen deep learning models as a dominant feature pushing for improvements in, anomaly discovery, enhancement of business processes and prognosis with regards to maintenance. However, these models' very sophistication often raises issues of their 'black box' nature, posing challenges to increasing interpretability and trust in some of the most critical corporate decisions. In Fig. 3 the process of how the explainable algorithm erases the black box and makes it a white box is depicted.

2 The Imperative for Explainability in Industry 4.0

Industry 4.0 employs well-designed models of Artificial Intelligence (AI), which can make decisions to convert everyday manufacturing with strategic plans to increase productivity. These tools can also make decisions in executing business with the previous data, even in case of critical circumstances. In general, the decision models

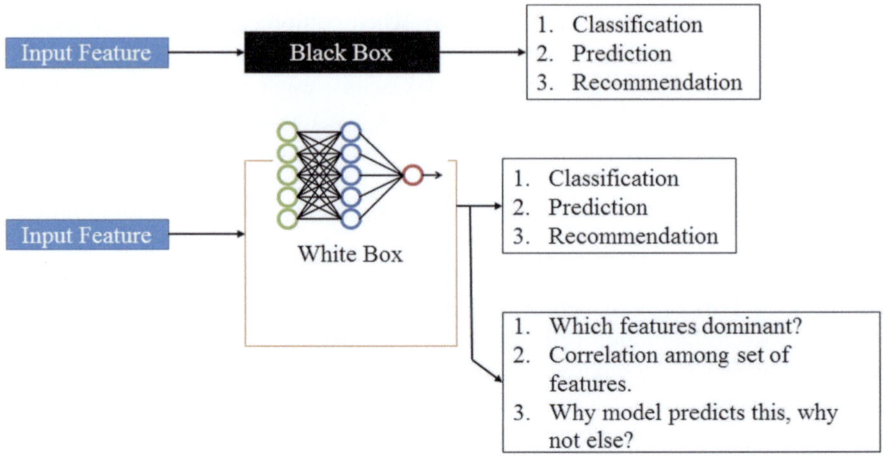

Fig. 3 AI versus XAI

by the individual can be extended to be collaborative and resilient decision infrastructures that will be more efficient in emergency circumstances. Innovative technologies like machine learning, especially the concept of deep learning, facilitate processes inside the industry in the design of its product and system of manufacturing. Even though there is a drastic development in autonomous, artificial, computational capabilities, models, and tools, human play a vital role in decision-making even during the transformation from Industry 4.0 to Industry 5.0, which is considered an "Age of Augmentation".

A human remains a key player in decision-making during specific processes like the detection of fault, its sequence of clearance, and the management of alarm. This is due to the fact that the occurrence of a fault and causes for malfunction of the equipment will be different from one incident to another. The industrialists who used to predict the real-time process in factories equipped with AI models need to establish confidence in the decisions suggested by machine learning and its derived predictions. The nascent technology Explainable AI (XAI) allows humans to understand, have certainty about, and control the results of AI. Due to its potential to address the limitations of AI establishment, there is a high need for explicit declarative knowledge offered by XAI. As applied to industrial AI and machine learning this forward thinking technological approach addresses, transparency, auditability and trust [5]. However, most explainability problems in embedded AI and autonomous systems are mostly associated with deep learning models in the context of Cyber Physical Systems and Artificial Neural Networks. This is because it is extremely difficult to see the nature of the input/output mapping in the relevant systems.

2.1 Potential Biases and Discriminatory Outcomes

The designed AI algorithms, in general, acquire knowledge from past data, and if there are biases in the input data, the AI can propagate and enhance these biases. This will make potential bias in decision-making of the system process in manufacturing, operation, man management, and even in business. This bias in the input historical data set makes the AI algorithm produce an outcome which results in the discriminative decisions during distinct tasks like fault management and crises. While developing an algorithm for AI, there may be the possibility of the introduction of bias in the training data, either purposely or by mistake. Whether bias is encoded intentionally or unintentionally by the developers, it will lead to discriminatory outcomes. AI algorithms are commonly designed to operate as "black boxes," as their process of decision-making is not interpretable and transparent. This impenetrability leads to a tough situation of identification and mitigation of biases, which may result in unfair outcomes, sometimes even unnoticed by the practitioner till it creates problems. In practice, AI recommendations in the automation system of industry 4.0 are followed blindly and do not require further evaluation. Biased AI systems will outspread and even magnify existing inequalities in decision-making processes. It will even sustain existing social inequalities, especially in Human resource management, rebuffing opportunities to certain groups of people and also unfair resource allocation. The feedback loop with discriminatory outcomes from old biased decisions influences the closed-loop output decisions, further establishing biases over time. The efforts have to be made by the algorithm developers to overcome the bias found in the historical data set, algorithm and discriminatory outcomes. The team of people who develop an AI system should be ensuring transparency and accountability in their decisions and recommendations by auditing the system for bias on a regular basis. They can also follow regulatory protocols to establish fairness and equity in AI-based systems.

2.2 Debugging Difficulties and Production Delays

AI algorithm-based systems designed for the establishment of Industry 4.0 are always complex since it has to replicate human intelligence with tricky algorithms. So the developers will find debugging tasks to be challenging once. Disputes are difficult to identify since they are not always straightforward. It's challenging to understand deep learning models and their decisions. The opaqueness of the algorithm also makes debugging more difficult, and it cannot find the origin of the error. The quality of the past data taken as a training set contributes to recommendations from AI systems. Incomplete or inaccurate data feed to the algorithm will lead to specious decisions whose consequence will be the production delays. These are the serious issues that have to be identified and addressed. Production atmospheres in the industrial environment frequently change since it is dynamic. The models designed with AI should follow these changes and have to be adapted to environmental variations. This adaptability may acquaint with errors that need to be corrected. The concept of overfitting is also common in AI models, which perform well during the training phase but find it difficult to make decisions with new data or situations. This may end up in unpredicted behaviors in production environments. Interpreting the AI models' decision is important to identify the root cause of production delays. Not understanding the crucial decisions taken by the AI model in Industry 4.0 makes it harder for the debugger to address the issue. To overcome the above-discussed challenges, industries adapting AI for their company should be conscious of robust data collection and management practices. Explainable AI techniques are to be implemented as part of Industry 4.0 by following testing procedures and creating elaborate guidelines for troubleshooting AI systems in the production department.

2.3 Regulatory Hurdles and Compliance Issues

Since the AI systems for Industrie 4.0 mostly use big volume of data for training and testing, the critical issues are, therefore, the data security and privacy. The regulatory framework should have strict measures for personal data protection, guarantee transparency in data processing, and Explain the ability of the modeled AI algorithm. Bias in the training data used to perpetuate and transfer in the recommendation of the AI systems, care should be taken to mitigate the biased outcome so that social acts and laws are not violated on the premises. Ethical concerns have to take care, especially in the field like finance and medical. To ensure the responsible AI deployment well, well-defined ethical guidelines are to be framed and dissipated to the developer and user of the proposed model. In an industry which undergoes transition and revolution, a new challenge is to manage the IPR, copyright, and trading models driven by AI algorithms. Separate intellectual property laws should be formed to protect AI-related inventions and technologies. Accountability should be there for the liability that arises due to the error in the decisions made by the AI-based algorithm. Regulatory guidelines also include a component about accountability of liability caused by the bias or error in the systems. Regulatory hurdles and compliance issues require a comprehensive understanding of the legal and ethical landscape surrounding AI technologies, along with proactive measures to address emerging challenges. For the development of proper structures of legislation which may protect a society's interest but also allow people to come up with innovations that are viable, the legislators, the business partners, and the authorities have to come up with structures.

3 Benefits of Xai for Industry 4.0

XAI Eliminates the main problem of deep learning models in Industry 4.0 which is interpreting their decisions and gaining trust in this technology. As mentioned earlier explainability is the ability to produce models that present enough details to be trusted while at the same time delivering optimum system performance. Mathematically qualified results of deep learning models have deterministic decision rules of express frame methodology, and there is a better understanding of decisions among people. Transparency facilitates collaborative condition monitoring, diagnostics, and predictive maintenance of smart industrial assets and data and knowledge privacy protection. Therefore, the transformation of black-box models into explainable classifiers in XAI is a way to increase the performance of AI and simplify the learning process. Because it combines the best of deep neural networks and decision trees—namely, the key advantage of the former and the key advantage of the latter—the better generalization and greater classification accuracy. This makes XAI an indispensable tool for achieving the human-machine symbiosis of Industry 5.0.

3.1 Enhanced Trust and Collaboration Between Humans and AI

Truly trustworthy and explainable artificial intelligence can be built only by creating trust and collaboration between humans and AI. The Commission's regulation on AI has stressed the relationship between the interconnections of trust in AI and the need for robust and transparent AI technologies. As AI develops in various ways, the issues of bias, security, the black-box nature of AI, and many more must be addressed. Trustworthy AI combines trustworthiness components such as robustness, safety, privacy, and data governance with the complement of the AI assisting humans, making it work reliably and in a controlled fashion. A relationship that can be established based on a comprehensive understanding of sound AI, based on trust on the one hand, but paved the way for systems that could benefit society in the banking industry, the health industry, autonomous systems, the IoT, and more. By reviewing trustworthy and explainable AI, it is possible to close the gap between humans and AI, and create a stronger bond between transparent, accountable, and dependable partners [10].

3.2 Proactive Model Management and Performance Optimization

Regarding Proactive Model Management and Performance Optimization, the study illuminates complicated connections between human annotators and machine learning models. In particular, the study focuses on Explainable Active Learning as an essential field of machine instructors' interface. The study unveils complex empirical feedback loops, which are relied upon by human annotators because of their natural use of reason in their explanations of model behavior and predictions using combinations of empirical studies and surveys. The claim is concerned with future algorithmic improvements, the development of which should match the slow, uncertain, and diverse nature of human feedback.t is essential as a foundation for integrating more advanced feedback mechanisms into Active Learning algorithms to signal the type of feedback required by people as such and fosters the interpretability and trustworthiness of ML models overall. The experimental analysis considers the annotators' subjective feelings. These include annotators' confidence in using machine learning models and job satisfaction in the annotation process. Further research uses five-point Likert Scale correlations to measure the participants' certainty about the applicability of their teaching models. Capability, benevolence, integrity, and predictability were the sections assessedIn addition, the survey assesses annotators' happiness and cognitive load during annotation tasks in the survey. This points to important characteristics of how ML models are experienced at work. It is noteworthy how the research emphasizes the need to create human consistent interfaces not only for the improvement of the various model's performance but also for catering to different needs and preferences of artificial intelligence educators. Such initiatives should encourage the people and AI frameworks to work with equal relations, and think about the variations amidst people and the influence of the environment on the results of the AI system.

3.3 Responsible AI Development and Ethical Alignment

In the last few years, there have been a number of advancements in AI that have brought changes to the concepts of explainability commonly referred to as XAI, which makes it possible to explain AI models. This presents a brief on technical and ethical issues relevant to XAI and observes that their dutiful application must be made to suit the requirements of stakeholders who seek explanations. Under the AI control of the essential infrastructures that directly influence the human well-being and fundamental decision-making process primarily powered by the machine learning (ML), the ethical aspect is something to ponder on while defining such systems. The presented analysis can be considered as a proper attempt to bring the gap between the technical advancements in XAI and ethical issues, connected with practical application of AI systems. Besides, it notes that xai should be incorporated into a large accountability system where human decision-makers are still required to justify in the normative for ML models.By bringing out the ethical nature of stakeholder needs like consent, fairness and risk assessment, this paper calls for a shift from a limited to a broader understanding of explainability's role in ensuring responsible and safe machine learning deployment while at the same time basing itself on stakeholder requirements. Finally, through this article, different ways designers can choose XAI methods will be revealed, which are intended to help safety engineers make better choices during the selection and implementation process, thus promoting a culture of ethical alignment within AI development based on evidence acceptance by regulators and service providers.

4 Xai Techniques for Demystifying Deep Learning in Industry 4.0

Explainable Artificial Intelligence plays a crucial role in understanding the importance of deep learning models in Industry 4.0. Deep learning models are highly effective in many decisions, and often, they are criticized for the lack of transparency, which can hinder their trust while making critical decisions. Explainable artificial intelligence provides insights into decision-making by providing the appropriate explanation. There are two explainable artificial intelligence, namely model-based XAI and postHoc XAI. The model base depends on the structure of the model itself, and the models like linear regression, logistic regression, and decision tree are model-based XAI. In these, the model structure is understood, and it has given such results because they are predictable by design. If the structure of the model is not predictable by nature, then it is referred to as PostHoc XAI.In posthoc XAI Techniques like feature mapping, model distillation, Local Interpredictable model agnostic explanation, Shapley Addictive Explanation, Concept Activation Vector, counterfactual Explanation, and Attention Mechanisms can demystify deep learning models in Industry 4.0. Figure 4 shows the classification of XAI Techniques.

Transparent methods, in turn, are built with models, whose decisions are understandable without almost any explanation. Some of the transparent methods include; Bayesian models, decision trees, and linear regression as well as fuzzy inference systems. Linear relationships of internal features are better explained by transparent models if the relationships between them are not very complex. Post-hoc methods are focused solely on explaining efforts made in the model that has been trained by a black-box. These methods seek to identify the factors contributing to the model's performance without changing its original work. Besides, in the process of considering methods and techniques of XAI for different data kinds, including text, image, audio, and video, the text often states that each media differs not only in essence but also in its character and that, therefore, it is necessary to employ diverse strategies and approaches in every case. If this classification scheme can be achieved, it will possibly help point out the direction on how to identify the suitable XAI methods and at the same time explain the various advantages and limitations of different approaches as well as present the viable multi-modal application scenarios. Furthermore, when discussing the ways and approaches for applying the XAI methods and techniques to the different data kinds, including the text, images, audio, and video, the text often mentions their individuality and states, therefore, that it is necessary to

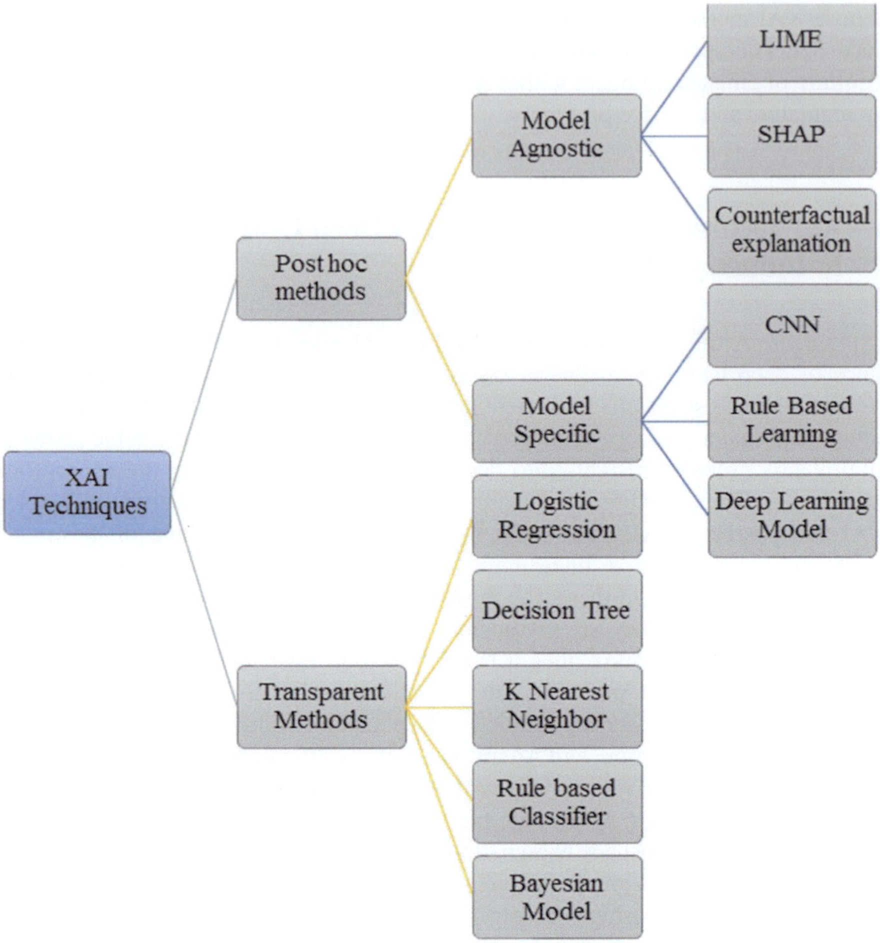

Fig. 4 Classification of XAI techniques

apply various approaches and techniques each time. Thus, it is wished that this classification framework can help in identifying the right XAI techniques, and explain their advantages and disadvantages at the same time, besides presenting potential multi-modal use cases.

4.1 Model-Agnostic Methods

The importance of model-agnostic interpretability methods cannot be overestimated. Such techniques as Shapley values and surrogate models help to explain the activities of models like neural networks and ensemble models. Shapley values help attribute

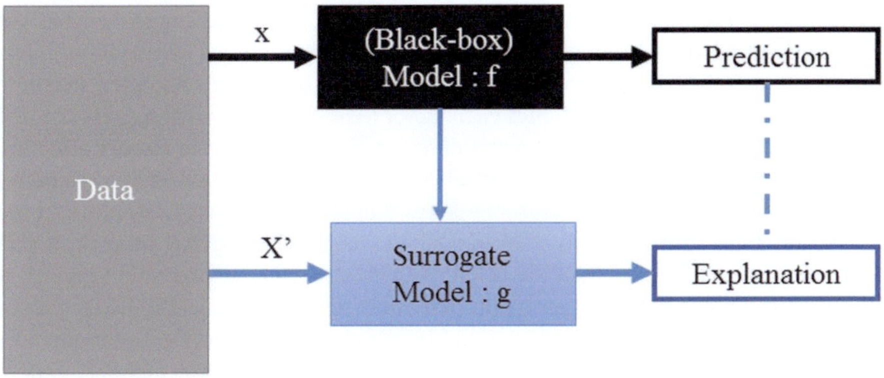

Fig. 5 Generation of explanation

each feature's importance to an actual prediction. Their application is especially accurate in the case of models which are based on trees. Surrogate models are alternative and interpretable black-box models. They explain fully and effortlessly but with some degree of loss of full fidelity. The balance between model-agnostic and explanatory power can be done by combining both methods. It may enhance machine learning systems' trust, transparency, and accountability and make them more accessible and understandable for people in various industries.In model agnostic, the data x is given to black box model f; then it will have a prediction as x(f), and then the model with data x is given to the explainable algorithm, which is in surrogate model g and gives the explanation for the predicted output as shown in Fig. 5.

LIME—can be interpreted as Local Interpretable Model-agnostic Explanations. LIME in its basic form can interpret any machine learning model because it is using each prediction to train a local interpretable surrogate model. It creates a new artificial dataset that corresponds closely to the original dataset but contains modified attribute values for the interpreted instance. The data points are weighted based on their proximity, and then features are sampled to obtain a prediction to simplify the black-box model's behavior.LIME can play a significant role in Industry 4.0 by improving transparency and interpretability in complex machine learning approaches in multiple industrial platforms. It generates local explanations for specific predictions, and people can easily evaluate how automatically developed black-box models arrive at decisions, promoting trust, validation mechanisms, compliance with Industry 4.0 regulatory frameworks, and conduct during use. Furthermore, LIME can generate interpretable surrogate models, assisting in detecting bias, errors, alerts, and other anomalies, which help to improve industrial decisions and outcomes.

SHAPley Additive explanations, known as SHAP, is a unified value-based model-agnostic methodology used to clarify the impact result of machine learning with an importance score assigned to each feature. SHAP values are derived from basic cooperative game models even in relation to the Shapley value and for example, it indicates the degree of the contribution of each feature towards the prediction. Concerning the application of the concept of Industry 4. 0, Besides, it can be suggested that the application of SHAP methodology is most useful when it comes to providing interpretative and fully comprehensible explanations of exceedingly complex machine learning decision-making. It enables each factor to be evaluated with regard to model predictions, contributing to the stakeholders' understanding of the factors affecting the AI system's decisions in an industrial context. This feature plays a very central role in Industry 4. 0 process because it assures that all the processes involving AI are safe, secure, and conform to the required standards. Further, SHAP values can also offer avenues such as classification of parameters indicative of potential features of significance, detection of outliers, and potential improvement of industrial procedures by pointing out factors that are highly influential in determining the model's results. Such information facilitates the enhancement of decision-making, system functioning, as well as application of AI-based system solutions in multiple spheres of Industry 4. 0, which helped in incrementing the effectiveness level of industrial operations.

4.2 Model-Specific Methods

XAI methods for a specific model of a deep learning system are unique approaches for elucidating the model built specifically depending on the algorithm of the given system and its structure. They are the possibilities to fulfill all the requirement regarding the explainability by scrutinizing the inputs of the model and the rationale of the determined output in order to learn. Several model-specific techniques which fall under the umbrella of XAI are especially useful in the attainment of the desired increase in transparency and interpretability when it comes to the employment of AI within the 4. 0 era. They are essential in understanding how complex industrial AI models applied in manufacturing, predictive maintenance, and quality control, make decisions. The outcomes generated by these models can be explained through explainable Artificial Intelligence (XAI) techniques, which help obtain interpretability behind the black-box models and enhance the accountability and credibility of AI models on the stakeholders and public. In XAI, saliency maps for Convolutional Neural Networks and layer-wise decomposition for Recurrent Neural Networks are two explainability techniques through which the undertaker understands which parts of the input image are most influential and what parts significantly impact distinguishing between classes. As shown in Fig. 4, saliency maps of an image indicate the parts of an input image that have impacted the classification process. On the other hand, layer-wise decomposition for RNNs refers to decomposing the sequential processing in RNNs into its components—for example, hidden states or memory cells—to understand how the information gets into the network over time. When these two techniques are applied to CNNs and RNNs and their layers are compared, the user can analyze how the layers have interacted to develop the final image and how the prediction quality has changed [11]. Therefore, learning more about how the AI model processes the sequential data and gives predictions is possible. In the Industry 4.0 context, saliency maps for CNNs and layer-wise decomposition for RNNs allow an understanding of the AI models used in manufacturing, predictive maintenance, supply chain, and quality control work.

In summary, XAI approaches are essential in enabling industry professionals and other stakeholders to understand the operations in complicated neural networks, guarantee the credibility of AI system decisions, and build and maintain transparency and accountability within automatic processes. Therefore, when mere organizations implement these explainability methods, they can consequently get reliable and trustworthy AI technologies that are valuable in their various operations and suitable for Industry 4.0, leading to improved efficiency and decision-making mechanisms.

5 Considerations for Xai Implementation in Industry 4.0

Industry 4.0 aims at higher levels of output through application of new technologies such as artificial intelligence, cyber-physical systems, IoT and cloud computing among others. In today's Industry, AI plays a prime role in automation with intelligent controllers that can monitor the operation parameters, interpret the decision recommended by AI, diagnose the malfunction of the equipment, and analyze its performance. The decision made by the proposed AI systems should be 'explainable' to experts to deploy and integrate intelligent systems to Industry 4.0. The XAI will be facilitated by developing an algorithm that gives results that make the human to understand decisions proposed by AI-modelled systems.

5.1 Trade-Off Between Explainability and Accuracy

To extract the importance of AI in Industry 4.0, it is essential to increase trust in predictions, which can be achieved by improving the accuracy and developing complex models which are usually considered black boxes. Explainable methods should be developed to understand the prediction and its logic by finding the important characteristics features from the given data set. These methods also facilitate understanding model logic with more clarity [12].

AI/ML depends on present and future data glaringly to ensure the reliability and accuracy. The main challenge in the development of processor based solutions is the processing of huge data volume in centralized manner which poses severe privacy concerns. Industrial AI systems should possess explainability in data collection and its training process and be free from bias. The datasets used for training should be diverse to avoid algorithms being biased to certain types of data to get predictions with utmost accuracy [13].

The accuracy of detecting defects in the manufacturing process and removal the corresponding parts may reach 100%, which may not be possible by a human operator. Thus, AI-trained model will remove differences between the users inside and outside the industry, which increases the robustness of identifying malfunction during the manufacturing process. Thus, the manufacturer can outsource the creation of the AI model to the company that will receive a large set of images of appropriate and inappropriate parts. Sometimes, even the manufacturer may not so embarrassed to display images of wrong part. Before sharing the photos with an outside party one should make sure they encrypt it so that the privacy setting may not allow the party to manipulate the photos. The AI model would be trained on the encrypted images, and the same would be done during the inference so that the model is not fed with out-of-distribution data. The explainable AI algorithm comprises a separate program to give the explanation to the human about the prediction it proposes so that any error occurs with its decision due to bias or any error in its data set.

5.2 Computational Cost of XAI Techniques

The cost that the industry has to spend for developing algorithms for XAI techniques will vary from one application to another. It is influenced by the complexity of the model to be explained, its data set size, resources available and finally, its method of interpretability. Simple Techniques like LIME (Local Interpretable Model-agnostic Explanations), SHAP (Shapley additive explanations), and tree-based methods (e.g. decision trees) are efficient particularly for small datasets and simple models. The computational cost may increase for large datasets and complex models.

Saliency map generation requires accurate input features for model prediction using deep learning methods. It needs intensive computational Techniques to obtain a precise output model. Grad-CAM (Gradient-weighted Class Activation Mapping) and guided backpropagation are techniques used to reduce computational costs for saliency map generation. Prototype-based explanations, such as exemplar-based methods, are the prototypes-based explanation model which derives the explanation from the data and have their own computational cost. The cost depends on the selected prototype count and its feature space complexity. Developing rule models that are interpretable by humans from complex models and large data sets requires extra computational expense. In practice, the computational cost of XAI techniques gets increased, and its value vary based on factors like the method of computation, the feature size of the data and model, and the required level of interpretability. Care should be taken that there is a balance between computational cost and accuracy of explanations for interpretability.

5.3 Integration of Domain Knowledge and Collaboration

Domain knowledge and related information in its association with the process are the key components of Explainable Artificial Intelligence (XAI). It helps in improving the credibility of prediction and its reason by the AI systems. Experts will have lots of trend-specific knowledge about the features, relationships and restrictions of a certain problem sphere. They can define the related characteristics of the problem domain. Domain experts will also encode the rule base and verify the output explanations provided by the XAI systems. Data Scientists and AI researchers together with the domain experts specify the requirements and goals of the XAI system. They can also give back the form of output regarding the performance of a model and the reasons behind it, which eventually improves the performance of the XAI system. Communication and documentation related to the process of XAI should be clear to trust and explain the result of XAI. Incorporating domain knowledge and fostering the interaction between disciplines as well as stakeholders make the XAI systems provide more accurate, understandable and trustworthy explanations to enhance decision making and acceptance.

6 Applications of Xai in Industry

Indeed, it can be posited that Explainable Artificial Intelligence (XAI) has a prominent place in Industry 4. 0 since it is critical to make features of such techniques transparent and interpretable when implementing them in complex manufacturing systems. Abilities of employing XAI in Industry 4. These values are depicted in Fig. 6.

6.1 Predictive Maintenance

By integrating explanatory AI (XAI), it is possible to change traditional practices of preventive and predictive maintenance totally. The introduction of XAI into Predictive Maintenance systems provide businesses with the ability to predict when and what maintenance is needed based on trends and or anomalies in data. Also, these technologies enable organisations to explain in an open manner the maintenance advice afforded by the AI models. They like transparency because where safety, reliability and trust are concerned this type of practice make sense since it is easier for such stakeholders to comprehend why forecasts were made about what was looming concerning the upkeep needs of their properties, therefore enhancing the decision making process for both them and enhancing the credibility of maintenance teams and such systems. The other advantage of applying XAI in Predictive Maintenance is due to its potential to help the staff members that are involved with carrying out maintenance operations to be trained on what the parts that lead to failures are, the root causes behind probable problems, and how the time not in service could be reduced while at the same time enhancing the performance of the assets. Therefore, integrating XAI together with another key concept of Industry 4.0 such as Predictive Maintenance might lead to general improvement of the operations, reduced costs in terms of service, and constant functionality of vital assets. 0 environments [14].

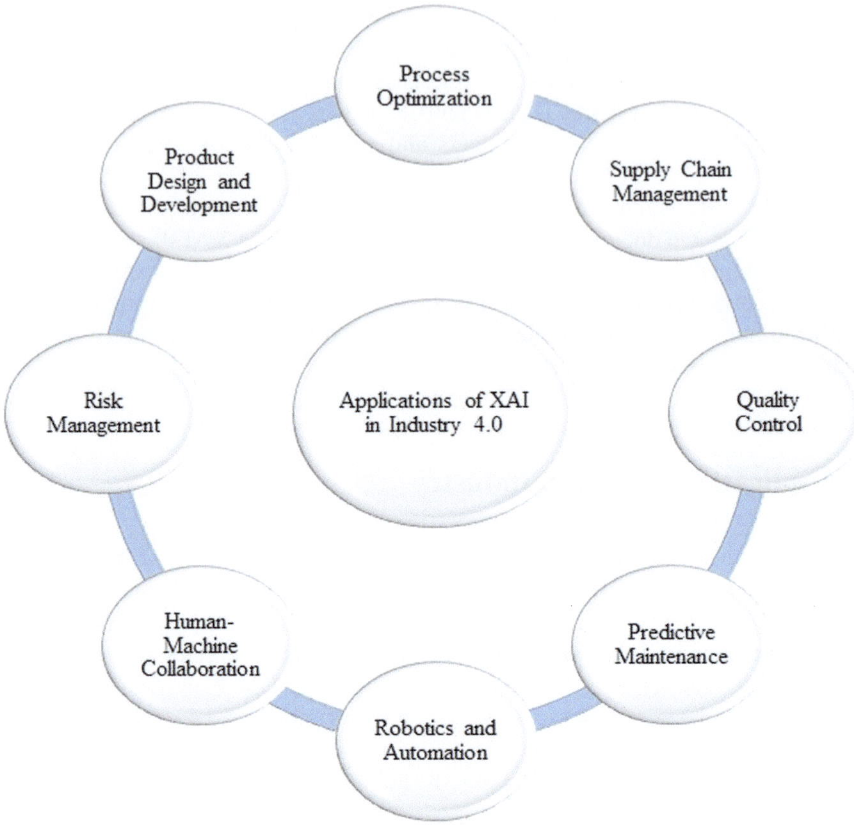

Fig. 6 Applications of XAI in industry 4.0

6.2 Quality Control

If Quality Control complicated the way organizations manage the quality of a product during manufacturing, then it can be said that when Explainable Artificial Intelligence (XAI) is applied to Quality Control, it alters this manner significantly. Apart from adopting XAI in the quality assurance systems of companies, organizations can also name failures and deviations in products, and then explain how and why such decisions were made by the AI systems in the quality assessment processes. By applying XAI, it becomes easier for the stakeholders to be informed on the features that impact on the quality control aspects like the aspects that are considered by the AI model for identifying the defects or the anomalies [15]. It enhances the belief when assessing these standards and provides the human inspectors to collaborate better with Artificial Intelligence systems. Furthermore, through RCA, production system enhancement, and efficient use of big data, enhanced overall product quality within industry 4. It is pointed out that: 0 settings, among others, can be realized by applying XAI in Quality Control frameworks. In Industry 4. In 0 settings, XAI offers understandable prepositions concerning the QC decision-making process that helps the establishments to attain more homogeneous products with minimal defects to cater to customers' needs that requires high-quality products [16].

6.3 Supply Chain Management

Supply Chain Management when combined with Explainable Artificial Intelligence elevates the visibility and operability of an organisation's supply chain processes to a level not previously possible. When applied to supply chain operations, there are benefits in improving inventory management, demand forecasting and logistics planning while at the same time giving an explanation of the actions taken by the respective AI throughout the supply chain. XAI can help stakeholders have a means of understanding why particular recommendation on the supply chain, be it inventory replenishment, transportation route, or supplier, need to be developed and executed. This makes it easier for organizations to put their trust on the outcomes of supply chain decisions made by the artificial intelligence mainly because of the insights and recommendations which can be explained easily. Furthermore, XAI in SCM can assist to potential risks and the decrease of disruptions, enhance the supply chain redundancy by means of explaining risk analyses and backup strategies. Enabling clear understanding of decision-making in complexity of supply chain optimisation, XAI can assist members of the organisation community to cut supply chain costs and improve overall customer satisfaction with the supply chain management practices, when Industry 4. 0.

6.4 Process Optimization

Combining Explainable Artificial Intelligence (XAI) and Process Optimization alters the expectation of how organizations improve efficiency, effectiveness, and performances of various activities. XAI's integration into process optimization initiatives can help establishments determine the areas that need improvement in terms of resource use or the areas that could be eliminated because, in one way or the other, they act as hindrances. In addition to this, firms can also provide high level justifications of why the certain optimization decisions made by the algorithms should be accepted. By utilizing XAI, the stakeholders get insight into all the variables involved in the AI engine and the limitations and trade-offs the engine makes during the identification of process improvement recommendations. This openness ensures people have faith in the outcomes of optimization hence the human operators can effectively combine with data-fostered enhancements that AI systems apply successfully. Besides, Explainable AI not only gives an understanding of how the process is going, but also indicates why specific actions were made during the decision-making processes and reveal potentially hidden trends that may open up new improvements for efficiency enhancement in firms' processes. Thus, such like businesses can gain on operational excellence through a combination of innovation founded in agility combined with transparency precipitated by the use of XAI for spiriting projects relating to process optimization in dynamic business contexts [11].

6.5 Robotics and Automation

Robotics and Automation, that recently gained attention as the powerful tool for Explainable Artificial Intelligence (XAI) can be regarded as the new generation of intelligent machines and self-managed systems. Some other things that robots can do if we teach XAI along with Robotics and Automation is as follows Overcoming of complicated decision-making, carrying out tasks with aptitude as well as the capability to perform a task based on the environment. To elaborate this further, it must also be stressed that any function executed by an AI driven system based on robotics can also be explained in the open. The justification of certain path planning decisions made by robots or any other decisions made by these systems while performing tasks such as object recognition task prioritization, among others, will be understandable thanks to XAI in Robotics and Automation, which will unveil algorithms used or data fed into these programs when performing tasks or the process through which particular decisions were made from such inputs. It is very crucial in such knowledge sharing because when people are collaborating with the modern engineering technologies such as the artificial intelligence used in various operations, then they require such information a lot. It makes them sure about the functions of those gadgets since they understand why some options were selected over others: in addition to trusting these devices, they also begin to admire the features that are owned by them. By mapping out all possible algorithms in the robotic work involving various units such as data processing units that have been employed then all areas where an error could have been made are revealed to the stakeholders building trust between human beings who use equipment containing AI systems and human beings doing similar tasks will enhance efficiency in the automated works. Moreover, for safety reliability performance, interpretability additional information XAI in Robotics and Automation provides organizations with robot behavior anomaly identification error, correction, and prevention mechanisms.Thus, by applying XAI to improve robotics and automation across various fields, companies can achieve increased efficiency and scope of innovations in various industries, which will contribute to the mass implementation of intelligent robotic systems as one of the key subjects of the Industry 4. 0 [17].

In robotics, XAI is vital in ensuring that robots can explain the actions they are taking or the decisions they are making to human handlers or other AI systems. This is important in creating trust that humans have with robots especially when the robots deal with human workers in detail. This makes robots capable of explaining their operations and conditions to the operators through XAI models and thus enabling operators to predict the robot's behavior. Having openness about how a decision was made can improve the safety of an autonomous system by showing where there is an error, a certain kind of bias, or uncertainty. Incorporating explanations into such a system, as done by XAI models, allows the operators to step in and alter the robots' actions in case of mistakes or to avoid mishaps. Such a level of interpretability is important in addressing the social justice issue of responsible use of autonomous systems in various sectors including production, health, transport and farming [18].

6.6 Human–Machine Collaboration

Introducing XAI in the human and artificial intelligence interaction alters the dynamics of how people relate to intelligent systems and make a symbiotic relationship. Thus, organizations can enhance decision-making, problem-solving, and ideas and innovation generation when incorporating XAI in situations where people interact with machines; moreover, it helps them explain to others the basis of the recommendations that AI systems provide during collaborative activities. Decisions or suggestions made by an AI system must be justified by XAI as to why it recommended or took specific actions based on the insights derived from such areas, and thus making it possible for humans to have confidence in using the full potential of smart systems. This openness fosters the bringing of humans and machines into a common ground where it becomes rather easier to cooperate, to share information, as well as learn from each other. Furthermore, it can be stated that the existing task allocation within organizations can be improved and at the same time, observing the performance along the feedback channels that have been established can be benefited if there are understandable indications about where each person or device contributed most when they were working hand in hand to accomplish organizational goals—hence, using Interpretable Insights from both Human and Machines: The Human-Machine Partners for Enhancing Task Allocation Performance Monitoring Feedback Systems [19]

6.7 Risk Management

Risk Management combined with Explainable Artificial Intelligence (XAI) provides institutions with a powerful means of optimising decisions, addressing unknowns, and optimising the general risk analysis activity. Risk treatment practices are enhanced when companies use XAI in risk management practices to easily notice, assess and prioritize risk factors while it also enables an organization to explain how, based on the AI algorithm, risk management strategies promote risk mitigation. With regards to XAI, all stakeholders have to understand why the system arrived at such risk assessments, i. e., XAI of data inputs that were received as well as of the assumption made by the model and of the insights about the risk indicators that were used by the system to arrive at potential threats. Such openness helps in the establishment of credibility on managing of hazards hence provides the decision makers with understandable information on which they can reason as they manage the hazards. Further, it also paints out the patterns or trends in Risk Management which are otherwise highly concealed in and around the scenario simulations in which organizations' exposure levels in risks are transformed with variations of some variable set during the analysis phase but this is stated in terms of what could be experienced if such incidents occurred that would lead to emergence or identification of fresh risks/dangers.

6.8 Product Design and Development

Product Design and Development with Explainable AI brings a new way of inventing, creating, and introducing new products into the market through integrating Product Design and Development into Explainable Artificial Intelligence technologies. It is established that with the help of Explainable Artificial Intelligence (XAI) integrated in the product design and development process of various enterprises, one is capable of enhancing ideation, efficiency, and satisfaction at the end-user level. This is done by providing straightforward rationales for chosen designs and plausible alterations indicated by, and based on, AI algorithms. XAI helps the designers and engineers to understand the thought process behind designing features and prioritizing them along with interpreting the user feedback by detailing the simple data, trends and analytics that the AI system studies during the course of designing. This much openness fosters cross-functional teamwork and increases the sharing of information and ideas between designers of multiple functionalities, leading to the development of more innovative products that meet the user's needs.

Explainable Artificial Intelligence (XAI) application in Product Design and Development can be helpful to firms in improving the process of design iteration, prototype, and verifying its value in the market. This is done through inclusive and easy to comprehend information that an organization receives concerning its users and the market, designs and trends and competitors respectively. Using the XAI approach in the product design and development, companies can reduce the flow of getting new product designs and features, design risks, and enhance the performance of the products. This in turn results in achieving competitive advantage and enhanced customer satisfaction under the prevailing conditions of high and growing market volatility. When incorporated into the PDLC, XAI enables organizations to design and develop products that are novel and satisfactory to users. In easing network complexity, XAI helps in product design since it explains the outcome of AI-based decisions made. XAI models can be utilised by designers and developers where the logic behind the output of the AI algorithms has to be explained to the creation of recommendations or forecasts. This makes it easy for the designers to point out the possible bias, mistake, or weakness of the AI models, making the products created much better and reliable. Furthermore, XAI can assist in the decisions regarding the appropriate product features or functionalities by considering the users' feedback. From the analysis of the frequencies of interactions with the product, XAI models can help the designers to better understand user behaviour, their requests or issues, and thus, design the product that will be more relevant to the users' needs. Thus, based on the described iterative design process and XAI facilities, it is possible to develop products that are more comprehensible, more user-oriented, and more engaging for consumers.

7 Applying Explainable AI (Xai) Techniques in Industry 4.0

The systematic approach used for the implementation of Explainable AI in Industry 4.0 is to first look at the problem based on an industry and second is to collect and preprocess data from sensors and IoT devices. Discrete patterns are built and trained with the data, which can be anything from decision trees or linear regression models. SHAP or LIME technique is used to make an explanation for a particular prediction of the model applied. Several local and global explanations are shown using SHAP summary plots. Claring representations using some of the form of visualizations such as SHAP summary plots. The final step involves either validation by other stakeholders within the particular domain or interpretation by the expert in a feedback loop to enhance stringency [20]. The following step by step method to implement explainable AI (XAI) techniques in Industry 4.0 was described in Fig. 7.

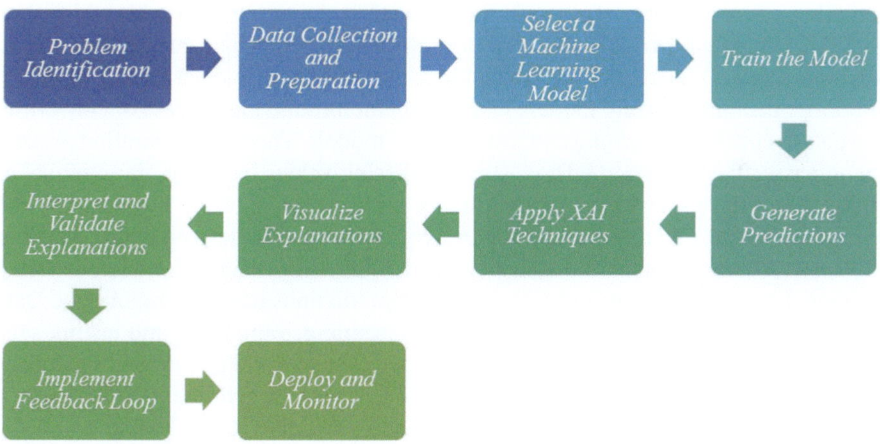

Fig. 7 Applying explainable AI (XAI) techniques in industry 4.0

7.1 Problem Identification

As part of the first step in applying Explainable AI techniques in Industry 4.0, problem identification is very crucial. A problem generally refers to the identification of a specific challenge or use case in an industrial setting where machine learning models are used to guide decision-making or predictive analytics. This requires deep knowledge of the operational environment and key goals of an organization. A well-formed problem statement directs stakeholders toward the goals and expectations of the XAI implementation. A well-formed problem statement gives a basis with which to decide upon sources of data, the machine learning models, and techniques in XAI to fit the purpose of the industrial application. Well-formed problem statements also give means by which success criteria might be defined and the effectiveness of the XAI in increasing transparency, interpretability, and trust in the machine learning models deployed in Industry 4.0 scenarios might be measured.

7.2 Data Collection and Preparation

The level of success attained in producing machine learning models for Industry 4. Starting from data collection, preparation, and analysis, zero applications are made. First of all, data is acquired and it is sensors, IoT devices, and operation systems that collect information necessary for training the model and to test it. Astonishingly, data quality and contents and correlation are one of the crucial ingredients among all. Data gathering concerns themselves with ETL, which is the process through which data goes through transform, cleaning, and loading so as to become usable by the selected ML algorithms. Data preparation also then comes into force in determining how well or poorly machine learning models would turn out through the process of converting raw data for use in the learning and testing of the models. The data preparation process encompasses the following efforts, such as missing values handling, the removal of outliers, and the preservation of the quality and reliability of the information that the model would use in training. Feature engineering includes selecting, creating, and transforming features that are relevant and informative for the model, enabling the model to learn patterns and make accurate predictions based on input data. Data preparation involves breaking the dataset into training, validation, and testing sets, whereby there is an estimate of the model's performance and generalization ability. On appropriate division of the data, the organization will give the model a chance to be trained on one of the subsets, a chance to validate its performance on the other sub-set, and a test on an independent, unseen dataset. This process would help prevent overfitting to assess the ability of the model to generalize to new data and optimize its performance for real-world applications. Data preparation is not a supplementary stage in industry 4.0, but instead a very important step toward model quality, reliability, and effectiveness; it puts in place accurate predictions, actionable insights, and informed decision-making based on AI-driven solutions.

7.3 Select a Machine Learning Model

Choosing a machine learning model is of great importance in the development of AI solutions for Industry 4.0 applications, as data-driven insights are the strong foundation on which many complex problems are solved and their implementation in practice. A critical aspect of this is the decision, which the organization must make in how to choose between different algorithms: linear regression, decision trees, support vector machines, neural networks, or ensemble methods, so that a model is chosen that best suits the prediction task, specific data characteristics, and performance requirements. All these algorithms are peculiar in themselves, and it is therefore essential for an organization to consider interpretability, scalability, complexity, and accuracy when choosing a machine learning model. The choice of a machine learning model depends upon the nature of prediction, the type of data, and the desired outcome to be obtained. Organizations must decide whether prediction has to be made through regression, a prediction of continuous values; classification, where the outcome is discretized, usually via a classification tree or similar algorithms; or clustering, where unsupervised learning algorithms are used, such as k-means. There should also be scalability, computational performance, and ability to process large volumes of data in time-sensitive manners. Keeping all these factors in view, it is then decided by the organization which model suits it best, which fulfills the needs and objectives of the organization in Industry 4.0 environments.

Therefore, the organizations in the process of choosing a machine learning model for their predictive analytics have to seek for a balance between a higher model sophistication and model interpretability. Even such a simple model as linear regression gives reasonable prospects for the prediction but provides very low interpretability, and it could be challenging to elicit through how a model of higher complexity such as deep learning neural nets for example, deriving its ability to predict. On the opposite end, we have models with lower interpretability index such as linear regression, decision trees and so on, though they yield lower accuracy. From this, an organization may combine the relative complexity of the model and the interpretability of such a model for stronger prediction capability and additional understanding of features of the data for relevant decisions within the Industry 4.0 context. The primary processes of creating AI solutions for Industry 4.0 include identification of the machine learning model. That is because data analysis can be useful for increasing the efficiency of processes and better decision-making. Concerning the choice of the model, there are several principles, such as the critical evaluation of different machine learning algorithms, the nature of the prediction task, and a ration between model complexity and interpretability that guide the organization in selecting the best model based on its goals and objectives. Selecting a specific machine learning model enables the organization to leverage on various AI technologies, rationalize and enhance on processes, and transform in the rate at which Industry 4.0 is advancing in order to realize sustainable, competitive future growth.

7.4 Train the Model

An important step in the development of Industry 4.0-driven machine learning-driven machine learning applications is training of the machine learning model. Such models learn from history to identify patterns, establish relationships, and trends, enabling the model to take the right prediction or classification of new data. Algorithms like linear regression, decision trees, support vector machines, or neural networks are used in organizations where the model is trained using labeled datasets, so input features are assumed to map to the target output. The organization will iteratively tune the model's parameters and optimize performance metrics, so that the model is more predictive and adaptive in relation to real-world scenarios. Sets of training will be divided into training and validation sets for capturing and avoiding overfitting while at the same time enhancing the generalization capabilities of the model. Techniques such as cross-validation, hyperparameter tuning, and regularization are used to tune the model parameters, making them more precise and robust to unseen data. The underlying patterns in the data are captured through the training process, learn from previous experience, and make the right prediction on new observations. This process is normally iterative and usually calls for monitoring, evaluation, and refinement for the right amount of performance and reliability in modeling to generate the right prediction.

The most important steps to pay heed to during data preprocessing, feature engineering, and model selection include organizational emphasis on these aspects. As an equal measure, preprocessing of the data is achieved through handling missing values, scaling features, and encoding categorical variables in order to assure quality and consistency of data. It includes the selection of relevant features and creating new variables and their transformation to enhance power to predict and generalize. The entirety of feature engineering includes the selection of model: in other words, the best suited algorithm, architecture, or ensemble method that best fits the predictive task and the datal characteristics, coupled with the requirements for performance. This means that preparation of data and the right model, will enhance a training process in order to enhance the rate of accuracy and success of a model in relation to the industrial application of the 4.0. That is, model training is one of the crucial activities through which organizations can leverage the power of AI in predictive analytics, optimization, and decision making in the context of industry 4.0 settings. However, proper investment not only in the quality but also in robust process of training results in the construction of exceptionally efficient, accurate, reliable, and most of all scalable models of machine learning that function towards the core objectives of data-driven organizations, promoting evidence Training of the machine learning models strategically allows for organizations to harness the potential that comes with them and support the AI solutions, improve industrial processes and attain long-term successes within the complex digital environment of Industry 4.0.

7.5 Generate Predictions

Generating predictions is one of the significant steps in using machine learning models in Industry 4.0. The information systems generate predictions based on the data-driven insight in support of decision-making and operational efficiency. Machine learning models predict future outcomes based on historical data, patterns, and relationships. This helps one predict trends, optimize processes, and avoid risks. An organization will be able to get predictions on any given parameter, for instance, equipment performance or quality control metrics, at its disposal with new data fed into the trained model. Such predictions would be useful as input for the strategic planning, resource allocation, and performance optimization process in Industry 4.0 environments. Predictions involving running actual data, either real-time or in batches, through a trained machine learning model and generating forecasts or classifications for input features. Different types of machine learning models could be used by an organization depending on the nature of the prediction task and the simplicity or complexity of the data: regression, classification, clustering, and deep learning. The predictive power of the model allows an organization insights into future trends, anomalies, or potential problems, so proactive decisions, risk management, and performance optimization in industrial processes can be conducted.

These will supplement the benefits and values that applied AI-driven solutions can deliver within Industry 4.0 by affecting the precision and credibility of the forecasts offered by the model. The actual quantitative evaluation of the model with respect to its ability to make predictions or performance indicators such as accuracy, precision, recall or F1 score is still missing in the organization. The monitoring of the model's performance will enable the organizations to understand their shortcomings, fine-tune the model parameters, and optimize its predictive accuracy for its real-world applications. Industry 4.0 operations are facilitated as organizations can make data-based decisions, optimize their processes, and drive a continuous improvement process with reliable predictions. It thus follows that the validation of making the right and accurate predictions using the machine learning model is consequential in tapping the potential of AI technologies in Industry 4.0. Companies can take advantage of predictive analytics; they will be able to forecast future outcomes and find potential opportunities for optimization, and thus avoid risks in industrial processes. The strategic use of machine learning models to generate the predictions will enable the organizations to make informed decisions and improve operational efficiency, thus fostering innovation in the era of Industry 4.0, encourages sustainable growth, and gives competitive advantage in the digital age.

7.6 Apply XAI Techniques

It is of utmost importance that apply explainable AI techniques to increase transparency and interpretability of machine learning models in Industry 4.0. XAI makes it possible for stakeholders to understand how AI models make predictions, what drives the results, and the line of reasoning in the making of any prediction SHAP, LIME, or feature importance analysis application can make it possible for an organization to generate explanations that reveal the guts of a complex machine learning model. Those justifications will help stakeholders, who are often not ML experts, to trust the model's outputs, validate the predictions made, and make decisions accordingly based on the evidence brought by the model. XAI techniques fill the gap between black-box machine learning models and human stakeholders by intuitive and interpretable explanations of model predictions. For example, visualizations, such as SHAP summary plots or feature importance charts, would help make rationality transparent about the decisions made by the model and the importance of all features, also conveying the importance of the different variables that are used to predict the outcome. These illustrative explanations may help the stakeholders to understand the outputs of the model, justify its accuracy, and thus develop confidence in AI-based solutions deployed in the Industry 4.0 context. Deploying means of XAI helps the organizations in enhancing the presence of transparency, accountability, and trust in their machine learning models and empower the stakeholders to take data-driven decisions with trust.

Moreover, the applications of XAI allow the organizations to come up with the compliance with regulatory requirements, ethical standards, and industry best practices in transparency and accountability to AI. Through giving interpretable explanations about the model's predictions, the organizations will be able to demonstrate the fairness, reliability, and bias mitigation strategies implemented in their AI systems. The application of the XAI helps multi-disciplinary teams to work together efficiently; domain experts, data scientists, and decision-makers can interpret the model outputs and use AI-driven insights for optimizing processes, improving efficiency, and driving innovation into Industry 4.0 applications. The final issue is related to trust, understanding, and acceptance of the machine learning models within Industry 4.0 environments. Utilizing the XAI techniques that make interpretable explanations, the organizations are also able to provide transparency, reliability, and usability of their AI solutions. Stakeholders can also make more informed decisions and create value, which will enable the parties to harness the full potential of AI technologies to support them in the decision-making process. A good strategic application of XAI techniques enables organizations to derive advantages from AI-derived insights, driving innovation, and ensuring sustainable growth in the era of Industry 4.0.

7.7 Visualize Explanations

For Industry 4.0, it is important to explain why the particular machine learning models are chosen and how the information can be understood by others. For instance, instead of using SHAP Plots, using force plots or summary plots will be useful as they reveal the exact working of the model and what is in the output can therefore be understood by the stakeholders in terms of the factors which they can understand. For the case of visualization, this will help stakeholders to be aware of the correlation between the input characteristics and output predictions. From this, whether there is a specific input feature/ output function for a specific stakeholder, pattern, fashion, or some outliers in the data set can easily be identified. Thus the stakeholders will be able to see how the model arrived to the decision they made. Visualizing explanations allows machine learning models to be made more accessible to organizations, and, therefore, makes stakeholders able to make informed decisions based on the insights derived from the visualizations. Explanations serve as a visual aid for organizations to explain how the model explains the prediction and show the influence of the most important features, and how changing the variables influences the results. Visualizations are a very strong tool for presenting complex information in an easily digestible way. For example, the understanding of the most valuable information will be performed much faster and effectively by the stakeholders. Thus, using such elements will enhance the responsibility and interpretability of machine learning models to other stakeholders, which ultimately will boost the collaboration of multidisciplinary teams and timely decisions in Industry 4. 0 applications. Besides, it assists in establishing the trust and confidence by the stakeholders in the machine learning models by providing them the understandable and interpretable output of the decision-making of the model. Several visualizations enable them to comprehend the algorithm's functioning, credibility of outcomes predicted by the model, presence of bias/error, and the dependability of explanations obtained. By visualizing the explanation of the results, organizations can improve the audibility of their machine learning model whilst the stakeholder, depending on the type of model, will be able to verify if the machine learning model is actually in the right course, verify whether the prediction is correct or otherwise and ensure that the model conforms with the domain knowledge as well as expectations. This transparency and interpretability are the success factors of confidence in the results generated by the models to facilitate the transformation through the adoption of the Industry 4. 0 scenario.

7.8 Interpret and Validate Explanations

The other two fundamental processes that assist in the development of the machines learning based models in the Industry 4. Thus, explanation and validation of these explanations are the primary meanings for the term and its derivatives 0 domain transparent, reliable, and effective. There is a necessity to into explanations and comprehend them to be able to analyze and understand the causes of the model's output. Thus, it is possible to receive information on the rationale behind decisions that are being made. The fact that it is possible to understand how single variables are connected to the outcome helps to reveal patterns, trends, and relations in a dataset for decision-making purposes. This may enable stakeholders to search for possible mistakes and/or bias in the model and verify crispness. The validation of the explanations as well as the modeling is always bound to ensure the last variability of the output of a model as per the defined criteria and benchmarks. The legitimacy of the reasons for the results is as significant as the interpretations to ground up the compliance, coherence, and conformity of the developed outputs with the knowledge of the domain and expectations to achieve that. Validation enables the identification of potential mistakes and / or prejudices in the clozes and to compare for reliability. The variability of the model's output relative to the criteria and milestones set also helped to maintain an additional level of dispersion. Thus, the Industry 4.0 application is more confident with the machine learning model to operate efficiently for the intended results and goals within the outlined framework. 0 applications. By applying the discussed interpretations and validation, organisations will continue developing their models, state potential problems, and adjust their decision-making with appropriate and validated visions. Interpreting the forecast and validating it in this cycle is a very favorable climate that encourages feedback practice, research based on quantitative data, and optimization of the use of the machine learning model in Industry 4. 0 environments.

7.9 Implement Feedback Loop

Applying a feedback loop helps even at this basic level to fine-tune and optimise the machine learning algorithms within Industry 4.0. Organization normally sets feedback mechanism and get to know about the performance and results of models used in environments. The kind of feedback loop thus kept an eye on this prediction and assesses how effective and precise the model is in accepting and identifying flaws and gaps. The gathering of feedback from users, domain experts, and systems outputs provides the organization a continuous improvement of the performance of the machine learning model and most likely facilitates issues that might arise and improve reliability and efficiency. Such a feedback loop is a non-stop, real-time learning and adaptation mechanism that allows an organization to fine-tune its machine learning model using real-time data and user experiences and, consequently, constantly tune and optimize Industry 4.0 applications. The mechanism of a feedback loop in an organization also fosters a culture of continuous improvement and innovation by incorporating feedback mechanisms in the process of deploying the machine learning models. Organizations proactively address challenges, respond to changing requirements, and adjust to evolving business needs by incorporating feedback mechanisms into the model deployment process. This feedback loop identifies areas of concern but also offers organizations with an effective means of making data-driven decisions on which improvements to focus and drives strategic initiatives based on actionable insights derived from user feedback and model performance evaluations. The cyclical characteristic of the matrices gives a way to gauge the relevance, efficiency, and applicability of an ML model to growing requirements throughout Industry 4. 0 environments and maintain growth and competitive advantage.

7.10 Deploy and Monitor

Model deployment and model monitoring are certainly crucial stages of the AI solution's life cycle within the Industry 4. 0 context. In evaluation, the trained model is used in production system where it is deployed for real time prediction. While using the deployment of the system, certain features like integration into the already existing structures should be seamless, flexibility when it comes to capacity to handle the varying volumes of work and stability in the diverse working conditions. Proper deployment puts organizations in a place to leverage the forecast capability of the model, that is, optimise process and ascertain the efficiency of the same, and in a broader sense obtain value from Industry 4. 0 applications. Production models should be supervised to ensure they continue to perform reliably and accurately even in the production environment. In addition, it is required to pay attention to changing behavior of the model and assessing change and shift of data as well as problems in the prediction continuously. This means that accuracy, precision as well as recall when adopted as measurements depicts how effective the model is, and where there is a probability of some form of error so that action can be taken in order to improve the model's performance not to be compromised in the future is taken by organizations. Supervision also allows the organisations to check that it has the right outputs, is in keeping with the regulation and to address new issues or shifts in the distribution of data. The deployment and the monitoring of the machine learning model are therefore systematic to guarantee the efficiency and the dependability of the model in Industry 4. 0. Organizations must, therefore, establish ways of monitoring, raising an alarm once deviation from regular behavior path is observed, providing feedback which can improve the model's performance. The monitoring tools and frameworks will help the organizations to identify problems at an early stage, fine tune parameters of models and improve their dependability and credibility. The deployment and monitoring of the models will reduce risks, maintain the accuracy, and enhance the organisation's AI processes in Industry 4. 0 applications.

8 Case Studies

8.1 Unveiling Operational Insights in Combined Cycle Gas Turbine (CCGT) Data

Combined Cycle Gas Turbines (CCGT) present an interesting and complex interplay of operating parameters affecting power generation efficiency. This paper attempts to illuminate the underlying operating dynamics of CCGT data, hoping to uncover the relationships between the environmental conditions and power output with advanced machine learning and Explainable AI techniques. This would facilitate the foundation of the model using the publicly available dataset with six years of operational measurements on a CCGT generator. In the prediction of the net hourly electrical energy output, the study uses operational data on ambient pressure, exhaust vacuum, ambient temperature, and relative humidity. Each of the models—linear regression, random forest, and XGBoost—is carefully preprocessed in the process of this study, an endeavor for a comparison of their performance in predicting power output. From the result analysis, it can be noticed that environmental conditions are highly correlated with the power output in the CCGT system, hence illustrating the importance of knowing how these correlations influence the working of the system. The fact that SHAP provides explanations for these predictive results stems from the machine learning algorithms used, hence allowing stakeholders to understand why such results are obtained. The disparity in performance, as shown from the metrics used, such as R2 score, MSE, and RMSE, is evidently high for the XGBoost Regression Algorithm to achieve accurate prediction of power output. The application of XAI tools and machine learning models in this case study brings value to elucidating the operational dynamics of CCGT systems. Through complex data analysis, this methodology would give stakeholders actionable insights to promote better decision-making, proactive maintenance strategies, and deep understanding of the interplay of environmental conditions with power generation in industrial settings [20].

8.2 Exploring Operational Insights in Boiler Feed Pump Gearbox Data

In mechanical systems, the gearbox is instrumental in controlling the relationship between three factors with rotational implications, for example, speed. A critical factor is its direct implication on the performance functionality of a machine. This article presents a case study for analyzing a data set from a boiler feed pump gearbox. Notably, it examines the nature of the relationships between control parameters, measured in stop-valve position values, and vibration measured in rms levels. Moreover, the use of XAI technology is intended for stakeholder outcomes on system performance dynamics; such outcomes are to be gathered through the following methodology. A data set is first obtained that contains parameters of operations in the gearbox, including stop-valve positions and rms-vibration. Through various preprocessing efforts, the retrieved data is then used to uncover the impact of stop-valve positions on vibration levels and its performance implications. Use machine learning models and XAI techniques to produce human-readable explanations of the predictive results. The study identifies the statistically significant relationship between stop-valve positions and rms-vibration levels, which clarifies the operational impact of the control parameters on the gearbox performance levels [20]. Based on the transformation of complex machine learning output to easily understandable text-based explanations, stakeholders obtain insights into the most important factors affecting the predictive results. Finally, stakeholders' insights will enable them to implement better-informed decisions and improve the operational performance by using the predictive model operators. In conclusion, the case study showcases the importance of transparent operational dynamics of boiler feed pump gearboxes through advanced machine learning techniques and XAI tools. Through data analysis which provides stakeholders with real-time and situational reports, the methodology increases transparency of operations, enables smart decision-making, and allows industrial facilities to operate on a preventive maintenance approach.

8.3 Predicting Thrust Bearing Wear in Feed Water Pumps

Anticipating thrust bearing wear is indeed critical to maintaining long-term operational effectiveness and preventing catastrophic failures. This case study leverages actual data collected via a monitoring and reveals data condition methodology to attempt to predict the wear of the feedwater pump's thrust bearing based on influencing parameters such as flow and head. Using machine learning models and Explainable AI approaches, each of them offers a measure of the predictability of the model that will be made available to all stakeholders. The methodology originated from the evaluation of a dataset that consisted of different flow values and head characteristics, which are vital in predicting thrust bearing wear. Following data preprocessing procedures, the dataset was converted into a supervised learning problem to create input-output sets suitable for the predictive model. The performances of the thrust bearing wear data set were validated using several machine learning models such as linear regression, Random Forest, and XGBoost, and an ensemble model that includes all of the methods was applied for the performance comparison. The assessment was performed to measure the effectiveness of the generated machine learning models to predict flow and head for the computation of thrust bearing wear. This provides useful insights to stakeholders by using XAI tools to explain the predictive outcomes. Using the Swamp dataset, it is evident that the model performances showed that in the current predictive task, linear regression performs the best as compared to other models. Specifically, it proves that if models are highly to give accurate predictions, interpretability is also essential for one to comprehend the operational dynamics. This case study demonstrates that it is possible to maximize the feedwater pumps's life span by predicting when thrust bearings are most likely to wear and relaxing the rest of the time. This methodology also enables stakeholders in operational decision-making, planning for maintenance schedules, and better understanding of implicated factors on equipment wear in industrial settings [20].

9 Conclusion

XAI is the answer to the deep learning's potential also in industry 4. But the 0 and beyond, for that matter, must not only be developed, but also unlocked. As a result, the techniques applied in XAI help people understand the confusing actions of these strong but nontranslatable models, which in return provides them an opportunity to attain the goal of AI's sensible advancement through knowledge and trust. Among the numerous benefits that this article demonstrates as resulting from explainability of AI systems, is that with the help of explainable AI, numerous models are actively managed by people, as well as innovation is encouraged, and at the same time, the latter is done in compliance with certain rules and requirements that are required for safe working conditions under the existing standards and legal frameworks set by various regulatory authorities such as for instance, OSHA and so on et c Different methods are explored here: LIME (Local Interpretable Model-Ageless Explanations), SHAP (Shapley Additive Explanations), saliency maps and layer-wise decomposition are some of the model specific Xai techniques employed across different industrial domains in this chapter.

For XAI to be successfully incorporated into Industry 4. 0, it has to have a broad view that compare transparency with accuracy, solve the problem of a massive volume of computational resources used by XAI methods, and enable collaboration between the developers of AI and specialists from other fields. By achieving this fusion, enterprises will be in a position to actually harness AI's capability to generate structural change while at the same time becoming responsible, fair, and people-focused processes. In these new frontiers of the Fourth Industrial Revolution, Explainable AI is that compass which charts the path to the world where technology and people cooperate more synergistically, augmenting each other's strengths as a means to promote knowledge, value creation, and progress for all.

References

1. Lasi, H., Fettke, P., Kemper, H.G., Feld, T., Hoffmann, M.: Industry 4.0. Bus. Inform. Syst. Eng. **6**(4), 239–242 (2014). https://doi.org/10.1007/s12599-014-0334-4
2. Qin, J., Liu, Y., Grosvenor, R.: A categorical framework of manufacturing for industry 4.0 and beyond. Proced. CIRP **81**, 625–632 (2020). https://doi.org/10.1016/j.procir.2016.04.038
3. Karpatne, A., Atluri, G., Faghmous, J.H., Steinbach, M., Banerjee, A., Ganguly, A., Kumar, V.: Theory-guided data science: a new paradigm for scientific discovery from data. IEEE Trans. Knowl. Data Eng. **29**(10), 2318–2331 (2017). https://doi.org/10.1109/TKDE.2017.2720168
4. Zhang, Y., Zhu, X.: Explainable AI framework for image analysis. (2018) arXiv preprint arXiv: 1806.07306
5. Lara-Benitez, P., Carranza-García, M., Ramos-Román, J.: Knowledge extraction from machine learning models: towards explainable AI systems in the manufacturing industry. Knowl. Based Syst. **226**, 107146 (2021)

6. Arrieta, A.B., Díaz-Rodríguez, N., Del Ser, J., Bennetot, A., Tabik, S., Barbado, A., Herrera, F.: Explainable artificial intelligence (XAI): concepts, taxonomies, opportunities and challenges toward responsible AI. Inform. Fusion **58**, 82–115 (2020). https://doi.org/10.1016/j.inffus.2019.12.012
7. Murdoch, W.J., Singh, C., Kumbier, K., Abbasi-Asl, R., Yu, B.: Definitions, methods, and applications in interpretable machine learning. Proc. Natl. Acad. Sci. **116**(44), 22071–22080 (2019)
8. Bhatt, U., Xiang, A., Sharma, S., Weller, A., Taly, A., Jia, Y., Eckersley, P.: Explainable machine learning in deployment. In: Proceedings of the 2020 Conference on Fairness, Accountability, and Transparency, pp. 648–657. (2020)
9. Adadi, A., Berrada, M.: Explainable AI for healthcare: from black box to interpretable machine learning models. In Towards Integrative Machine Learning and Knowledge Extraction (pp. 327–347). Springer, Cham. (2018)
10. Chamola, V., Hassija, V., Sulthana, A.R., Ghosh, D., Dhingra, D., Sikdar, B.: A review of trustworthy and explainable artificial intelligence (Xai). IEEE Access. (2023)
11. Messalas, A., Aridas, C., Kanellopoulos, Y.: Evaluating MASHAP as a faster alternative to LIME for model-agnostic machine learning interpretability. In: 2020 IEEE International Conference on Big Data (Big Data) (pp. 5777–5779). IEEE. (2020)
12. Ogrezeanu, I., Vizitiu, A., Ciusdel, C., Puiu, A., Coman, S., Boldis, C., Itu, A., Demeter, R., Moldoveanu, F., Suciu, C., Itu, L.: Privacy-preserving and explainable AI in industrial applications. Appl. Sci. **12**, 6395 (2022). https://doi.org/10.3390/app12136395
13. Abirami, T., Mapari, S., Jayadharshini, P., Kavipriya, M., Kavin, T., Kanagasubramaniyan, V.S.: A machine learning techniques for early autism spectrum disorder detection through comparative analysis of feature engineering and classification models, ICAICCIT-23, IEEE Delhi section, (2023)
14. Kamalam, G.K., Krishnasamy, L., Rajasekar, V., Fathima Kadhoon, M.: Comparative analysis of maize leaf disease detection using convolutional neural networks, In: 2023 IEEE International Conference on ICT in Business Industry & Government (ICTBIG), (2023)
15. Arumugam, J., Lalitha, K., Supreetha, S.M., Shrinithi, R.T., Tamilarasan, S.: Machine learning for detecting twitter bot. Fifth Int. Conf. Comput. Intell. Commun. Technol. (CCICT) **2022**, 278–282 (2022)
16. Singh, A., Dhanaraj, R.K., Sharma, A.K.: Personalized device authentication scheme using Q-learning-based decision-making with the aid of transfer fuzzy learning for IIoT devices in zero trust network (PDA-QLTFL). Comput. Elect. Eng. **118**, 109435 (2024)
17. Preuveneers, D., Ilie-Zudor, E.: The intelligent industry of the future: a survey on emerging trends, research challenges and opportunities in industry 4.0. J. Ambient Intell. Smart Environ. **9**(3), 287–298 (2017). https://doi.org/10.3233/AIS-170432b
18. Jagatheesaperumal, S.K., Pham, Q.V., Ruby, R., Yang, Z., Xu, C., Zhang, Z.: Explainable AI over the Internet of Things (IoT): Overview, state-of-the-art and future directions. IEEE Open J. Commun. Soc. **3**, 2106–2136 (2022)
19. Ghai, B., Liao, Q.V., Zhang, Y., Bellamy, R., Mueller, K.: Explainable active learning (XAL) toward AI explanations as interfaces for machine teachers. Proceed. ACM Human-Comput. Interact. **4**(CSCW3), 1–28 (2021)
20. Amin, O., Brown, B., Stephen, B., McArthur, S.: A case-study led investigation of explainable AI (XAI) to support deployment of prognostics in industry. In: Proceedings of the European Conference of the PHM Society 2022. pp. 9–20. (2022)

N. Sanjana Sanjana N is a highly skilled Data Science professional with an MTech from Kumaraguru College of Technology, where she maintained an exceptional CGPA of 9.68. She completed her B.E. in Information Technology from Bannari Amman Institute of Technology with a CGPA of 7.91. Her technical expertise spans Python, C, C++, and data visualization tools including Tableau and Power BI. She has demonstrated strong research capabilities through multiple publications in prestigious IEEE conferences, focusing on innovative applications of machine learning and image processing in agricultural and healthcare domains. Her notable projects include work on edge detection techniques for tumor detection and disease detection in crops using advanced algorithms. Her research contributions show a particular focus on computer vision and AI applications in real-world problem-solving, particularly in agricultural technology and medical imaging.

R. Immanual Immanual is an assistant professor with seven years of academic experience. He has a master's in Energy Engineering and has published several research papers. He has expertise in thermal systems, renewable energy, and engineering design. Immanual has filed 25 intellectual property applications, with six published and five granted. He has received multiple awards for his inventions at international exhibitions. He has secured external funding worth over INR 10 lakhs for various projects. As a passionate educator, Immanual has organized 12 workshops and trained over 300 students on emerging technologies. He established labs for engines and 3D printing to enable multidisciplinary projects. His students have won numerous national and international competitions under his mentorship. Immanual aims to foster entrepreneurship and innovation among students. He serves as an ASME leadership team member, promoting mechanical engineering education. With his academic and research experience, he aspires to inspire visionary leaders.

K. M. Kirthika Kirthika KM is an Assistant Professor at Sri Ramakrishna Institute of Technology, specializing in Data Mining. She holds both her M.E. and B.E. degrees in Computer Science and Engineering from Avinashilingam Institute of Home Science and Higher Education for Women, Coimbatore. As of February 2016, she had contributed 8 months of academic experience at SRIT, having joined the institution on June 26, 2015. Her educational background and focus on data mining demonstrates her commitment to the field of computer science education and research.

S. Sangeetha Dr. S. Sangeetha is an Associate Professor specializing in Power Systems, with over 10 years of academic experience. She completed her B.E. in Electrical and Electronics Engineering from Bharath Institute of Technology, University of Madras, followed by an M.E. in Power Systems from Sona College of Technology, Salem, and has earned her Ph.D. in Electrical Engineering from Anna University, Chennai. Her career progression includes positions at various institutions, with her most significant tenure at Sri Ramakrishna Institute of Technology, where she has advanced from Lecturer to Associate Professor. Her research contributions include notable publications in international journals focusing on power systems and renewable energy, particularly in solar PV systems and wind energy conversion. She has demonstrated practical expertise through consultancy work with Foretec Engineers and has been instrumental in organizing national-level workshops. As NAAC coordinator and through various institutional roles, she has made substantial contributions to academic administration while maintaining active membership in ISTE. Dr. Sangeetha has also delivered expert lectures on power quality improvement and energy conservation, showcasing her commitment to both academic excellence and practical industry applications.

Transformative Healthcare: Industry 4.0 Integration of Distributed Deep Learning and XAI

S. Keerthika, Hassan Oukhouya, S. Priyanka, P. Jayadharshini, J. Vaitheeshwari, and G. Roshini

Abstract Now, the sectors of health care do not have much to think about in today's time facing gigantic obstacles such as millions of data, diagnostic time delays, and inefficient therapies in the present Industry 4.0 era but gives new solutions with a merger of distributed deep learning and explainable AI. With distributed DL, it accelerates processing countless and large amounts of medical data to build highly accurate models for applications such as medical imaging, remote monitoring, and tailormade treatment. This speeds up the process of diagnoses, ensures better patient outcomes, and so on. XAI plays a crucial role in making AI-driven judgments transparent and understandable, thus increasing the confidence between healthcare providers and patients while knowing how those decisions are made. Together, DL and XAI are

---

S. Keerthika (✉) · S. Priyanka · P. Jayadharshini
Assistant Professor, Department of Artificial Intelligence and Data Science, Kongu Engineering College, Erode, India
e-mail: keerthivss97@gmail.com

S. Priyanka
e-mail: priyasubramani93@gmail.com

P. Jayadharshini
e-mail: jayadharshini.ai@kongu.edu

J. Vaitheeshwari · G. Roshini
Student, Department of Artificial Intelligence and Machine Learning, Kongu Engineering College, Erode, India

H. Oukhouya
LaMSD, MSASE, Department of Economics, FSJES, Mohammed First University of Oujda, BV Mohammed VI B.P.724, 60000, Oujda, Morocco

MAEGE, Department of Statistics and Applied Mathematics, FSJES Ain Sebaa, Hassan II University of Casablanca, BP2634, Route Des Chaux et Ciments Beausite, 20254, Casablanca, Morocco

J. Vaitheeshwari
e-mail: vaitheeshwarij.22aim@kongu.edu

G. Roshini
e-mail: roshinig.22aim@kongu.edu

L. Krishnasamy et al. (eds.), *Distributed Deep Learning and Explainable AI (XAI) in Industry 4.0*, Information Systems Engineering and Management 55,
https://doi.org/10.1007/978-3-031-94637-0_3

transforming health care from a reactive approach to a proactive approach; predictive analytics help clinicians detect health problems before these become critical. These tools also foster better collaboration among healthcare providers and researchers, thus accelerating developments in areas like drug discovery and precision medicine. Importantly, they ensure that the judgments made by AI are fair, responsible, and respect the privacy of the patient while ensuring strict legal compliances. By taking DL and XAI together, healthcare can become efficient, more transparent, and, above all, more patient-centric, and hence it is likely to yield healthier results for everyone.

Keywords Healthcare innovation · Data overload in healthcare · Diagnostic accuracy · Remote patient monitoring · Personalized treatment plans · Healthcare efficiency · Technological transformation in healthcare

Overview

Industry 4.0 will transform old ways to respond to the new demands of the healthcare sector. The integration of technologies like XAI and DDL offers capabilities and insights never heard or seen previously in the healthcare delivery domain. This integration enhances the accuracy, efficiency, and patient-centeredness of healthcare systems. This article discusses the effects of technological convergence and promotes collaborative models in the healthcare landscape by detailing advances in treatment effectiveness, diagnostic precision, and health management strategies. In this paper, we hope to clear the path for a future where advanced technologies empower healthcare providers, inspire confidence in stakeholders, and redefine what is thought possible from medical potential by clarifying the synergistic power of XAI and DDL.

The workflow for our project

See Fig. 1.

Industry 4.0 challenges in healthcare

Under Industry 4.0, the health sector is strongly challenged in its need for specialist-based knowledge expertise as a way of advancing sustainably. Some of the main challenges would include strict adherence to HIPAA regulation requirements that enforce the confidentiality of the patient, his/her confidentiality, and confidentiality of patient communications. Another obstacle to adapting to such firm demands is the evolving nature of digital technologies. Non-adherence does not only risk one's legal standing but also destroys patient trust in any healthcare delivery system.

As more interconnected health infrastructures tap into the power of big data analytics, the potential exposure to cyber threats increases. The protection of sensitive patient information from various data breaches and ransomware attacks, as well as unauthorized access, thus requires the implementation of broad cybersecurity

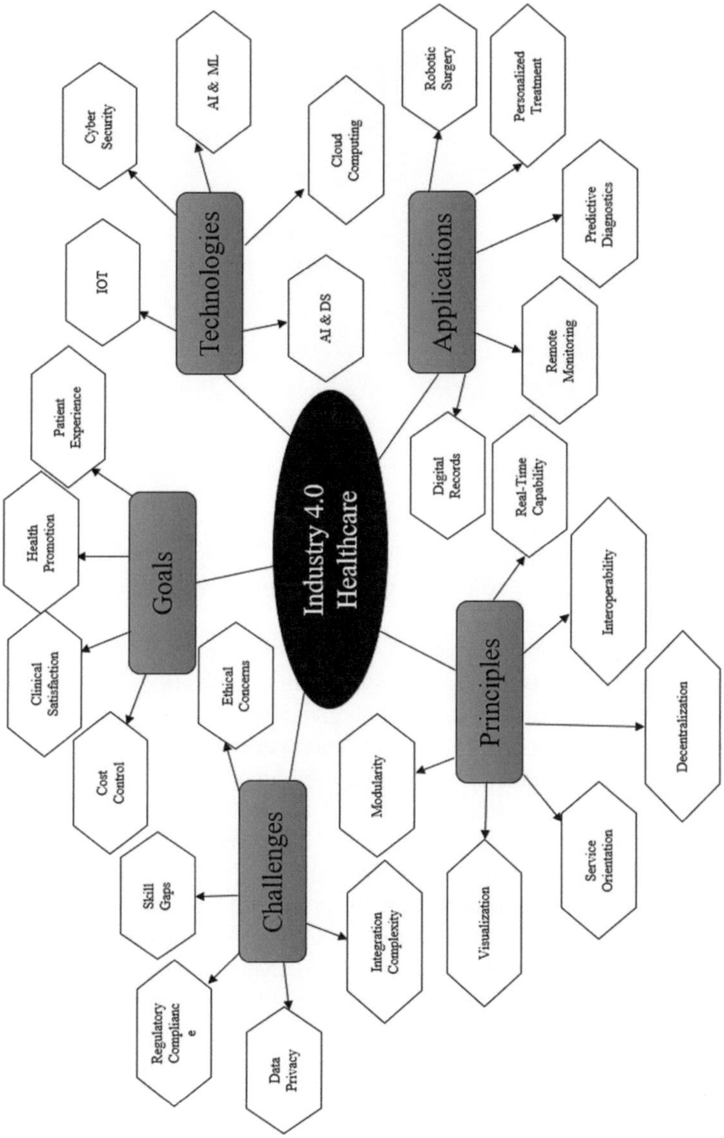

Fig. 1 Industry 4.0 workflow in the healthcare sector

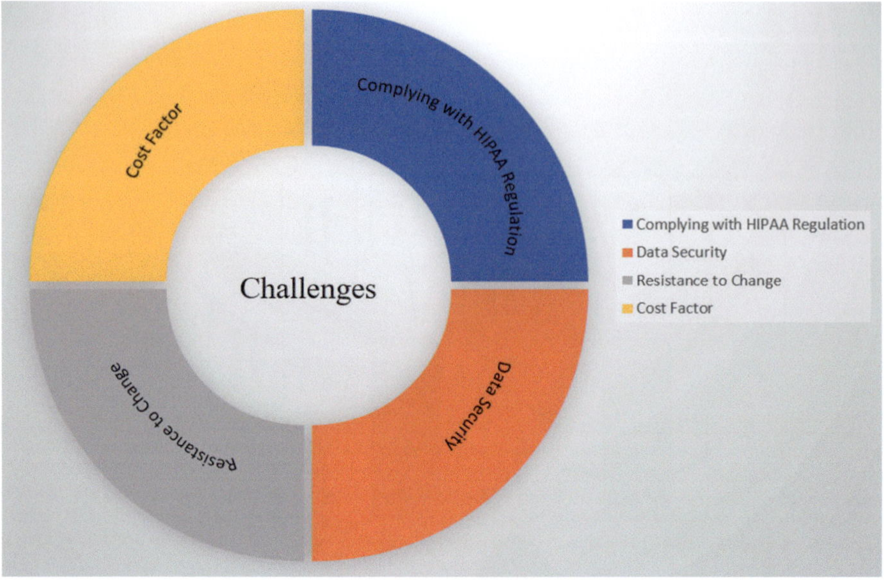

Fig. 2 Challenges of health care in industry 4.0

measures combined with vigilant monitoring. Only a strong cybersecurity framework will maintain the integrity of patient data while preserving stakeholders' trust amid changing digital landscapes.

It is evident that the adverse impact of a data breach extends far beyond just monetary loss to patient safety and reputation of the organization. Physician aversion toward new technologies coupled with organizational inertia prevents a new solution meant to enhance the care for patients and also efficiency from getting into actual practice. Such resistance can be overcome by an effective process of change management, along with leadership support and requisite training and support to the stakeholders is no longer an afterthought for most organizations. High investments in modern technology, laying down infrastructures, and training personnel are essentials. The discussion will focus on how these technological advancements will transform the way healthcare services would be delivered, improve the outcome of patients' lives, and help organizations function effectively in healthcare. Balancing the need for technological advancement with financial knowledge is a critical issue, especially where financial constraints and competition exist in healthcare (Fig. 2). In this era, the healthcare industry faces many challenges, including regulatory compliance and data security, protection against changes, and pricing decisions. Solving these challenges requires an integrated approach that includes strict compliance, cybersecurity procedures, effective change management strategies, and sound financial management. Only by recognizing these challenges can healthcare organizations harness the transformative potential of Industry 4.0 to improve patient outcomes and achieve better business.

1 Healthcare and Industry 4.0

1.1 Knowing What Industry 4.0 is and How It Affects Healthcare

"Industry 4.0" denotes the fourth industrial revolution, which is distinguished by the amalgamation of cutting-edge technologies and conventional industries. This is a paradigm shift that includes the widespread use of automation, digitalization, and connection to improve personalization, flexibility, and productivity in a variety of industries, including healthcare. The fundamental ideas and components of Industry 4.0 will be thoroughly covered in this chapter, along with a thorough examination of any possible ramifications for healthcare systems. Now, the discussion is on how the above technological advances will transform the practice of healthcare delivery and the improvement of patients' conditions besides making health care organizations operate more efficiently.

1.2 Data Management, Healthcare Service, Resource Allocation, and Right Diagnosis Are Some of the Opportunities and Challenges in the Healthcare Industry

Opportunities and Difficulties

- **Precise Diagnosis and Therapy**: Deep learning algorithms, big data analytics, and new imaging modalities constituted of Industry 4.0 advancement ultimately provide improved diagnosis accuracy and treatment plans. Vast datasets with sophisticated analytical methods enable medical practitioners to create customized treatment regimens, further enhancing the patient's outcome. These developments further assist in promoting the general quality of care provision to the patients since they also help in the implementation of individually devised therapy methods and in maximizing the precision of diagnosis.
- **Resource utilization**: This is the greatest problem in the delivery of healthcare regarding the proper distribution of medical staff, facilities, and equipment. Predictive analytics and optimization algorithms are some of the typical Industry 4.0 technologies that can give maximum resource usage. Those solutions can minimize costs and facilitate greater access to treatment by providing assurance that healthcare services are administered effectively and efficiently. Industry 4.0 leads to patient-centred care that is individualized, proactive, and seamless. The ability of wearables, linked health platforms, and remote monitoring to continuously collect data will enable health service providers to give their patients personalized recommendations and timely interventions to improve their well-being.

- **Patient Care**: Industry 4.0 has the potential to hold for individualized proactive and seamless patient care. Wearables, linked health platforms, and remote monitoring offer a continuous gathering of data. That will further help healthcare professionals provide patients with suggestions and appropriate timely action thus enhancing their well-being.
- **Data management**: The healthcare industry generates gigantic amounts of data. Such data encompass genetic data and electronic health records (EHRs) along with medical imaging. Advanced analytical methodologies, secure storage solutions, and distributed data processing capabilities of Industry 4.0 technologies make using this data easier. The use of these technologies can enhance the research capability and the decision-making capability of healthcare organizations. This integration of data-driven insights encourages advances in healthcare research and innovation besides improvements in clinical outcomes [3].

2 In Distributed Deep Learning (DDL)

2.1 Intro to Deep Learning and Its Applications in Medicine

This chapter dives into the exciting field of Distributed Deep Learning and how it may be the key tool for changing the healthcare sector, especially with regard to addressing the opportunities and challenges created by Industry 4.0. DDL is a critical method to process large volumes of medical data in real-time because it enables us to leverage the power of cooperation and parallelism by breaking computing work over many nodes or devices.

Foundations of Deep Learning: In this section of the book, we cover some of the fundamentals of deep learning DL before we progress to detail on distributed methods. Giant datasets can be leveraged by computers to learn from and make predictions thanks to the subset of machine learning referred to as deep learning. To this end, deep neural networks-artificial neural networks with multiple layers-must be trained to recognize complex patterns and relationships in datasets. These networks automatically extract relevant features and make inferences with very little intervention from humans, thereby making them more accurate at their forecasts and improving generalizability to different contexts.

2.2 State Advantages of Scattered Methods: Scalability, Fault Tolerance, Data Parallelism

- **Data Parallelism**: DDL splits enormous datasets into smaller subsets and distributed them across different nodes. This allows for processing huge datasets in parallel, thus making possible faster handling of large-scale medical datasets, such as those usually used in genomics or medical imaging research by acceleration of training and inference [5].

- **Redundancy and Fault Tolerance**: Distributed systems have consistent computation and fault-tolerant performance even in the event of failure at a single node. This feature is very important for healthcare applications, wherein uninterrupted access to computational resources is required to provide patients with fast and precise care.
- **Scalability**: Since nodes may be added or deleted in order to fulfill dynamic computing requirements, DDL supports smooth scalability. Due to this flexibility, business healthcare undertakes growing volumes of data and shifting workloads quite effectively.

2.3 Methods for Dispersed Inference and Training: Model Parallelism, All Reduce, and Parameter Server

- **Parameter Server**: In this method, worker nodes compute on their individual data subsets, while a central server keeps track of the model parameters. The server receives updated parameters from the nodes regularly, and it aggregates them to update the global model.
- **All Reduce**: This method eliminates the need for a central server by enabling nodes to directly share and aggregate model changes. It guarantees consistent model updates across all nodes and facilitates effective synchronization (Fig. 3).
- **Model Parallelism**: This technique divides a model into numerous nodes by distributing its various components, especially for very large models that are too big to fit into the memory of a single node. Every node processes a subset of the model, and communication takes place to keep the computations synchronized.

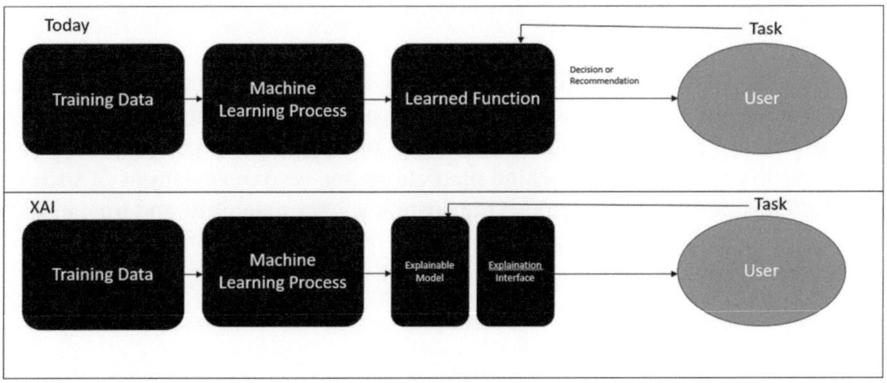

Fig. 3 Doing a task with AI versus eXplainable AI

2.4 Practical Uses in the Healthcare Industry

- **Medical Imaging Analysis**: By training large-scale models on large-scale medical imaging datasets, DDL improves the precision of tasks like organ segmentation, tumor identification, and illness diagnosis.
- **Remote Patient Monitoring**: Real-time anomaly identification and preemptive healthcare treatments are made possible by distributed algorithms' ability to examine data from several wearables or sensors.
- **Personalized Treatment Planning**: DDL makes it easier to create individualized treatment plans that take medical history, lifestyle choices, and genetic variants into account by processing patient-specific data quickly and effectively.

3 Distributed Deep Learning for Large-Scale Medical Data Management

In the last chapter, we discussed distributed deep learning for large-scale medical data management. As AI systems become more complex and crucial in healthcare decisions, it is also important to understand the decision-making practices of such systems. In this light, eXplainable Artificial Intelligence (XAI) stands as a critical aid because it provides such transparency of an AI model's workings in order to build greater confidence between the physician and the patient. This chapter explores the role played by XAI in the healthcare industry, describes the approaches currently accessible, and underlines benefits XAI brings to the table in Industry 4.0.

3.1 Conceptual Background for Explainability in AI for Healthcare

Explanability in Healthcare AI is Triggered by High Consequences: Its Decisions Directly Affected Patients' Health. With AI systems increasingly becoming an integral part of diagnosis, treatment planning, and monitoring patients, there is a huge necessity to have insight behind predictions and recommendations of such AI systems. Thus, XAI seeks to achieve transparency, interpretability, and trust in such AI systems so that healthcare professionals can rely confidently on AI decisions and explain them to the patients.

3.2 Techniques Use for XAI Are Rule-Based Systems, Decision Trees, LIME, SHAP, and Counterfactual Explanations

Several techniques have been developed to explain the decisions taken by AI models. Some of them are explained below.

- **Local Interpretable Model-agnostic Explanations (LIME)**: This explains individual predictions by the following approach: it modifies the input data to determine what changes in the model output follows and which features are most impactful on a particular prediction through generating an interpretable surrogate model that can approximate complex models locally.
- **SHapley Additive exPlanations (SHAP)**: SHAP values can be attributed based on game-theoretic principles by providing a framework for explaining the importance of every feature within a specific prediction. This isn't only measuring individual contributions from features but gives an overall sense of global behavior, increasing interpretability of lots of predictions in a model.
- **Decision Trees and Rule-based Systems**: This kind of model is centered around interpretability. Decision Trees mimic human decision making by organizing information into a hierarchical structure of straightforward rules that are easy to follow and understand. Similarly, rule-based systems rely on very simple IF–THEN rules to explain what conclusions the model has made to the healthcare practitioner and stakeholders.
- **Counterfactual Explanations**: Counterfactual explanations give insight into how changes in the input data may drive alternative decisions. They enable users to see in what ways the model would have needed to behave otherwise.
- **Evaluating Explainability**: To guarantee the efficacy and reliability of explanations, it is essential to evaluate their quality. These are a few standards for evaluation:
- **Fidelity**: Fidelity gauges how well the model's actual behavior matches the explanation. The underlying decision-making process is faithfully represented by an explanation.
- **Interpretability**: This is how much people can understand and feel an explanation; interpretability is strengthened with short explanations that are easy to understand.
- **Usability**: Usability assesses how well explanations aid in decision-making. Practical, implementable insights ought to be offered by explanations.

3.3 Benefits of XAI in Healthcare

Two most important benefits of interpretable AI for the healthcare industry are increased confidence and transparency. These systems allow patients, healthcare providers, and others to better trust them because they make clear the logic and variables underlying decisions generated by AI. This clarity nurtures an environment friendly to creative ideas flourishing in clinical practice while, at the same time, it encourages greater acceptance and implementation of such AI technologies.

Systems based on XAI provide clear explanations that complement the knowledge of healthcare professionals with detailed insights, thus enabling them to make informed decisions and hence better patient outcomes 1. It also makes the understanding of rationale behind AI-based judgments easier for auditors and regulators, thus aiding ethical norms, accountability, and fairness in the field of regulatory compliance. Furthermore, XAI reduces the probable biasing by detecting it and making it transparency. This is one of the ways through which it avoids the ethical dilemmas in AI, leading to bias as well as prejudice. The other major advantage of XAI is that it empowers patients, allowing them to understand their diagnostic and treatment options, thus taking charge of their care and making informed decisions.

3.4 Real-World Applications

XAI has been applied in healthcare for several purposes.

- **Interpretable Medical Imaging Analysis**: XAI techniques increase the interpretability of radiologists to support the validation of AI-driven conclusions, enabling the understanding them of which areas or features within medical images contribute to a diagnosis.
- **Personalized Treatment Planning**: XAI enables healthcare providers to formulate custom treatment plans for every patient according to his or her unique requirements, preferences, and characteristics. This is accomplished by defining factors that contribute to recommendations.
- **Precision Medical**: Such genetic or molecular markers linked with diseases can be identified, and then XAI can help in the development of precision medicine techniques along with drugs that can be tailored (Table 1).

4 Healthcare Integration of DDL and XAI

In the previous chapters, we discussed DDL and XAI in isolation. It is now time to merge these two concepts and look at how revolutionary they would be if they are applied to the healthcare sector. We can address difficult problems in healthcare and bring forth a new age of effective, transparent, and patient-centric care by putting together the powers of DDL and XAI.

Table 1 Use cases of explainable AI

S. No.	Category	Description
1	Customer retention	Improving customer loyalty and retention
2	Claims management	Enhancing efficiency in managing claims
3	Insurance pricing	Using AI for better pricing models
4	Fraud detection	Identifying and preventing fraudulent acts
5	Payments exceptions	Managing and resolving payment issues
6	Collections	Optimizing collection processes
7	Robo-advisors	Providing automated financial advice
8	Banking customer engagement	Enhancing customer interaction in banking
9	AI-assisted drug design	Leveraging AI for drug development
10	AI-integrated healthcare conditions prediction	Predicting healthcare conditions using AI

4.1 Benefits of Integration: Trust, Accuracy and Openness Are Benefits of Integration

Benefits of Integration: There are several advantages of integrating DDL and XAI in healthcare (Fig. 4).

- **Accuracy and Transparency**: DDL also facilitates the processing of vast medical datasets with much higher accuracy, than AI models. Explainable AI further expands on the reasoning for predictions, hence encouraging openness in the

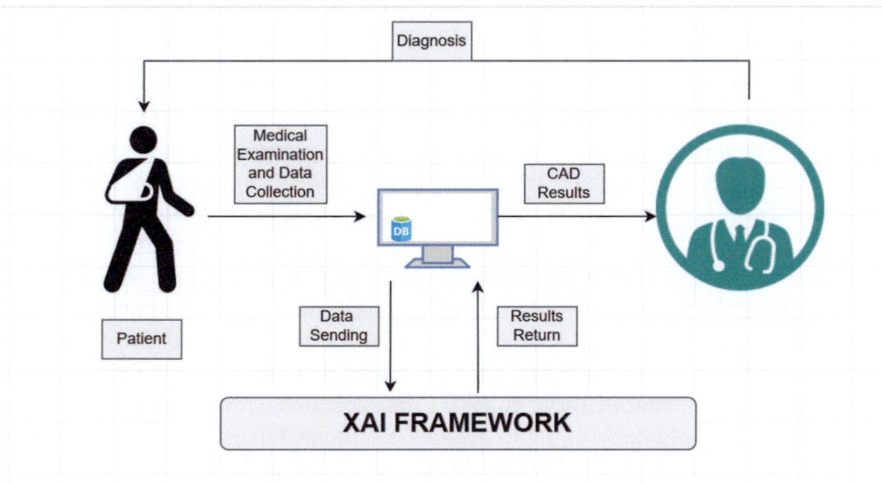

Fig. 4 XAI in healthcare

decision-making process. When AI technologies are accurate and transparent, they become more available to use within health procedures due to better trust from patients' perspectives and healthcare practitioners' perspectives.

- **Proactive Healthcare Interventions**: With DDL, the chance to process and analyze the data well can help in detecting early patterns, trends, and potential hazards for healthcare professionals. Explainable AI allows proactive therapies and prevention of care through the interpretation of outcomes. Change from reactive to proactive health care in which immediate intervention of a patient's condition may make a huge difference in terms of outcome.
- **Individualized treatment Planning**: XAI explains the motivators of recommendations for treatment, whereas DDL contributes to the development of very accurate prediction models [20]. That is why doctors are most probably going to compose special programs for every patient's treatment, concerning the peculiarity, preferences, and risks that can arise in this case.
- **Enhanced Regulatory Compliance**: Hence, this combination of DDL and XAI will support both ethical norms and regulations. While interpretability provides justice, accountability, and transparency in the AI-driven decision, it is DDL that allows safe handling and effective processing of sensitive healthcare data [2].
- **Seamless Collaboration**: It was possible for data scientists, researchers, and healthcare professionals to operate remotely through the use of DDL. Various teams could cooperate with each other toward the rapid production of innovative drug development and precision medicine, also optimizing healthcare delivery, through swapping and analysis of data across different nodes.

4.2 Use Cases Range from Personalized Treatment Planning to Remote Monitoring of Patients and Medical Image Analysis

- **Personalized therapy Planning**: XAI Elucidates on those factors driving the pushing of therapy recommendations, and DDL simplifies developing highly accurate prediction models. By such a link, medical practitioners are now in a position to personalize the treatment programs for every patient through the analysis of their preferences, special features, and potential hazards.
- **Improved Regulatory Compliance**: DDL and XAI together support both the norms of ethics and regulatory compliance. While explanations and interpretability-guaranteed justice, accountability, and transparency to AI-made decisions-DDL enables safe and efficient processing of sensitive health data [2].
- **Rubber-Band Collaboration**: In DDL, data scientists, researchers, and healthcare workers could be working in dispersed settings. Diverse groups can collaborate perfectly well in accelerating innovations to come directly into drug development, precision medicine, and optimization of delivery in health care through the sharing and analysis of data coming from various nodes.

4.3 Predictive Analytics-Driven Proactive Health Care Interventions

Health Care Providers Shall Forecast and Reduce Vulnerability to Health Hazards Using Predictive Analytics: DDL + XAI enables such possibilities of predictive analytics.

- **Predictive modelling**: DDL can train models that will be able to predict the chances of certain health-related outcomes or events, such as the onset of a specific illness or readmissions [20]. It is able to find trends in past and current data indicating such risks in the future.
- **Risk stratification**: Based on the factors that dictate expected risks, XAI systems help healthcare professionals allocate resources and strategize therapies. Based on this risk classification, the patients with the most extensive needs are immediately treated.
- **Patient-specific preventive measures**: Healthcare professionals can develop patient-specific preventive measures through the use of XAI explanations to understand each patient's risk factors [3].

5 Function of IoT Devices in Data Collection for Healthcare

Internet of Things has influenced the way data collecting is carried out in healthcare through the capability of fetching data from hundreds of sources and monitoring things continuously. It allows real-time analysis of a patient's health situation. This includes wearables, smart gadgets, and connected medical equipment that track vital indicators such as blood pressure and blood sugar levels, and heart rate. This non-invasive data collection provides better insight as well as an instant awareness of the patient's health, which is easy to notice irregularities sooner and, hence, to provide quick treatment that enhances the accuracy of diagnoses and therapies.

Other than monitoring sleep and physical activity, these technologies are also crucial data in managing chronic diseases, like diabetes and hypertension. Smart home technology can also care for elderly people through the tracking of movement patterns and falling, early medical intervention for older citizens' health and safety, and their quality of life. These qualities are highly beneficial to underserved or rural areas with few health facilities.

With EHRs, IoT data can be integrated in such a manner that it gives a more holistic view of patient history and their current health conditions. However, the big challenge with organizing and interpreting these massive amounts of data from Internet of Things devices arise. This makes explainable artificial intelligence relevant. Finally, these technologies serve as enablers that help the healthcare industry make more intelligent, strategic decisions by providing robust tools that can efficiently process and yield insights on both structured and unstructured IoT data.

5.1 Joining Distributed Deep Learning and XAI with IoT

IoT and deep learning, with artificial intelligence, show an overwhelming leap in healthcare through the power of comprehensive data and advanced metrics. It is a step toward patient care improvement. Since deep learning can process big data coming from IoT devices on multiple nodes or servers, it enhances data performance and capabilities in terms of analysis. This deployment enhances the computing performance and the system's computing resources, by which healthcare systems can manage the huge amount of data according to the usage of IoT.

Deep learning of IoT data assists medical practitioners identify patterns and links that are usually hidden beyond traditional analysis. A simple example is a deep learning model that analyzes wearable devices to predict problems such as heart diseases or respiratory conditions before they happen. This then leads to timely effective interventions that improve patient outcomes and costs incurred in healthcare.

It means that these systems must be trustworthy to provide advice and that judgments guided by AI must also be understandable. This technology on XAI makes this easier to understand because it tends to make the decision-making process transparent and understandable for artificial models. As such, this kind of transparency between patients and healthcare practitioners enables more acceptance and confidence in AI-enhanced healthcare solutions.

The combination of XAI and deep learning fosters the progression of tailored medicine. AI models significantly increase intervention and therapy effectiveness by providing personalized treatment plans according to a patient's exclusive medical profile. For example, AI-based systems can offer patients personalized drug proposals according to patients' real-time responses, that also simplifies adverse effect minimization and optimization of therapeutic outcomes. This integration also allows for continuous training and skill set building.

The output from XAI can be fed back into the deep learning processes in order to enhance and improve the model's accuracy and reliability. Deep learning is essentially incompatibility with IoT integration, and XAI has potential that would revolutionize healthcare through real, transparent, and personalized care. This approach not only enhances personal health outcomes but also helps in increasing the efficiency and sustainability of healthcare. The Flowchart of Integration of IoT with Distributed Deep Learning and XAI is shown in Fig. 5.

5.2 Privacy and Security Concerns

Finally, healthcare will definitely be revolutionized in terms of integration of IoT devices with deep learning and XAI; however, there is related security and privacy risk with such devices mainly because they are also being hacked at a very high rate. IoT devices themselves happen to be prone to hacks since they are being used

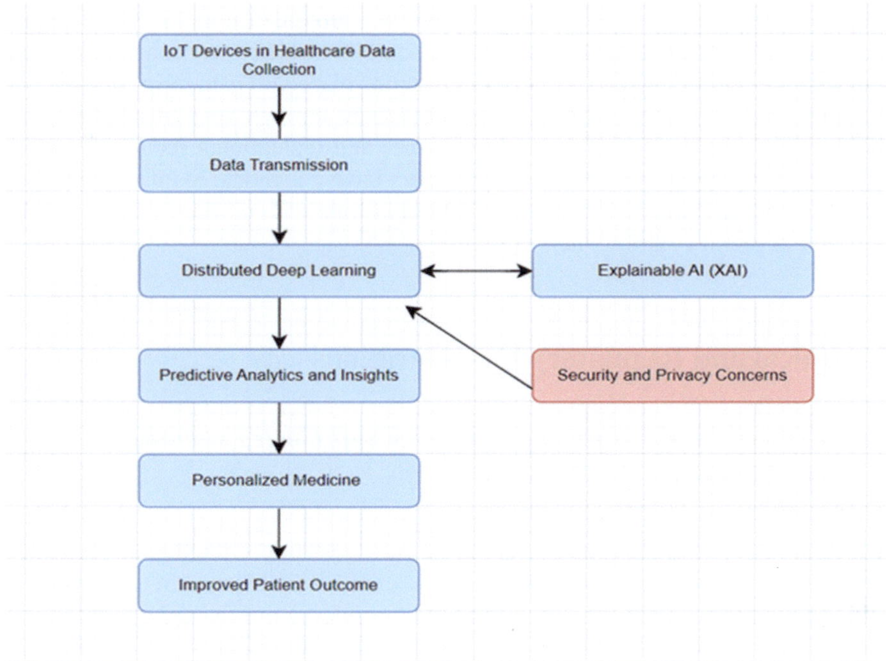

Fig. 5 Workflow of IoT integration with DDL and XAI

for gathering and transmitting health information. The privacy of these devices, as well as the information generated by these devices, must be safeguarded to protect patients' privacy and the quality of healthcare service delivery. Information breaches may compromise the anonymity of a patient; hence, it can be very serious. Between the IoT devices and medical systems, strong encryption methods and secure communication protocols need to be established to minimize such risks. Current security vulnerabilities present in most IoT devices require periodic security updates and patches to remodel the system. Although decentralized systems may prove efficient and effective, they increase the attack surface and make the system vulnerable to cyber-attacks. Unwanted access to private data can also be avoided through strict controls over access, robust authentication techniques, and regular security audits. Patient information management and protection are also included in ensuring openness in the system. This reflects the openness of decision-making processes. This transparency will be at the core on which building confidence between patients and healthcare providers depends so that AI-powered health solutions are trusted and safe. XAI may well prove to be very effective in dissolving the fears of data privacy and security through clear explanations of how data processing procedures work and what measures have been taken in place for security. Even in healthcare, the effectiveness and safety of such breakthrough medical technology can be amplified to better satisfy patient conditions and foster more confidence in AI-driven healthcare solutions, provided that comprehensive security measures are considered along with data protection regulations. Strict limitations placed on gathering, storing, and utilizing personal health information are stipulated through laws like HIPAA and GDPR. Non-compliance will bring heavy fines and reputational risks to the institution. In that respect, the healthcare providers have a mandate to ensure that their application of IoT, deep learning, and XAI conforms with all these regulations by inclusion of privacy rules in their systems.

6 Ethical and Regulatory Considerations

6.1 Ethical Implications of Using AI and Deep Learning in Healthcare

I use AI and deep learning in healthcare. Integration of artificial intelligence in healthcare How do you feel about the integration of artificial intelligence in the healthcare field? I personally consider it a revolution that promises breakthroughs in diagnosis, personalization, and performance. Along with these promises, however, come important ethical issues that require careful evaluation and mitigation strategies. Patient privacy and data security become the most important issues in the age of artificial intelligence-supported healthcare services. Since AI systems heavily rely on sensitive patient data to train algorithms in an effective manner, any breach thereof will put their confidentiality and integrity at great risk-the very things that are highly needed to avoid breaching patients' trust and leaving them independently vulnerable. Hence, such risks can only be alleviated with strong data protection control measures, including encryption and stringent security controls. Deep learning models know patterns in large data sets that are biased through thought processes of race, gender, or health and other information. This, therefore, calls for vigil and reform of the system to ensure that health services are delivered fairly and with no prejudice. Some of the key steps toward bringing bias-free artificial applications in medicine would be through technologies like bias detection and management of variable data. Doctors and patients should be aware of the way AI forms decisions because, by definition, normally AI-based recommendations in almost all healthcare settings determine instant decisions that influence patient care. For this reason, Explainable Artificial Intelligence (XAI) technology is developed to demystify AI decision-making processes, thus enabling clinicians to communicate AI-driven insights to patients in an effective way, thereby building trust and informed decision-making. Informed consent is a very fundamental factor in the ethical use of medical technology. Patients should have a right to know when AI is going to be used in their care and what information is going to be gathered, as well as how it will be used and protected. The importance of clear open communication with patients concerning the benefits, risks, and limitations of AI-driven interventions is necessary to treat them in accordance with their rights and to seek their informed and voluntary consent [8]. Apart from the above, accountability, justice, and equitable distribution are parts of social expectations. Around the world, there are many challenges regarding how to adapt existing systems to the fast pace of AI innovation, or how to keep patient health and rules safe. Finally, establishing clear guidelines and standards for the development, implementation, and use of AI technology in healthcare is important to address all of the ethical issues. In other words, the participants must share responsibility for the use of flexible resources when designing, implementing, and managing. This approach not only enhances the care of the patients but also maintains standards of practice while safeguarding the dignity, privacy, and independence of the patient while they age in AI-driven medicine. While the scenario persists, ongoing cooperation among researchers, policymakers, clinicians, and patient advocacy groups will be critical to help achieve equity for the future of the medical specialty.

6.2 Regulatory Challenges and Processes for AI in Healthcare

Today, artificial intelligence advances in healthcare at a phenomenal rate, thus bringing about challenging regulatory issues. It has become, therefore, an issue that policymakers and regulators across the world need to undertake steps that will help simplify complex issues that ensure safety, privacy, and ethics when the AI is being applied in the health sector. Distribution and maintenance of smart medical devices and software will be the focus of centralized management models while conventional systems of management will be unable to match this shifting nature of such complex AI algorithms that learn continuously on fresh data. The development of extensive technical standards will be essential in healthcare for the dependability, safety, and efficacy of AI-powered medical devices while driving innovation. Such regulatory bodies as the U.S. Food and Drug Administration (FDA) are actively developing these recommendations with the aim of finding a balance between such growth as technological one, on the one hand, and maintenance of patient privacy and health standards, on the other. Cooperation and information exchange among diverse healthcare systems also lie at the center of effective crisis management [23]. To provide the best healthcare services, AI has to integrate information from a multiple source. It means setting up strong data sharing protocols between various medical facilities without compromising their patients' privacy. The application of AI support in decision-making related to the clinical practice also makes it difficult to give blame for the mistake or an unfavourable outcome in hospitals. The ethical and safe development of AI requires standards that succinctly define the responsibilities of the developers, implementers, and managers of AI. These standards should include protocols for data breach handling, conflict handling, and keeping track of the performance of AI. It goes without saying that AI technology has the capacity for improving patient care outcomes, treatment plans, and diagnostics in a customized manner with predictive

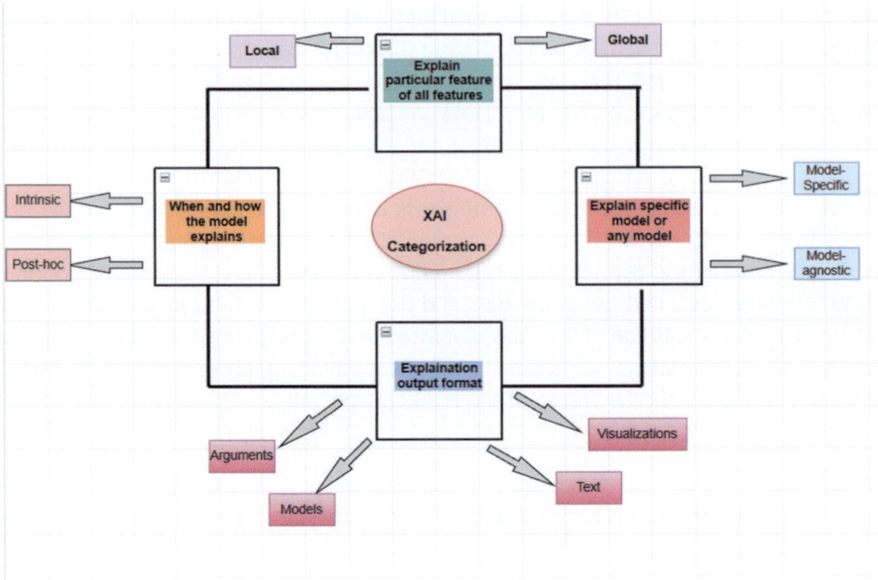

Fig. 6 Categorization of XAI

analytics. It is their ability to contribute, as both academic and private sector stakeholders, to the production of best practices and standards that protect those ethical norms of patient care through intellectual knowledge. With innovation as a shared responsibility and through strong leadership models, the healthcare industry should well be able to use artificial intelligence to alter the fortunes of health outcomes for patients while preserving their rights and welfare for future generations (Fig. 6).

7 Data Management and Infrastructure

7.1 Data Collection, Storage, and Preprocessing for Scientific Research

See Table 2.

Table 2 Healthcare data insights

Aspect	Description
Decentralized deep learning in healthcare	DDL is the ability to share large amounts of health data between different systems while carrying out various tasks. This will give more time and capacity in the processing, hence it's important in healthcare due to complexity and volume of the data [11]
Medical data collection	Medical data exists through various sources such as EHRs, traditional medical records, and PGHD. These are of various formats, with varying amounts and qualities; hence, a robust strategy in collecting data seems to be necessary. Due to the emergence of Industry 4.0, real-time data from IoT devices and wearables have been significantly helpful for patient monitoring and treatment improvement
Medical large data storage solution	Cloud computing offers a scalable option about the safe storage of big medical data files for DDL. This approach ensures compliance with US norms of privacy regulations, like HIPAA. Edge computing collaborative coexistence with cloud systems brings the data processing closer to where the data is going to be collected, reduces latency, and enhances the real-time response considerably, particularly in remote regions
Importance of preprocessing	It is essential to preprocess the raw medical data so it is ready to be fed into a deep learning model. The process involves cleaning noisy data, filling in incomplete data, and further securing data on lines of patient anonymity. Moreover, feature selection and extraction lead to cleaning data for better performance in healthcare predictions and treatment decisions. This, in turn, enables stabilization of the models, enhancing quality of data, and contributing to superior health care results

7.2 Infrastructure Requirements for Deploying Industry 4.0 Technologies in Healthcare Settings

- To interpret artificial intelligence (XAI), robust hardware is essential due to the intense data processing demands. High-performance computing (HPC) clusters, GPU-powered servers, and dedicated AI hardware (like GPUs and TPUs) enable fast model training and decision-making, allowing AI systems to handle complex tasks such as image processing, medical diagnoses, and treatment recommendations.
- Software architectures and platforms form the backbone of Industry 4.0 applications in healthcare. Open-source frameworks such as TensorFlow and PyTorch provide libraries for building and deploying deep learning models, while healthcare providers offer solutions for EHR integration, decision support, and telemedicine [22]. Containerization technologies such as Docker and Kubernetes simplify software deployment and management, increase capacity, and increase the ability to integrate disparate medical IT systems.

7.3 Cloud and Edge Computing's Place in Healthcare Data Management

- Effective data exchange and communication among IoT devices is crucial in distributed healthcare systems, including those utilizing edge computing and cloud servers. Low-latency networks enhance the ability to process data in real-time, supporting critical applications like remote patient monitoring, robotic surgeries, and telemedicine consultations [24]. Secure communication protocols (e.g., TLS, VPN) ensure the protection of sensitive medical data during transmission, minimizing the risk of data breaches and unauthorized access.
- Safeguarding patient data in healthcare management environments is a key component of stringent administrative practices. Infrastructure must adhere to industry standards and protocols to maintain data confidentiality, privacy, and integrity [11]. Methods such as access control and continuous monitoring help mitigate cybersecurity risks while ensuring compliance with regulations such as GDPR and FDA guidelines.
- Healthcare technology for healthcare to ensure the security and integrity of AI-powered devices. Features provide unique benefits to meet the changing healthcare needs of physicians, patients, and researchers. Enter when necessary. In healthcare, cloud platforms provide centralized information for electronic medical records, medical records, and research data, supporting collaborative research and public health review.
- This cloud service will enable flawless interoperability and integration of healthcare data in conjunction with AI-driven insights that would support predictive analytics, disease prevention, personalized treatment planning, and public health surveillance. It decentralizes computing and data processing resources through cloud platforms to efficiently manage data across various healthcare systems. On the other hand, edge devices, including IoT sensors, wearables, and medical equipment, could collect, monitor, and make decisions in real-time without requiring cloud central servers. This edge computing approach minimizes latency while it harnesses augmented capabilities in critical applications like telemedicine for remote patient care and emergency medicine and especially robotic surgery in remote locales where connectivity might be either very spotty or not very stable.

7.4 Hybrid Approach

Hybrid cloud edge architecture combines the best of the worlds of cloud and edge computing to optimize that kind of data management strategies that really perform well in healthcare. Edge tools favor local data to deliver instant insight and action, but cloud services support incorporating tasks requiring big data collection, long-term storage, and review. This allows health IT infrastructure to be flexible, adaptable, and functional in almost any application from the smallest scale of personalized medicine to the large scale of population management into health and medical research.

8 Economic Impact and Sustainability

8.1 Economic Benefits of Integrating Advanced XAI Technologies in Healthcare

- Business Benefits of Integrating Advanced XAI Technology in Healthcare

Health care should implement the deployment of XAI technology in order to reap considerable economic benefits for changing the essence of medical decision-making, business efficiency, and patient outcomes. XAI moves beyond the traditional information process and grants the transparency and disclosure necessary for building doctors' and patients' confidence and acceptance.

So, the healthcare XAI is also expected to optimize procedures for treatment better, possibly increasing efficiency with resources and lowering overall costs.

XAI supports the smooth flow of decisions by having transparent and interpretable insight concerning AI-driven insights in the diagnostic process, reducing errors, and creating personalized plans for therapy to improve better outcomes of care with decreased unnecessary expenditure. The use of XAI can reduce the possibility of repeat testing, shorten the stay of hospital patients, and enhance the use of resources with optimized diagnostic procedures in relation to the prediction of the outcomes of patients, thus providing an optimized plan for treatment. For example, predictive analytics initiated by AI can help doctors identify the high-risk patients prior to such costly complications and readmissions being allowed to happen [8].

Therefore, XAI implantation in clinical practice not only enhances the accuracy of the diagnosis but also tends towards more personalized approaches in medical approach and treatment. XAI algorithms analyze complex data, namely genetic, biomarkers, and imaging studies with a pattern better than the traditional methodologies. Thus, this technique can enhance the clinical outcome while also minimizing some trial-and-error therapy approaches, simultaneously enhancing drug efficacy and response from the patient.

XAI enables patients to gain a better understanding of their health and treatment, thereby creating trust between the patients and doctors based on clear documentation and explanation of medical decisions toward increasing patient satisfaction and compliance rate with the treatment plans [14]. This collaboration will finally give better health outcomes and reduce healthcare costs incurred by noncompliance and redundant visits.

The adoption of XAI in the healthcare arena would drive business growth and spur innovation across the healthcare ecosystem. Ever-increasing demand for AI expertise, software development, and hardware supports job growth and sustains the competitive focus on innovation and quality improvement [5].

- Sustainability of XAI-Driven Healthcare Solutions

Some of the economic benefits are vast for explainable artificial intelligence in healthcare; it should hence be aligned toward long-term success and ethical adoptions of this technology. Sustainability here encompasses different dimensions-ethical, social, and environmental as well as aspects of economics that have a highly significant impact on how such XAI-based solutions of healthcare are adopted and effective. Besides these dimensions, addressing these guarantees the applications of XAI not only make clinical services more efficient and cost-effective but also promote access equitably, patient trust is maintained, and rules and regulations are followed to foster an all-rounded approach to healthcare innovation.

- Ethical and Regulatory Compliance:

The safe application of XAI in a clinical setting depends on the ethical concerns [18]. Among these obligations, maintaining confidentiality, protecting patient privacy, and ensuring informed consent are the critical steps in maintaining ethical norms and public confidence in the healthcare industry. For example, some regulatory frameworks impose strict data protection and governance procedures, such as HIPPA in the United States and GDPR in Europe. Such policies enforce the protection of patient rights and thus reduce the possibilities of bias and discrimination as well as potential misuse of information from AI-generated data, thereby enhancing the ethical application of AI in health care settings [25].

- Scalability and Interoperability:

Scalability is required to support XAI-driven healthcare solutions across various healthcare settings and populations. Scalable AI models improve efficiency and reliability across clinical environments in real-world settings by their ability to adapt to the volumes and complexities of data being captured 12. Since XAI-driven healthcare solutions require a seamless interaction with existing healthcare IT systems, interoperability of such systems becomes essential for the exchange and integration of data to ensure effective coordination and support of decision-making across clinics and specialties.

- Environmental Impact:

There are three critical environmental sustainability considerations relating to XAI technology in healthcare: energy consumption and carbon footprint. Optimal tuning of AI algorithms and infrastructures to minimize energy usage in model training and inference may help ease the deleterious effects on health care from these causes of environmental degradation. The use of renewable energy sources for data centres and healthcare facilities ensures increased sustainability in the AI-driven healthcare ecosystem.

- Long-Term Value and Return on Investment (ROI):

The benefits of XAI-driven healthcare solutions-long-term value and ROI-have to be drawn both through the benefits they bring to the business and social impact. Economic analyses compare the costs saved-from improvements in health outcomes, lower costs for healthcare providers, and greater efficiency in operations-that are undertaken [13]. In addition, quantifying social outcomes in higher quality of life, equitable access to care, and reduced healthcare disparities make for a balanced approach for which XAI promises to be transformative in healthcare reform.

9 Practical Applications

9.1 Successful Implementation of XAI and DDL in Healthcare

9.1.1 Precision Radiology with XAI-Augmented Imaging

Medical imaging analysis was transformed forever with the combination of Distributed Deep Learning and explainable Artificial Intelligence at a top-tier health care facility. The DDL inclusion of vast datasets ensured training of algorithms to detect anomalies in X-ray images, thereby providing precision in early lung cancer diagnoses [16]. Giving radiologists heatmaps that identify questionable areas in pictures enhances the power of XAI so much to boost their trust in analyzing results and devising focused treatment plans. With this integration, early and accurate diagnosis is achieved for lung cancer, which later raises patient survival rates and reduces the need for intrusive operations.

9.1.2 Remote Patient Monitoring and Proactive Interventions

A new remote patient monitoring application with Artificial Intelligence that Can Be Explained (XAI) and Distributed Deep Learning (DDL) developed by a health-care business will allow the discovery of the very minute data trends from the wearable devices and possibly hinting upon the patient's deteriorating health. XAI will explain such patterns and support proactive treatment. The patients are provided with recommendations tailored to their cases. This indicates that the patients are in charge of their healthcare. As expected, this website has experienced a reduction in the readmissions by 20%, better outcomes from the condition for the patients and a higher participation from the patients in care.

9.1.3 Personalized Medicine Using Genomic Information

As an emerging objective at a research centre, it has enabled the incorporation of Distributed Deep Learning (DDL) and eXplainable Artificial Intelligence (XAI) into extracting insights from genetic data that could later tailor the medicines for uncommon diseases. In other words, the integration helps the DDL algorithms process and analyse huge genomic datasets that may eventually search for genetic variants and biomarkers that might be related to the rare conditions. The meaning and consequences of identified genetic markers are explained so that XAI makes this step more efficient and provides researchers studying the causes of the disease with valuable new information. Such a combination makes all the difference since it accelerates the search for targeted therapies and gives patients with rare genetic disorders personalized treatment options.

9.2 Discuss Privacy Protection, Ethical Issues and Compliance to the Laws

9.2.1 Impact on Healthcare Delivery

Healthcare delivery has been improved with the employment of Distributed Deep Learning (DDL) and Explainable Artificial Intelligence (XAI) [19]. DDL learning algorithms enhance the accuracy of diagnosis because they make fewer errors and missed detections due to the significant datasets on which they are trained. DDL and XAI help the shift towards proactive care; with DDL, trends are detected and hazards are predicted beforehand, but it has a focus on prevention. In conclusion, interpreting XAI gives health care providers with valuable information relating to the peculiar needs of a patient, and hence, tailor-made treatment to result in better outcomes and higher patient satisfaction. Furthermore, DDL optimizes the use of resources and the health care system through simplification, lowering costs, and thereby improving access to care. Moreover, XAI-driven explanations foster a team-based approach to healthcare as it empowers patients to own the care plan and understand their health condition.

9.2.2 Regulatory Compliance and Ethical Considerations

While integrating DDL with eXplainable Artificial Intelligence (XAI) in healthcare is anchored on the principles of fairness, responsibility, and transparency, by examining variables that impact AI-driven decisions, XAI must be maintained; otherwise, one will not spot or mitigate any biases in AI. Since DDL is scattered with data processing, it becomes more important to protect the patients' privacy by ensuring safe transport and storage of data [9]. This integration enhances transparency, which in turn increases confidence between regulators, healthcare providers, and patients. Ethical frameworks offer crucial direction to make sure that the creation and application of DDL and XAI technologies are aligned with ethical standards and social norms. These factors collectively enhance responsible and trustworthy AI integration in healthcare for rightful decision-making and safeguarding the confidentiality of patients.

9.2.3 Collaborative Platforms and Knowledge Sharing

Distributed Deep Learning (DDL) and eXplainable Artificial Intelligence (XAI also facilitate cooperation and information sharing, besides enhancing health care delivery. DDL enables collaboration between academics, data scientists, and experts in healthcare from many sectors by the use of distributed computing that relies on their respective areas of expertise [18]. DDL systems also allow for safe sharing of data among medical facilities, thus promoting cooperative research and better patient care. Additionally, the DDL platforms facilitate availability of big-scale medical data sets through open data initiatives, which energize new innovation and research in the health sector. DDL-XAI synthesis will promote collaboration for the health delivery service, ensure efficiency, and effectiveness while inspiring innovation and advancement in the field.

9.3 Future Prospects

Looking into an integration of DDL and XAI, the future of healthcare seems promising:

- **Advanced Imaging and Diagnostics**: By combining DDL and XAI with more sophisticated imaging modalities like PET and MRI, it will be possible to improve diagnostic precision and get a better understanding of complicated disorders [16].
- **Drug Discovery and Development**: By predicting possible side effects, optimizing dose schedules, and identifying viable drug candidates more quickly, DDL and XAI help streamline the drug development process.
- **Healthcare Delivery Optimization**: Operational efficiency may be increased, improving patient flow and resource allocation, by combining DDL and XAI with electronic health records and healthcare management systems.

10 Conclusion

In a nut shell, Distributed Deep Learning holds huge potential for industry transformation in healthcare though some barriers and restraints are still there. The challenges relative to data availability and quality are well vindicated, with some concerns over privacy, security, and other ethical issues. The interpretability-accuracy trade-off needs to be resolved, which calls for the advancement of XAI techniques. It is essential to quickly develop and deploy resources for distributed computing due to scalability and complexity issues. Finally, the decision-making by AI has to be done ethically and lawfully, requiring responsibility, equity, accountability, and fairness. Some of the promising future studies include multi-modal data integration, hybrid models, more architectures of .clf, and law and judiciary.

References

1. Duan, Y., Edwards, J.S., Dwivedi, Y.K.: Artificial intelligence for decision making in the era of Big Data—evolution, challenges and research agenda. Int. J. Inf. Manage. **48**, 63–71 (2019). https://doi.org/10.1016/j.ijinfomgt.2019.01.021
2. Rehse, J.R., Mehdiyev, N., Fettke, P.: Towards explainable process predictions for Industry 4.0 in the DFKI-Smart-LegoFactory. KI-Künstliche Intelligenz **33**(2), 181–187 (2019). https://doi.org/10.1007/s13218-019-00586-1
3. Nguyen, T.D., Kasmarik, K.E., Abbass, H.A.: Towards Interpretable Deep Neural Networks: An Exact Transformation to MultiClass Multivariate Decision Trees. arXiv preprint arXiv: 2003.04675. https://arxiv.org/pdf/2003.04675.pdf (2021)
4. Kowshika, P., Mousika, S., Divya, P., Lalitha, K., Jeevanantham, A., Muthukrishnan, H.: Enhancing the automated diagnosis system of soft tissue tumors with machine learning techniques. In: Udgata, S.K., Sethi, S., Gao, X.Z. (eds.) Intelligent Systems. Lecture Notes in Networks and Systems, vol. 431 (2022)
5. Shen, W., Yang, C., Gao, L.: Address business crisis caused by COVID-19 with collaborative intelligent manufacturing technologies. IET Collab. Intell. Manuf. **2**(2), 96–99 (2020). https://doi.org/10.1049/iet-cim.2020.0041
6. Holzinger, A.: From machine learning to explainable AI. In: Proceedings of the 1st World Symposium on Digital Intelligence for Systems and Machines (pp. 55–66). IEEE (2018). https://doi.org/10.1109/DISA.2018.8490530
7. Abirami, T., Mapari, S., Jayadharshini, P., Krishnasamy, L., Kavin, T., Kanagasubramaniyan, V.S.: A machine learning techniques for early autism spectrum disorder detection through comparative analysis of feature engineering and classification models. In: ICAICCIT-23, IEEE Delhi section, Nov 2023
8. Monteath, I., Sheh, R.: Assisted and incremental medical diagnosis using explainable artificial intelligence. In: Proceedings of the 2nd Workshop on Explainable Artificial Intelligence, pp. 104–108 (2018)
9. Ahmed, I., Jeon, G., Piccialli, F.: From artificial intelligence to explainable artificial intelligence in industry 4.0: a survey on what, how, and where. IEEE Trans. Ind. Informat. **18**(8), 5031–5042 (2022)
10. Santhiya, S., Mapari, S., Abinaya, N., Jayadharshini, P., Priyanka, S., Krishnasamy, L.: Early detection of cervical cancer using machine learning classifiers for improved diagnosis in underserved regions. In: ICAICCIT-23, IEEE Delhi section, Nov 2023
11. Sehito, N., Shouyi, Y., Alshahrani, H.M., Alamgeer, M., Dutta, A.K., Alsubai, S., Nkenyereye, L., Dhanaraj, R.K.: Optimizing user association, power control and beamforming for 6G Multi-IRS Multi-UAV NOMA communications in smart cities. In: IEEE Transactions on Consumer Electronics, pp. 1–1. Institute of Electrical and Electronics Engineers (IEEE) (2024)
12. Nascita, A., Montieri, A., Aceto, G., Ciuonzo, D., Persico, V., Pescapè, A.: Unveiling mimetic: interpreting deep learning traffic classifiers via XAI techniques. In: Proceedings of the IEEE International Conference on Cyber Security and Resilience, Rhodes, Greece, pp. 455–460 (2021)

13. Chaudhary, S., Joshi, P., Bhattacharya, P., Prasad, V.K., Shah, R., Tanwar, S.: Untangling explainable AI in applicative domains: Taxonomy, tools, and open challenges. In: Proceedings of the 4th International Conference on Computing, Communications, and Cyber-Security, pp. 857–872 (2023)
14. Clinciu, M., Hastie, H.: A survey of explainable AI terminology. In: Proceedings of the 1st Workshop on Interactive Natural Language Technology for Explainable Artificial Intelligence, pp. 8–13 (2019)
15. Science and technology for the explanation of AI decision makingerc-2018-adg grant [Online]. Available at: https://xai-project.eu/. Accessed 28 Jan 2023
16. Explainable artificial intelligence for defense advanced research projects agency [Online]. Available at: https://www.darpa.mil/program/explainable-artificial-intelligence. Accessed 28 Jan 2023
17. Terziyan, V., Vitko, O.: Explainable AI for industry 4.0: semantic representation of deep learning models. Procedia Comput. Sci. **200**, 216–226 (2022)
18. Daglarli, E.: Explainable Artificial Intelligence (xAI) approaches and deep meta-learning models for cyber-physical systems. In: Artificial Intelligence Paradigms for Smart Cyber-Physical Systems, pp. 42–67. IGI Global (2021). https://doi.org/10.4018/978-1-7998-5101-1.ch003
19. Prasad, V.K., Tanwar, S., Bhavsar, M.D.: Advance Cloud Data Analytics for 5G Enabled IoT, pp. 159–180. Springer, Cham, Switzerland (2021)
20. Velasquez, N., Estevez, E., Pesado, P.: Cloud computing, big data and the industry 4.0 reference architectures. J. Comput. Sci. Technol. **18**, Art. no. e29 (2018)
21. Binns, R., Kleek, M.V., Veale, M., Lyngs, U., Zhao, J., Shadbolt, N.: 'it's reducing a human being to a percentage'; perceptions of justice in algorithmic decisions. In: Proceedings of the CHI Conference on Human Factors in Computing Systems, pp. 1–14 (2018)
22. Singh, A., Dhanaraj, R.K., Sharma, A.K.: Personalized device authentication scheme using Q-learning-based decision-making with the aid of transfer fuzzy learning for IIoT devices in zero trust network (PDA-QLTFL). In: Computers and Electrical Engineering, vol. 118, p. 109435. Elsevier BV (2024)
23. John, S., Han, W., Rajesh, K.: Data-centric AI: bridging the gap between data engineering and machine learning. IEEE Trans. Knowl. Data Eng. **35**(7), 1902–1915 (2022). https://doi.org/10.1109/TKDE.2022.3141045
24. Chen, W., Lin, Z., Tan, P.: Advancing predictive analytics in Industry 4.0 with interpretable deep learning. In: Proceedings of the International Conference on Smart Manufacturing, pp. 101–110 (2023). https://doi.org/10.1145/3579987
25. Saurabh, G., Ankit, D., Bikram, R.: Explainable AI in smart cities: applications, challenges, and future directions. Int. J. Urban Comput. **19**(2), 87–99 (2023). https://doi.org/10.1016/j.ijurban.2023.04.001

S. Keerthika

S. Priyanka

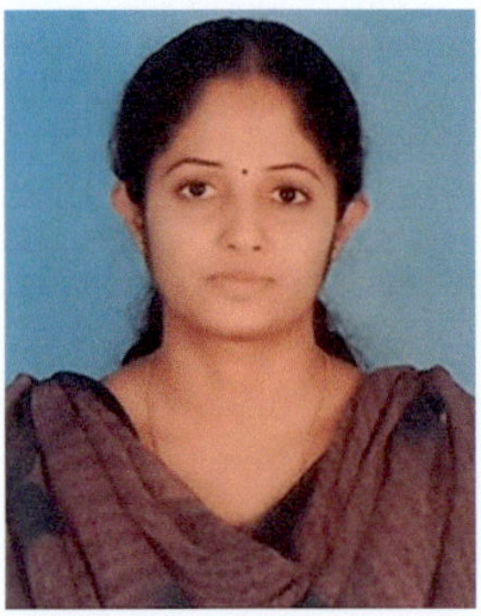

P. Jayadharshini

J. Vaitheeshwari

G. Roshini

Impact of XAI and Integrated Distributed Deep Learning in Industry 4.0

M. Dhurgadevi, N. Naveena, V. E. Sathishkumar, A. Sugitha, and A. Banupriya

Abstract The modern world is advancing along a digital revolution. Internet is everywhere and everyone is dependent on it. It's obvious that industries should also be digitalized. Such a transition is represented by Industry 4.0. A lot of new technologies play a vital role in Industry 4.0. The main advantages of industry 4.0 include flexibility, sustainability, efficiency, increased productivity, and so on. The biggest problem in implementing Industry 4.0 is the skillset of employees. As the employees are trained to work in a physical environment, a sudden transformation to digital requires significant skill upgrading. Skills can be upgraded by providing proper training, support, and encouragement. As many technologies stand as pillars for industry 4.0, this chapter focuses the role of Artificial intelligence (AI) and deep learning. Due to the increased amount of data, deep learning concepts are often too demanding to be implemented on a large-scale industry as they also consume more time. To address this issue, distributed deep learning is necessary, which enhances the stability and utilizes the resources efficiently. With an increase in distributed deep learning methods, it is inevitable to design an AI model to clearly elucidate these methods to users, that is, an eXplainable AI (XAI) for the deep learning algorithms. By integrating these two technologies and implementing them in Industry

M. Dhurgadevi (✉)
Department of Computer Science and Engineering, Rathinam Technical Campus, Coimbatore, India
e-mail: devi.durga@gmail.com

N. Naveena
Department of Computer Science and Engineering, Sri Krishna College of Technology, Coimbatore, India

V. E. Sathishkumar
Department of Computing and Information Systems, Sunway University, Bandar Sunway, Malaysia

A. Sugitha
Department of CSE (Cyber Security), Sri Krishna College of Technology, Coimbatore, India
e-mail: sugitha.a@skct.edu.in

A. Banupriya
Department of Information Technology, C.S.I. College of Engineering, Ketti, India

L. Krishnasamy et al. (eds.), *Distributed Deep Learning and Explainable AI (XAI) in Industry 4.0*, Information Systems Engineering and Management 55,
https://doi.org/10.1007/978-3-031-94637-0_4

4.0, creating an innovative approach to increase productivity and digitize industries becomes a hassle-free process. This chapter proceeds with summarizing the deep learning concepts accompanied with its eXplainable AI (XAI) model in a distributed environment, along with its highlights and challenges.

Keywords Industry 4.0 · eXplainable AI · Distributed deep learning · LIME · SHAP

1 Introduction

The modern world is advancing along a digital revolution. Internet is everywhere and everyone is dependent on it. It's obvious that industries should also be digitalized. Such a transition is represented by Industry 4.0. In the context of Industry 4.0, numerous computers, machines and smart devices communicate with each other and make decisions [1]. A lot of new technologies play a major role in Industry 4.0. Primary benefits of Industry 4.0 are enhanced production, sustainability, adaptability, and efficiency. The biggest problem in implementing Industry 4.0 is the skillset of employees. As the employees are trained to work in a physical environment, a sudden transformation to digital requires significant skill upgrading. Skills can be upgraded by providing proper training, support, and encouragement. As many technologies stand as pillars for industry 4.0, this chapter focuses on the role of Artificial Intelligence (AI) and deep learning. Deep learning principles are frequently too time-consuming and labor-intensive to be applied on a wide scale in industry because to the rising amount of data. To address this issue, distributed deep learning is necessary, which enhances the stability and utilizes the resources efficiently. With an increase in distributed deep learning methods, it is inevitable to design an AI model to clearly elucidate these methods to users, that is, an eXplainable AI (XAI) for the deep learning algorithms. By integrating these two technologies and implementing them in Industry 4.0, creating an innovative approach to increase productivity and digitize industries becomes a hassle-free process. This chapter proceeds with summarizing the deep learning concepts accompanied with its eXplainable AI (XAI) model in a distributed environment, along with its highlights and challenges.

The next phase of the industrial sector's digitization is Industry 4.0, or 4IR. Disruptive trends include the expansion of data and networking, analytics, robot improvements, and human–machine interaction are driving it. Manufacturing is undergoing a change to Industry 4.0 technologies, which automate procedures, connect devices via IoT, and use big data for analytics-based decision-making.

Industry 4.0 has completely changed the way companies create, produce, and market their goods. A few of the technologies that are being incorporated into production processes more and more are artificial intelligence (AI), machine learning, cloud computing, and the Industrial Internet of Things (IIoT). Coherent and integrated manufacturing produces items, facilities, and assets that are intelligent and

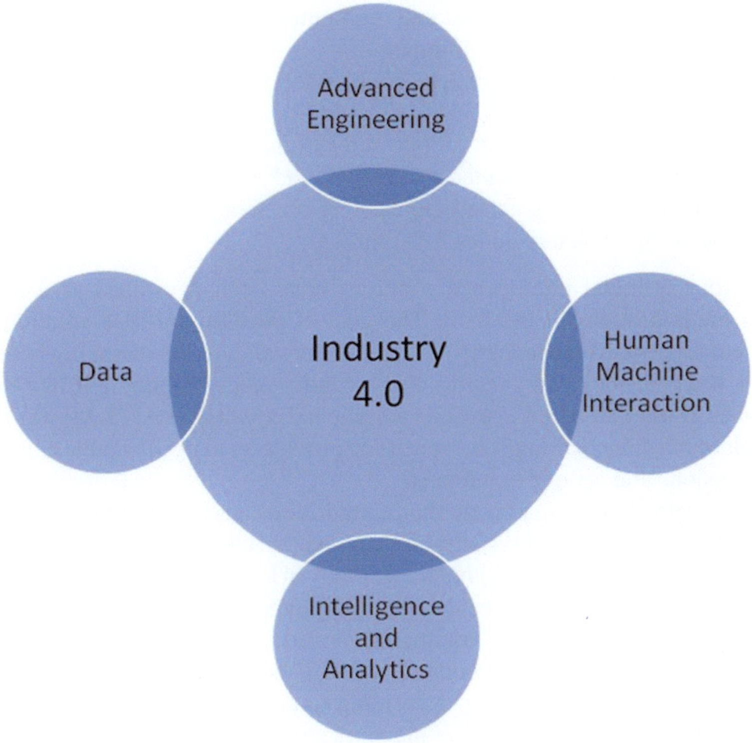

Fig. 1 New perspective of Industry 4.0

networked. The process of incorporating intelligent digital technologies into production processes is referred as "industry 4.0.". Figure 1 depicts the new perspective of Industry 4.0.

The First Industrial Revolution had begun by the early 1800s. With the development of the steam engine, the industrial sector became less dependent on the use of human and animal labor and entered a new era of precise engineering and manufacture. The assembly line and mass production techniques, many of which are still in use today, propelled the Second Industrial Revolution.

Robotics and factory automation saw their early beginnings during the Third Industrial Revolution. During this period, computerized business systems designed for data management and analysis were also used for the first time. The Fourth Industrial Revolution, or 4IR, is the term given to the next phase of production that follows Industry 4.0. Automation and intelligent technology that enable producers to make items more rapidly, cheaply, efficiently, and/or sustainably will define it.

Industry 4.0 elevates these innovations to new heights with the aid of four core disruptive technology categories.

1. Information, network access, and computational capacity: sensors, cloud services, blockchain, and the Internet
2. Analytics and intelligence: artificial intelligence, machine learning, and advanced analytics
3. Human–machine interaction: robotic and automated systems, self-driving automobiles, virtual and augmented reality (VR and AR).
4. Two instances of sophisticated engineering are renewable energy and additive manufacturing, which includes 3-D printing.

Numerous artificial intelligence solutions have been developed as a result of ongoing breakthroughs in the field. The goal of developing these solutions is to make them function independently. Without understanding the reasoning behind the decision, it would see, learn, decide, and act independently, putting it in a difficult situation as to whether or not to trust its judgment. Explainable AI (XAI) was created as a result of machine learning's incapacity to provide an understandable explanation for its decisions and actions to humans.

XAI is the process of creating methods, algorithms, and instruments that generate information, judgments, and systems based on AI that can be understood by humans. The significance of XAI is highlighted, especially in high-stakes commercial applications, for the moral and human-centered development of AI systems. Since artificial intelligence (xAI) goes beyond traditional performance measurements to directly understand the many learning models and their behavior, it has the potential to greatly enhance machine learning. According to taxonomy of interpretability techniques [2], the majority of XAI techniques were suggested for tasks using neural network models.

More number of researchers done research on XAI and its practical applications in various fields like medicine, agriculture, business etc. In health care the reviews include machine learning applications particularly cardiology [3], breast cancer diagnosis and surgery [4], medical image analysis [5], and radiology [6]. Providing trustworthy explanations backed by strong validations is one of the biggest issues in the XAI ecosystem [7]. The rational processes of such models are rendered more transparent and verifiable by an explainable artificial intelligence application, which offers comprehensive elements that elucidate the models' decision-making processes. This facilitates the understanding of the contributions of features to predictions and their impact on predictive performance [8]. In order to produce information that can be understood by humans, an XAI approach needs be created to check the many components of complex learning functions and break down opaque portions [9]. To turn black boxes into verifiable tools that support machine-learning judgments, XAI approaches need to offer trustworthy explanation components.

2 Industry 4.0 and Its Technologies

Industry 4.0 employs artificial intelligence (AI) to connect various technologies, enabling robots and software to see, comprehend, react, and gain knowledge from human behavior. This technology could lead to more efficient operations of the industrial production system. A few of the technologies that AI has enabled are cloud computing, IoT, and artificial intelligence. Because it enables intelligent robots to do self-regulation derivation, assessment, and analytics, it is the primary force revolutionizing industries. Deep learning and machine learning are very helpful in the manufacturing sector, helping businesses anticipate maintenance needs and reduce downtime. To guarantee the constant deployment and integration of AI systems, XAI must create human-readable algorithms from the outputs generated by AI expert systems. Explaining actions and outcomes is also important. Among the notable companies in this market are Intel Corporation, Nvidia Corporation, and International Business Machines Corporation (Quanta Storage Inc.). HP, IBM, Nvidia (Quanta Storage) Corporation and Intel are other notable companies. In addition, Robert Bosch GmbH, Fanuc Corporation, Cisco Systems Inc., DENSO Corporation, Stratasys Ltd., Schneider Electric SE, SAP SE, also have significant presences.

Nine technological pillars support Industry 4.0. These advancements allow for intelligent, self-governing systems and establish a connection between the digital and physical domains [10]. Table 1 lists the various applications of these pillars.

The first pillar of Industry 4.0 is Big Data along with AI analytics. Big Data is collected from various sources. The company's global network and other locations outside the plant floor are potential sources of data. Weather and traffic apps that enable more effective logistics can be among them, as can user input which then directs R&D and design. Real-time analytics driven by AI and machine learning,

Table 1 The nine technology pillars of Industry 4.0

S. No.	Technology	Applications
1	Big Data with AI analytics	Supply chain and industrial management sectors
2	Cloud computing	Remote monitoring, Data backup
3	Augmented Reality	Viewing digitalized parts, maintenance or assembly instructions, and training materials
4	Industrial Internet of Things	Maintain inventory and products, analyze client preferences, and streamline supply chains
5	3D printing	Quick prototyping, distributed manufacturing and mass customization
6	Autonomous robots	Mobile robots for pick-and-place jobs
7	Internet of Things	Digital twin
8	Simulation	Product development, Process optimization
9	Cyber security	Secure communication, risk mitigation

used in the supply chain and industrial management sectors, are utilized to analyze data and improve automation and decision-making in general.

The majority of cutting-edge technologies [11], including Internet of Things, AI and ML integration, are built on cloud technology, which is the second pillar of Industry 4.0. Modern cloud computing allows businesses to innovate. Industry 4.0 technology is powered by cloud-based data, and its cyber-physical systems depend on real-time coordination and communication.

Next technology is Augmented Reality (AR) [12]: AR is the process of overlaying digital content on top of the physical world. Workers can view digitalized parts, maintenance or assembly instructions, training materials, and real-time Internet of Things data via smart glasses or mobile devices while maintaining attention on a tangible object, such as a product or piece of equipment, when utilizing an augmented reality (AR) system.

The Industrial Internet of Things [13], or IIoT, is the fourth pillar and is so essential to Industry 4.0 that the terms are sometimes used synonymously. The majority of Industry 4.0 physical objects, including products, equipment, robots and machinery, use RFID tags and sensors to transmit real-time data on their performance, location, and state. Businesses can make use of technology to maintain inventory and products, manufacture and adapt goods fast, prevent equipment failure, analyze client preferences, and streamline supply chains, among many other things.

The fifth pillar is additive manufacturing, also known as 3D printing [14]. Originally intended as a tool for quick prototyping, it is increasingly employed for a wider variety of tasks, including distributed manufacturing and mass customization. 3D printing lowers costs and does away with the necessity for off-site/offshore production because it enables parts and products to be held as design files in virtual inventories and created on demand as needed.

Next is Autonomous robots pillar where Industry 4.0 will usher in a new breed of autonomous robots. Autonomous robots are designed to do tasks with as little human intervention as possible. They come in a range of sizes and forms, from autonomous mobile robots for pick-and-place jobs to inventory scanning drones. These robots can do intricate and challenging jobs because they are outfitted with sensors, state-of-the-art software, artificial intelligence (AI), and machine vision. Additionally, kids are able to identify, evaluate, and react to information they learn from their surroundings.

The seventh pillar is Internet of Things (IoT) sensor data is the foundation for digital twins, which are virtual representations of actual machinery, goods, processes, or systems. This fundamental element of Industry 4.0 enables businesses to analyze, comprehend, and enhance the operation and maintenance of industrial products and systems. A digital twin [15], for instance, can be used by an asset operator to pinpoint a specific broken component, anticipate future problems, and increase uptime.

The next pillar is simulation which helps creating digital models of the innovations in machines and processes. It is also useful in testing the system and achieving optimization in a virtual environment. By analyzing the models using simulation, the necessary corrections can be made and risks and costs can be reduced when implementing in real time.

The final foundation is cyber security [16]. As Industry 4.0 uses networking and Big Data more frequently, strong cybersecurity becomes essential. By automating threat detection, prevention, and reaction, businesses may lower the risk of data breaches and production delays across their networks by utilizing Zero Trust architecture and cutting-edge technologies like blockchain and machine intelligence. Producers can employ these technologies to create new, innovative products and services, automate processes, and boost productivity. But industry 4.0 is a paradigm shift that profoundly changes how we organize and carry out production processes, not only the incorporation of cutting-edge technology. It becomes imperative for firms to implement new Industry 4.0 best practices as they compete to remain competitive in this new era.

3 Explainable Artificial Intelligence (XAI)

Artificial intelligence is made available in this field to acquire cognitive skills. Another domain in which it can exhibit explainability is the human–machine interface; this is because explainable AI necessitates an extremely high degree of user involvement [17]. Figure 2 represents the xAI model with deep learning.

Artificial Intelligence based mechanisms are being used in most of the manufacturing companies in both product level and process level [18]. We can use AI in these companies for strong decision making and to increase the productivity. But this implementation seems complex as the employees are not trained to use AI. It is a big necessity to enhance the employees' digital skills which consumes time and money. Also, the quantity of data that is being handled by the companies is very large in velocity.

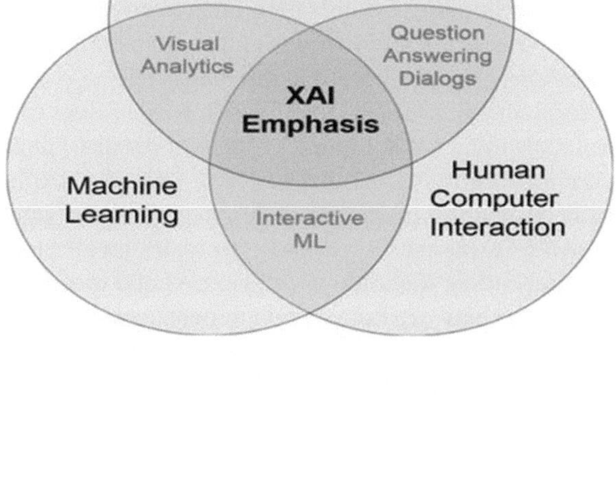

Fig. 2 Explainable Artificial Intelligence (xAI) [9]

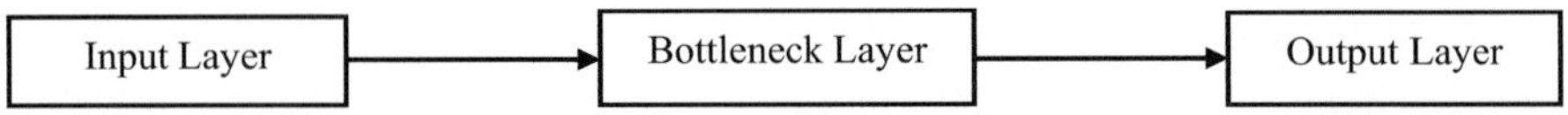

Fig. 3 Auto encoder architecture

To overcome most of these issues the concept of deep learning models and tools came into existence. Few of the models that can be used by companies are:

(a) **Auto encoders (AE)**

When higher dimensional data is given, Auto encoders will understand the encoding of the data [19]. The bottleneck layer will minimize the dimensionality of the input layer providing a compressed encoding. Again, the bottleneck layer expands the da ta back into original dimensionality and the same input is reconstructed in the output layer. Figure 3 represents the functional architecture of auto encoder.

(b) **Long Short-Term Memory (LSTM)**

LSTM is a feed forward architecture [20]. It recognizes the information from earlier data phases and builds the model. It is utilized in programs that call for memory. Three gates comprise it: input, forget, and output. The equation not needed in future is eliminated by the forget gate. The equation for the forget gate is given in Eq. 1.

$$Ft = (Wf[ht - 1, xt] + bf) \tag{1}$$

Other than deep learning models, there are also AI-based advanced tools for data analysis using which the employees make use of the resources in an effective way. Also, they reduce the waste and carbon emissions [21]. In most of the industries these tools do reliable analysis, assist in extracting text and provide support for coding. These tools also support renewable energy sources. Even food manufacturing industries, agriculture sector and healthcare also find more benefits in AI-base models and tools.

Currently the focus of the industries is turned towards automating the tasks with more efficiency by combining artificial intelligence and deep learning concepts which can result in better optimization [22]. It's well-known that machine learning based models are being used in designing, managing product life cycle, manufacturing operations and much more in the industries.

Explainable artificial intelligence (xAI) refers to a collection of techniques and protocols that enable human users to understand and rely on the output of machine learning algorithms. AI that has been trained to explain its goals, justifications, and decision-making process in a language that the general public can comprehend is known as explainable AI (xAI). To foster greater trust, XAI assists human users in comprehending the logic underlying AI and machine learning (ML) technologies.

The gap between human comprehension and artificial intelligence models' operations can be closed with the use of explainable AI. AI developments are now essential to improving industry performance and efficiency with the introduction of industry

4.0 [23]. Explainable artificial intelligence (xAI) facilitates transparent analysis of algorithms by businesses and developers, allowing them to evaluate and refine the exact process by which they arrive at answers. Now, a group of researchers that analyzed the current AI and explainable AI (xAI) based methodologies utilized in Industry 4.0 has brought attention to the need for XAI-based approaches to help construct efficient smart cities, manufacturing, healthcare, and human–computer interactions.

Generally, XAI uses either interpretable models or black-box analysis. The conventional approach, known as "black-box analysis," involves only opening the preexisting box of an algorithm and looking at the data within. Similar to a computer enclosed in a glass case, these models are meant to be examined. Although they might be somewhat complicated to make, user-friendly XAI encourages developers to keep exploring and creating new interpretable technologies. Industry 4.0 is primarily driven by artificial intelligence (AI), which automates intelligent devices to self-monitor, interpret, diagnose, and analyze on their own. Artificial intelligence (AI) techniques, including natural language processing (NLP) [24], computer vision (CV) [25], machine learning (ML), and deep learning (DL), assist industries in anticipating their maintenance requirements and reducing downtime.

XAI is following four principles of NIST standard framed by US which provides certain understanding of AI working models.

- The first principle is explanation, which explains how decisions are made with the necessary justifications and supporting data so that others may comprehend the process.
- Second one is Meaningful which states that first principle is satisfied by the end user. The explanation must to be simple enough for both individual and group users to understand.
- Third principle is Accuracy which is accuracy of explanation provided which indicates that logic is the logic is understood by all stakeholders requires 100% correctness.
- Last principle is Knowledge Limits which clarifies that the model can only be used in the particular scenarios for which it was designed.

These principles help us define the expected output from the XAI model and how an ideal XAI model should be. However, it doesn't indicate how the output has been achieved. Subdividing the XAI into three categories to better understand the rationale:

1. **Explainable data**: Which data does the model get its training from? Why was this specific set of data chosen? How much biased is the data?
2. **Explainable predictions**: What features did the model use that lead to the particular output?
3. **Explainable algorithms**: How is the model layered? How do these layers lead to the prediction?

Based on individual instances, the explainability may change. For example, the neural network can only be explained using the Explainable Data category. Research is ongoing that is focused on finding ways to explain the predictions and algorithms. At present there are two approaches:

a. Proxy Modeling—A different model from the original is used to approximate the actual model. This may result in different outcomes from the true model outcomes, as it is just an approximation.
b. Design for Interpretability—The real model's construction makes it simple to comprehend how it functions. However, this increases the risk of reduced predictive power and overall accuracy of the model.

The XAI is referred to as the White Box, as it explains the rationale behind its working. However, unlike the black box, its accuracy may decrease in order to provide an explainable reason for its outcome. Explainable techniques include Bayesian networks, decision trees etc. Hopefully, with the advancements in the field, new studies will come up to increase the accuracy of the explanations.

3.1 Critical Industries for XAI

XAI would be helpful in those industries where machines play a key part in decision-making. These use cases might also be useful in your industry, as the details may vary, but the core principles remain the same.

3.1.1 Healthcare in XAI

The decisions made by AI in healthcare impact humans in a very critical way. A machine with XAI would help the healthcare staff can save a lot of time, which they might use to focus on treating and attending to more patients. For example, diagnosing a cancerous area and explaining the reason in a matter of time helps the doctor to provide appropriate treatment [26].

3.1.2 Manufacturing in XAI

In the manufacturing industry, fixing or repairing equipment often depends on personnel expertise, which may vary. To ensure a consistent repair process, XAI can help provide ways to repair a machine type with an explanation, record the feedback from the worker, and continuously learn to find the best process to be followed [27]. The workers need to trust the decision made by the machine in order to risk working on the equipment repair, which is the reason XAI becomes useful.

3.1.3 Autonomous Vehicles in XAI

A self-driving car [28] seems great until and unless it has made a bad decision, which can be deadly. If an autonomous car faces an inevitable accident scenario, the decision it makes impacts greatly on its future use, whether it saves the driver or the pedestrians. Providing the rationale for each decision an autonomous car takes, helps to improve people's security on the road.

The XIA can be implemented using different techniques like Local Interpretable Model-agnostic Explanations (LIME) [29], XAI approach makes use of a local approximation of the model to offer comprehensible and interpretable insights into the variables that are most significant and pertinent to the model's predictions. Next is SHAP (SHapley Additive exPlanations) [30] that gives interpretable and comprehensible insights into the variables that are most pertinent and significant in the model's predictions by utilizing the Shapley value from game theory.

Third one is Explain Like I'm Five, or ELI5:ELI5, a XAI technique [31], which is easy to use and intuitive language to deliver interpretable and explainable insights into the aspects that are most relevant and influential in the prediction's models. These methods can be applied in a variety of fields and applications and offer varying degrees of interpretability and explainability.

4 Integrated Distributed Deep Learning and XAI Approaches in Industry 4.0

Recent developments in industries and due to the enormous amount of data, it is essential to automate the tasks in industries to compete and achieve success in the future world. Combining the concepts of XAI and Distributed Deep Learning (DDL) can bring a reliable model to operate the machines. However, there are numerous obstacles to overcome, some of which are listed below:

When the dataset is large, the XAI functions should be scalable with DDL.

1. To implement a distributed environment, it can be highly expensive.
2. It is also difficult to ensure the consistency throughout the system.

4.1 Integration Techniques

Model—Agnostic methods is one of the methods that can be used for integration. This method works by applying XAI methods which can work with any model architecture. Specifically, methods like LIME (Local Interpretable Model-Agnostic Model) [7] or SHAP (Shapely Additive exPlanations) can be used by making small modifications in explanation process.

The design of the integration architecture can include components as shown in Fig. 4.

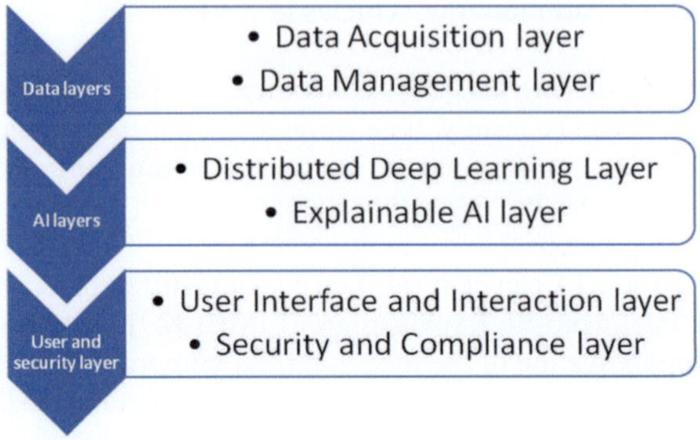

Fig. 4 Integration of XAI and DDL

4.2 Workflow

At first, a variety of sources, including sensors, Internet of Things devices, and data gateways, can provide data. The subsequent layer then handles data handling and preparation. Warehouses can be used to store and retrieve data. After processing the data, the data layers forward it to the AI layers. In DDL Layer the data is split and shared parallelly for processing. Proper training frameworks can be implemented in the DDL layer. After processing, the data is sent to next XAI layer for decision-making process. Explanations are refined and stored in this layer and a feedback layer is added to get more improvements. The next layer supports user interaction with the system. It consists of a dashboard to view the system performance, predictions and real-time monitoring. The security layer is applicable for full system for data protection. The compliance layer makes sure that the framework adheres to industry standards. The designed architecture should support efficient resource management and fault tolerance. The workflow is shown in Fig. 5.

4.3 Sample Application

The Energy Grid Management System [32, 33] efficiently integrates Explainable AI (XAI) with Distributed Deep Learning (DDL) to enhance grid operations and ensure reliable energy distribution.

Data collection is the first important step, where sensors and smart meters installed on the network collect real-time data on parameters such as energy flow, voltage levels, generation rates, and consumption patterns. Additionally, external data source such as weather forecasts and market demand forecasts are integrated to provide a comprehensive view of network conditions. After data collection, the data handling and preparation stage involves aggregating the collected data into a centralized repository, where the data is cleaned and normalized to remove noise and handle missing values. This ensures high-quality input for subsequent analysis.

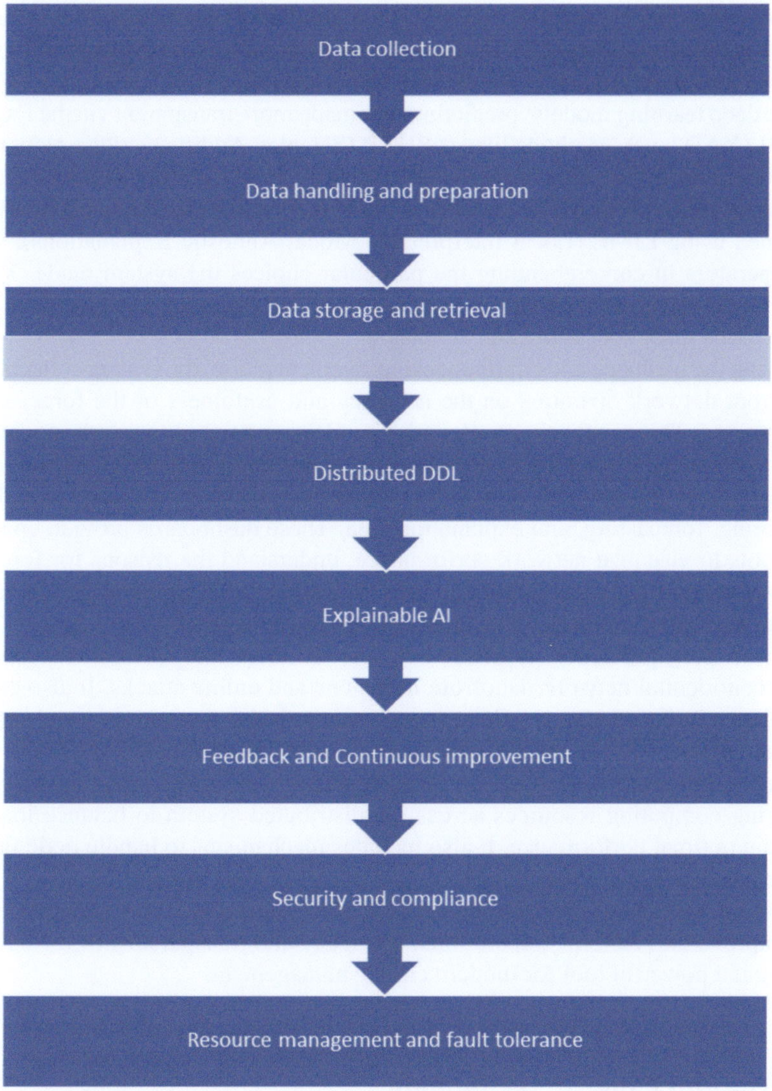

Fig. 5 Energy Grid Management System workflow diagram

The processed data is kept in distributed, scalable storage systems that enable high throughput and quick retrieval during the data storage and retrieval phase. Data indexing and caching mechanisms are used to facilitate fast access during model training and prediction.

The Distributed Deep Learning (DDL) layer leverages frameworks such as TensorFlow Distributed [34] or PyTorch Distributed [35] to train deep learning models across multiple nodes. Models such as Long Short-Term Memory (LSTM) networks and Convolutional Neural Networks (CNNs) are used to predict peak demand and analyze network topology. This distributed approach enables the system to efficiently process large-scale data and provide real-time insights.

The deep learning models' predictions are made more transparent via the Explainable AI (XAI) layer. Methods like as SHAP (SHapley Additive exPlanations) offer both local and global explanations by describing how different factors—like the weather or rates of electricity generation—affect forecasts. Individual forecasts are explained using LIME (Local Interpretable Model Agnostic Explanations), which aids operators in comprehending the particular choices the system made. Visualization tools like feature importance charts and heat maps are used to make these explanations more accessible and actionable.

During the feedback and continuous improvement phase, the system collects feedback from network operators on the accuracy and usefulness of the forecasts and explanations. This feedback, along with new data, is used to regularly update and refine the models, improving their performance and interpretability over time.

The User Engagement layer includes interactive dashboards that display real-time monitoring, forecasting, and explanation data. These dashboards provide operators with tools to visualize network performance, understand the reasons for forecasts, and receive alerts and recommendations to address potential issues such as overloads or outages.

Robust data protection is offered by the Security and Compliance layer to shield confidential network data from intrusions and online attacks. It also ensures compliance with industry regulations and standards for data privacy and network management.

Finally, the resource management and fault tolerance layer focus on efficiently allocating computing resources across the distributed system to balance load and maintain optimal performance. It also includes mechanisms to handle node failures and network failures, ensuring system continuity and data integrity.

Through this comprehensive integration, the Energy Grid Management System will improve the reliability, operational efficiency and transparency of the network, making it a powerful tool for modern energy management.

5 Pros and Cons of Technology

The Fourth Industrial Revolution may make it easier for businesses, clients, and stakeholders to access and transmit goods and services across the value chain. According to preliminary data, 4IR technology can be used to enhance supply chains, boost output during working hours, reduce waste in companies, and provide various other benefits for consumers, employees, and stakeholders. Considering the difficulties posed by the pandemic, utilizing Industry 4.0 technologies is very advantageous. XAI is anticipated to become more crucial in aiding AI systems' operational and monitoring responsibilities, particularly in order to improve AI maintainability, build confidence, and allay safety worries. One disadvantage is that while XAI offers tools for model-causal insights, which let users' pinpoint the variables influencing AI models' choices, it is unable to discern causal linkages within the data itself.

5.1 Advantages

The advantages of the integration are as follows:

- Enhanced Visibility—there will a clear understanding of the explanations given by XAI
- Reliability—the explanations given for automated decisions is useful, mainly in healthcare and finance companies.
- Strategic Decision making—As there is a visualization of the insights of decision making, technicians can understand the action to be done next.
- Flexible—Datasets in varying size, from small to large, can be handled by DDL.
- Productive—as the DDL works by distributing the process across many nodes, the resource utilization can be optimized.
- Real time Data Interpretation: As the data is processed in real-time, anomaly detection and instant decisions are possible.

5.2 Disadvantages

There are also few disadvantages as given below:

- Complexity—There are practical difficulties in integrating XAI with DDL. Therefore, architectural planning should be proper.
- Reduced Scalability—When large datasets are handled, there are many challenges in data management such as synchronization and consistency
- Resource Requirements—XAI methods can be requiring more power and memory which may not be available in some of the industries.
- Expensive—Creating and maintaining integrated infrastructure is costly and medium or small sized industries cannot afford.
- Data Privacy—In a distributed environment data privacy is a challenging task.
- User Adoption—Understand the explanations given by XAI is not possible for non-technical employees, which leads for additional training and education

Adoption of XAI may put controllers, developers, engineers, and data subjects at danger, even while it has the potential to increase openness and confidence in AI systems. The integration of Explainable AI and Distributed Deep Learning in Industry 4.0 provides many benefits. But it also presents challenges. To increase the advantages and minimize the drawbacks, careful planning, proper infrastructure, and ongoing training are essential. By addressing these challenges, industries can leverage AI technologies to create smarter, more adaptive, and transparent industrial systems.

6 Conclusion

Industry 4.0, which will transform production through intelligent, autonomous systems, is predicted to include digital technologies like artificial intelligence (AI), the Internet of Things (IoT), and robotics more deeply. Virtual reality and collaborative robotics will improve workforce capacities, and data analytics advancements will power predictive maintenance and resource efficiency. Blockchain and IoT will make supply networks more transparent and flexible. Industry 4.0 will keep redefining manufacturing procedures, promoting creativity, sustainability, and market competitiveness. End users' ability to make effective decisions may be impacted by the often-inexplicable outputs produced by AI-based DL and ML models. The use of XAI can clarify how and why the results of DL and ML models are produced. As a result, it is possible to gain knowledge of the operation, behavior, and results of models. Organizations seeking to handle the creation and application of AI models responsibly may find XAI to be helpful. XAI can assist developers in identifying possible problems like AI biases as well as in understanding the behavior of an AI model and how it arrived at a particular result.

References

1. Ghobakhloo, M.: Industry 4.0, digitization, and opportunities for sustainability. J. Clean. Prod. **252**, 119869 (2020)
2. Linardatos, P., Papastefanopoulos, V., Kotsiantis, S.: Explainable AI: a review of machine learning interpretability methods. Entropy **23**(1), 18 (2020)
3. Petch, J., Di, S., Nelson, W.: Opening the black box: the promise and limitations of explainable machine learning in cardiology. Can. J. Cardiol.Cardiol. **38**(2), 204–213 (2022)
4. Zhang, Y., Weng, Y., Lund, J.: Applications of explainable artificial intelligence in diagnosis and surgery. Diagnostics **12**(2), 237 (2022)
5. van der Velden, B.H.M., Kuijf, H.J., Gilhuijs, K.G.A., Viergever, M.A.: Explainable artificial intelligence (XAI) in deep learning-based medical image analysis. Med. Image Anal. **79**, Art. no. 102470 (2022)
6. Reyes, M., Meier, R., Pereira, S., Silva, C.A., Dahlweid, F.-M., Tengg-Kobligk, H.V., Summers, R.M., Wiest, R.: On the interpretability of artificial intelligence in radiology: challenges and opportunities. Radiol. Artif. Intell. **2**(3), Art. no. e1900443 (2020)
7. Khan, S., Yairi, T.: A review on the application of deep learning in system health management. Mech. Syst. Signal Process. **107**, 241–265 (2018)
8. Hochreiter, S., Schmidhuber, J.: Long short-term memory. Neural Comput.Comput. **9**(8), 1735–1780 (1997)
9. Ortigossa, E., Gonçalves, T., Nonato, L.: EXplainable Artificial Intelligence (XAI)—from theory to methods and applications. IEEE Access **12**, 80799–80846 (2024). https://doi.org/10.1109/ACCESS.2024.3409843
10. Jamwal, A., Agrawal, R., Sharma, M.: Deep learning for manufacturing sustainability: models, applications in Industry 4.0 and implications. Int. J. Inf. Manage. Data Insights **2**(2), 100107 (2022)
11. Shah, I.A., Jhanjhi, N.Z., Amsaad, F., Razaque, A.: The role of cutting-edge technologies in industry 4.0. In: Cyber Security Applications for Industry 4.0, pp. 97–109. Chapman and Hall/CRC (2022)

12. Malathy, S., Vanitha, C.N., Dhanaraj, R.K., Krishnasamy, L.: Augmented reality based medical education. In: ICCEBS, 2023, Sri Sairam Engineering College, Chennai
13. Dhanaraj, R.K., Singh, A., Nayyar, A.: Matyas–Meyer Oseas based device profiling for anomaly detection via deep reinforcement learning (MMODPAD-DRL) in zero trust security network. Computing **106**(6), 1933–1962 (2024)
14. Jandyal, A., Chaturvedi, I., Wazir, I., Raina, A., Haq, M.I.U.: 3D printing—a review of processes, materials and applications in industry 4.0. Sustain. Oper. Comput. **3**, 33–42 (2022)
15. Minerva, R., Lee, G.M., Crespi, N.: Digital twin in the IoT context: a survey on technical features, scenarios, and architectural models. Proc. IEEE **108**(10), 1785–1824 (2020)
16. Abirami, T., Mapari, S., Jayadharshini, P., Krishnasamy, L., Kavin, T., Kanagasubramaniyan, V.S.: A machine learning techniques for early autism spectrum disorder detection through comparative analysis of feature engineering and classification models. In: ICAICCIT-23, IEEE Delhi section (2023)
17. Ahmed, I., Jeon, G., Piccialli, F.: From artificial intelligence to explainable artificial intelligence in industry 4.0: a survey on what, how, and where. IEEE Trans. Industr. Inf.Industr. Inf. **18**(8), 5031–5042 (2022). https://doi.org/10.1109/TII.2022.3146552
18. Carvalho, T.P., Soares, F.A., Vita, R., Francisco, R.d.P., Basto,J.P., Alcalá, S.G.: A systematic literature review of machine learning methods applied to predictive maintenance. Comput. Ind. Eng. **137**, 106024 (2019)
19. Hansen, E.B., Bøgh, S.: Artificial intelligence and internet of things in small and medium-sized enterprises: a survey. J. Manuf. Syst. **58**, 362–372 (2021)
20. Ribeiro, M.T., Singh, S., Guestrin, C.: Model-agnostic interpretability of machine learning. arXiv preprint arXiv:1606.05386 (2016)
21. Dağlarli, E.: Explainable Artificial Intelligence (xAI) approaches and deep meta-learning models. In: Advances and Applications in Deep Learning. IntechOpen (2020). https://doi.org/10.5772/intechopen.92172
22. Hassija, V., Chamola, V., Mahapatra, A., et al.: Interpreting black-box models: a review on Explainable Artificial Intelligence. Cogn. Comput.. Comput. **16**, 45–74 (2024). https://doi.org/10.1007/s12559-023-10179-8
23. Peres, R.S., Jia, X., Lee, J., Sun, K., Colombo, A.W., Barata, J.: Industrial artificial intelligence in industry 4.0—systematic review, challenges and outlook. IEEE Access **8**, 220121–220139 (2020)
24. Santhiya, S., Mapari, S., Abinaya, N., Jayadharshini, P., Priyanka, S., Krishnasamy, L.: Early detection of cervical cancer using machine learning classifiers for improved diagnosis in underserved regions. In: ICAICCIT-23, IEEE Delhi section (2023)
25. Abirami, T., Mapari, S., Jayadharshini, P., Krishnasamy, L., Ragavendra Vigneshwaran, R.: Streamlined deployment and monitoring of cloud-native applications on AWS using Kubernetes, Keda, Argocd, Prometheus and Grafana. In: ICAICCIT-23, IEEE Delhi section (2023)
26. Faluyi, S.G., Chabchoub, Y., Togbe, M.U., Sublime, J.: Application of explainable AI to healthcare: a review⋆. In: IDDM'24: 7th International Conference on Informatics & Data-Driven Medicine (2024)
27. Sofianidis, G., Rožanec, J.M., Mladenic, D., Kyriazis, D.: A review of explainable artificial intelligence in manufacturing. Trust. Artif. Int. Manufactur **24**, 93–113 (2021)
28. Atakishiyev, S., Salameh, M., Yao, H., Goebel, R.: Explainable artificial intelligence for autonomous driving: a comprehensive overview and field guide for future research directions. IEEE Access (2024)
29. Mishra, S., Sturm, B.L., Dixon, S.: Local interpretable model-agnostic explanations for music content analysis. In: ISMIR, vol. 53, pp. 537–543 (2017)
30. Nohara, Y., Matsumoto, K., Soejima, H., Nakashima, N.: Explanation of machine learning models using improved Shapley additive explanation. In: Proceedings of the 10th ACM International Conference on Bioinformatics, Computational Biology and Health Informatics, pp. 546–546 (2019)

31. Kawakura, S., Hirafuji, M., Ninomiya, S., Shibasaki, R.: Adaptations of explainable artificial intelligence (XAI) to agricultural data models with ELI5, PDPbox, and skater using diverse agricultural worker data. Eur. J. Artif. Intell. Mach. Learn. **1**(3), 27–34 (2022)
32. Rathor, S.K., Saxena, D.: Energy management system for smart grid: an overview and key issues. Int. J. Energy Res. **44**(6), 4067–4109 (2020)
33. Byrne, R.H., Nguyen, T.A., Copp, D.A., Chalamala, B.R., Gyuk, I.: Energy management and optimization methods for grid energy storage systems. IEEE Access **6**, 13231–13260 (2017)
34. Dillon, J.V., Langmore, I., Tran, D., Brevdo, E., Vasudevan, S., Moore, D., Patton, B., Alemi, A., Hoffman, M., Saurous, R.A.: Tensorflow distributions. arXiv preprint arXiv:1711.10604 (2017)
35. Li, S., Zhao, Y., Varma, R., Salpekar, O., Noordhuis, P., Li, T., Paszke, A., Smith, J., Vaughan, B., Damania, P., Chintala, S.: Pytorch distributed: Experiences on accelerating data parallel training. arXiv preprint arXiv:2006.15704 (2020)

Dr. M. Dhurgadevi is working as Professor, Information Technology in Karpagam College of Engineering Coimbatore, Tamil Nadu. She did her Ph.D. from Anna University, Chennai in the area of Wireless sensor network. She did her post-graduation (M-E) in Computer Science and Engineering with first class and distinction, and did her B-Tech in Information Technology with first class. She has more than 15 years of experience in teaching graduate and postgraduate students. Areas of interests are Wireless Sensor Networks, Distributed systems, Cyber security, cloud computing and Data science. She also has four patents published to her credit. She acted as Resource Person for Various AICTE sponsored Workshop in IoT and acted as session chair for various TEQIP Sponsored and other International Conferences. She added her credits by NPTEL certification in Introduction to modern application development (Elite) and Wireless adhoc and sensor networks (Elite).She also certified SAP-ABAP and VM ware Data center Virtualized Trainer. She is a reviewer in Internal Journal of Business Intelligence and Data Mining. She has published 10 book chapters with Anuradha, Magnus, A.R Publications, Lambert publications and has more than 10 publications to his credit in reputed International Journals and 10 papers in International/National conferences.

Ms. N. Naveena is working as Assistant Professor, Information Technology in Nandha College of Technology, Erode, Tamil Nadu. She completed M.E. in Computer Science and Engineering with first class, and completed B.E. in Computer Science and Engineering with first class and distinction. She has more than 15 years of experience in teaching graduate and postgraduate students. Her areas of interests are Cyber security, Machine Learning, Deep Learning and Data science. She also has two patents published to her credit. She added her credits by Nptel certification in Cloud computing (Elite + Silver).

Mr. V. E. Sathishkumar is working as Lecturer at Department of Computing and Information Systems in Sunway University, Malaysia. Over the past ten years, his research has centered on constructing predictive models using real-time datasets from various domains, including Intelligent Transportation Systems, Smart Farming, Smart Grids, and Healthcare. Currently, he is dedicated to developing applications within the Healthcare domain. Prior to this, he was involved in the AI Convergence and Research project. His primary focus is on analyzing real-time datasets using Data Mining techniques. This involves the analysis of real-time data from multiple Korean data hubs to construct predictive models. His research interests encompass a wide range of topics, including Smart Farming, Cryptography, Biometric Technologies, Data Mining, Machine Learning, and Big Data Analytics. Parallel to his research, he is passionate about education. He is dedicated to imparting knowledge and technical skills to students, particularly in the domain of Data Science and related fields. In the classroom, he emphasizes the practical application of theoretical concepts, ensuring that students not only learn but also understand how these concepts are applied in real-world scenarios.

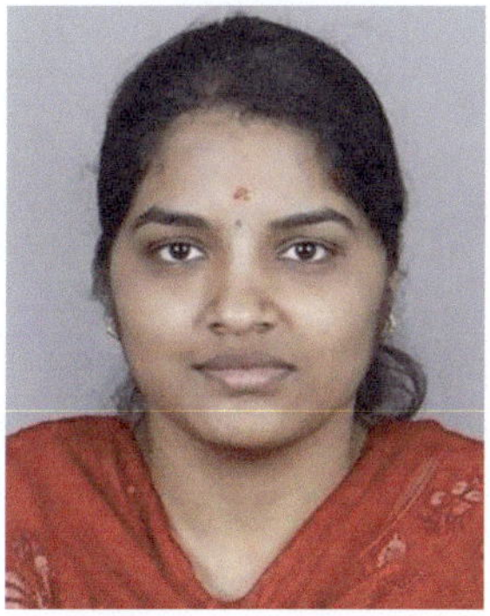

Ms. A. Sugitha is a seasoned academic professional with 8 years of teaching experience across diverse institutions. Currently, she is working at Sri Krishna College of Technology, since October 2023. With a strong passion for Cyber security, she is committed to advancing research and education in this rapidly evolving field, equipping students with the skills and knowledge required to excel in the digital era.

Ms. A. Banupriya is working as Assistant Professor, Information Technology in CSI College of Engineering, Ketti, Tamil Nadu. She completed her M.E in Computer Science and Engineering with first class, and completed B.Tech. in Information Technology with first class. She has more than 15 years of experience in teaching graduate and postgraduate students. Areas of interests are Data Mining, Machine Learning, Deep Learning and Data analytics.

Embracing Industry 4.0: Confronting Practical Realities and Navigating Complexities

C. Kishor Kumar Reddy, Mariam Fatima, R. Deepti, and S. Md. Shakir Ali

Abstract Modern industrial production systems have undergone a comprehensive paradigm change due to the Fourth Industrial Revolution, or Industry 4.0. The fundamental ideas of Industry 4.0 are explained in this chapter. These ideas include the smooth integration of cyber-physical systems (CPS), the Internet of Things' (IoT) continuous connectivity, and the application of advanced automation technologies powered by machine learning and data exchange. The chapter transverses the various opportunities and problems that industrial sectors face while implementing Industry 4.0 practices. It provides a thorough examination of the real-world problems that crop up during implementation, covering everything from infrastructural needs and technological complexity to organizational reorganization and personnel up skilling requirements. Additionally, a critical assessment of Industry 4.0's socioeconomic consequences is conducted, emphasizing how it will affect job dynamics, how skill demands will change, and how it will affect global competitiveness. Challenges include the substantial organizational capital investment required for integration, the uncertainty and unpredictable outcomes inherent in technological implementations, the necessity of supply–demand alignments and strategic analysis for decision-making, the identification of unprepared suppliers, and the wide time range between implementation and tangible results.

Keywords Industry 4.0 · Smart manufacturing · Cyber-physical systems · Internet of things · Automation · Digital transformation · Employment · Skills · Competitiveness · Organisational restructuring · Supply–demand alignments

C. K. K. Reddy (✉) · M. Fatima
Stanley College of Engineering and Technology for Women, Hyderabad, India
e-mail: ckishorkumar@stanley.edu.in

R. Deepti
Independent Researcher, Secunderabad, Telangana, India

S. Md. S. Ali
Department of Digital Business, Lithan Academy (eduCLaaSPvt Ltd), Singapore, Singapore

L. Krishnasamy et al. (eds.), *Distributed Deep Learning and Explainable AI (XAI) in Industry 4.0*, Information Systems Engineering and Management 55,
https://doi.org/10.1007/978-3-031-94637-0_5

1 Introduction

The arrival of Industry 4.0 signifies an extreme shift in the manufacturing landscape, catalyzed by the fusion of digital technologies with conventional industrial processes. This transformative paradigm, identified by the seamless integration of cyber-physical system, the Internet of Things, advanced automation, and information analytics, indicates a new age of unprecedented effectiveness, agility, and innovation [1–5]. The Historical narrative of the Industrial Revolution unfolds across several distinct phases: Industry 1.0 (1760–1820): This era marked the transition from manual labor to mechanized processes powered by steam and water, setting the groundwork for industrialization and the rise of factories. Industry 2.0 (1870–1914): Known as the technological revolution, this period witnessed the widespread adoption of electrical energy to power machines. Innovations such as the assembly line revolutionized mass production, driving unprecedented economic growth. Industry 3.0 (Late 20th century): This phase saw the integration of computer and communication technologies into manufacturing processes, leading to automation and digitalization. It brought about a new era of efficiency and productivity. Originating in Germany as "Industry 4.0," this concept represents the latest phase in the evolution of manufacturing. It emphasizes the digitization and interconnectivity of production systems, blurring the lines between the physical and virtual worlds [10]. Within this context, organizations worldwide are confronted with both chances and challenges as they embark on the journey of Industry 4.0 adoption [3, 6–10]. The promise of Industry 4.0 is that it has the potential to improve production processes, improve product quality, and transform supply chain management through the integration of digital technologies. Automation driven by artificial intelligence (AI) and machine learning algorithms allows companies to streamline routine operations, reduce errors, and improve resource utilization [1, 2, 5].

The IoT enables seamless connectivity and data exchange among interconnected devices, facilitating real-time monitoring, predictive maintenance, and responsive decision-making [4, 7, 8]. Moreover, data analytics leverages the huge troves of data brought by interconnected systems to extract actionable insights, enabling organization's to gain an edge in a drastically data-driven marketplace [2, 5, 8]. Alongside the transformative potential of Industry 4.0, organization's encounter a myriad of challenges that warrant careful consideration and strategic response [3, 6, 7, 9]. Foremost among these challenges is the looming spectre of job displacement as automation and digitization reshape traditional employment landscapes [3, 6, 9]. Concerns over cybersecurity vulnerabilities also loom large, underscoring the critical significance of robust security measures to shield sensitive data and critical infrastructure from cyber threats [3, 7, 9]. The imperative of up skilling the workforce to navigate the complexities of a digitized environment emerges as a pressing priority, necessitating comprehensive training and educational initiatives to bridge the skills gap and foster digital literacy [3, 6, 9].

The upfront capital investment required for the implementation of advanced technologies poses a significant financial barrier for many organizations, particularly

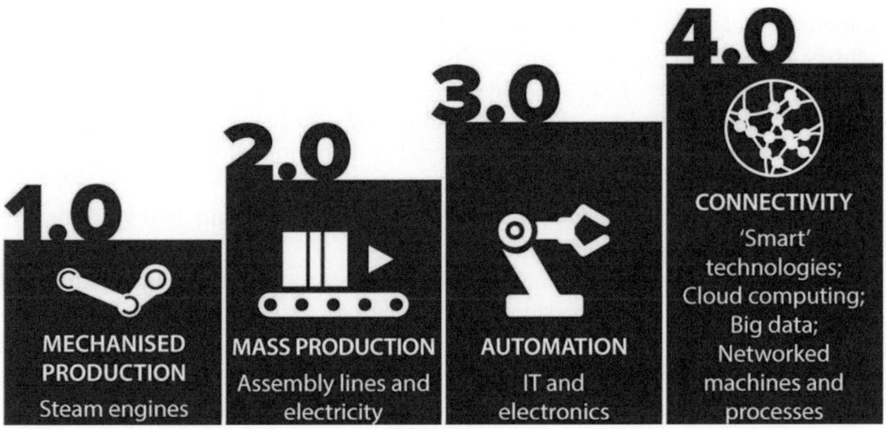

Fig. 1 Phases of industrial revolution

smaller enterprises [3, 6, 9]. While large corporations may possess the resources and expertise to invest in cutting-edge technologies, smaller businesses may face challenges in accessing capital and acquiring the necessary technical know-how [3, 6, 9]. Addressing these disparities requires a concerted effort to promote inclusivity and foster an environment conducive to digital innovation and adoption across all segments of the industrial ecosystem [3, 6, 9]. Against this backdrop of opportunities and challenges, this chapter embarks on an exploration of the practical issues and complexities inherent in the implementation of Industry 4.0 practices [3, 6, 7, 9, 10]. By delving into real-world scenarios and case studies, from infrastructure requirements to organisational restructuring and workforce development, this chapter aims to provide valuable perception and pragmatic solutions for organizations navigating the complexities of Industry 4.0 adoption [3, 6, 7, 9, 10]. Through a rigorous analysis of opportunities and obstacles, this chapter endeavors to empower organizations to utilize the changing power of Industry 4.0 while mitigating risks and maximizing benefits in an increasingly digitized industrial landscape [3, 6, 7, 9]. Figure 1 gives the phases of industrial revolution.

2 Background

Authors [10] propose to examine the problems and benefits of adopting business 4.0 technologies in manufacturing. Industry 4.0 represents the fourth industrial revolution and involves the integration of technological technologies such as the Internet of Things, big data analytics, cloud computing, and cyber-physical systems. Time management, optimization, and improved convenience and speed in production. With the creation of new value propositions and revenue streams, manufacturers can utilize customer data and feedback to tailor offerings and services to individual

Challenges include high costs and complexity, and implementing and integrating Industry 4.0 technologies necessitate substantial investments in hardware, software, and workforce training. Intuition and data security concerns, compatibility among various systems and devices, along with addressing data security and privacy issues, present significant challenges. Organisational and cultural shifts: The transition to Industry 4.0 requires fostering a collaborative, innovative, and data-driven mindset. Many manufacturers lack the requisite skills and expertise within their workforce to effectively manage and use technologies.

The authors [11], briefed the challenges and gaps encountered by SMEs (small and medium enterprises) in progressing nations when accepting Industry 4.0 technologies. SMEs in growing countries often lack the essential infrastructure, skills, and financial resources needed to embrace the latest manufacturing technologies associated with Industry 4.0. Barriers hindering the investment cost of small and medium-sized businesses to adopt Industry 4.0 are high, limited technical expertise, and resistance to change among SME owners and managers. The article proposes potential solutions to these challenges, such as government support initiatives, fostering collaborations with larger firms, and the development of customized Industry 4.0 solutions tailored specifically for SME's.

The authors [12], explained the analysis of the challenges confronting industrial firms, particularly amidst factors like intensified global competition, trade constraints, and shrinking profit margins, accentuated by the effect of the COVID-19 pandemic. It emphasizes the urgent need for industrial companies to fortify the resilience of their value chains and augment the flexibility and adaptability of their production processes in response to external disruptions and evolving market conditions. The study explores the conceptual impact and operational change that Industry 4.0 and smart manufacturing concepts can offer industrial enterprises aiming to navigate these challenges effectively. It delves into how these advanced technologies and methodologies can revolutionize conventional manufacturing practices, optimize operational efficiency, and foster innovation across diverse industry sectors. Addressing's include the concept of the smart factory, which integrates IoT, AI, and automation to establish interconnected, data-driven production environments that boost productivity and agility. The role of digitalization in streamlining processes, enhancing decision-making, and enabling real-time monitoring and control is complemented by lean production principles to eliminate waste and magnify operational efficiency. The significance of cultivating the art of innovation, collaboration, and steady improvement within industrial firms to foster sustainable growth, adaptability, and competitiveness in the evolving business landscape.

The authors [13] mentioned to integrate digital technology, data, and automation across devices, shifting from traditional processes to data-driven operations. Analyzing and sharing existing data to improve overall performance and adapt to business needs. Connect and optimize the supply chain. The aim is to improve productivity, flexibility, and responsiveness to industry changes. These elements collaborate to optimize operations, improve visibility, track assets, and streamline decision-making. Benefits include the adoption of digitalization and Industry 4.0 technologies offers various benefits, including enhanced efficiency, responsiveness, sustainability, and competitive advantage. Organizations embracing these technologies can lead the market, drive innovation, and achieve sustainable growth. Implementation implementing these technologies presents challenges such as cost, data security, interoperability, workforce skills, organisational culture, scalability, and integration with legacy systems. Overcoming these hurdles requires strategic planning, investment, cybersecurity measures, workforce training, change management strategies, and flexible, scalable systems. Highlights include the life-changing impact of digitalization and Industry 4.0 on supply chains, focusing on the need for organizations to adapt to the developing technological outlook to drive efficiency, sustainability, and competitiveness.

The authors [14] defined Industry 4.0 and bring significant advances and opportunities. Trace of the historical evolution of technology, starting from the Renaissance period in Western Europe, and highlighted the advancements in arts, science, and societal changes that paved the way for technological progress. The narrative progresses to the development of machinery, which transformed human labour by using various energy sources to streamline work processes and drive industrial economies. The concepts of 'industry' and 'industrialization' are crucial for grasping the evolution of technology and its societal impacts. Stress on the significance of digitalization, scientific knowledge, and technological innovation in advancing the fourth industrial revolution. It emphasizes the necessity for collaboration between natural and artificial intelligence in organizing labour and managing resources, both in public administration and organisational contexts. The challenges brought by the COVID-19 pandemic highlight the urgent need for innovative solutions to navigate global crises effectively. A detailed exploration of the 4th Industrial Revolution and Artificial Intelligence, shedding light on the complex interplay between technology, society, and economics, and AI's pivotal role in driving innovation, transforming industries, and shaping the future of business and society in the period of Industry 4.0

The authors [15], explained that Machine learning has become a pivotal element in the production sector, particularly within the framework of Industry 4.0, facilitating significant advancements in various operational domains. Process Optimization: By employing machine learning algorithms to examine extensive datasets derived from sensors and other sources, manufacturers can intelligently enhance production processes, leading to the realization of "Smart Manufacturing," where operational efficiency and adaptability are substantially improved. Automation of Tasks: Machine learning empowers automation by replacing manual labour in repetitive or hazardous tasks. This boosts productivity and enhances safety standards within manufacturing environments. Enhanced Quality Control: continuous monitoring of product quality is made feasible through the integration of machine learning algorithms with sensor technology. This approach ensures early detection of defects, minimizes waste, and elevates overall product quality. Proactive Maintenance Strategies By leveraging data from sensor-equipped machinery, machine learning algorithms can predict maintenance requirements before equipment failures occur, minimize downtime, and reduce maintenance costs. Demand forecasting: machine learning models are adept at predicting energy demand by analyzing historical data alongside external factors such as weather patterns. This capability enables manufacturers to optimize energy production and consumption, resulting in cost savings and operational efficiency improvements. AI-Powered Customer support chatbots driven by machine learning algorithms offer seamless customer service around the clock, diminishing the reliance on human intervention for handling routine inquiries. These chatbots continually learn from interactions, refining their responses over time to enhance customer satisfaction. The integration of machine learning technologies into manufacturing processes under Industry 4.0 is revolutionizing the sector, rendering operations smarter, more adaptable, and more efficient. This transformative shift benefits enterprises of all scales, empowering them to remain competitive amidst dynamic market dynamics.

The authors [16], note that Industry 4.0 marks a pivotal shift in manufacturing, driven by advancements in technology like IoT, cloud computing, and machine learning. This revolution promises profound impacts on the industry, bringing about increased efficiency, enhanced decision-making capabilities, and improved quality control Alongside these benefits, there are significant challenges that manufacturers must navigate to completely understand the power of Industry 4.0 positive effect of Industry 4.0 in manufacturing include enhanced productivity and efficiency, Automation and optimisation of processes lead to streamlined operations and higher productivity levels. Real-time data visibility and access to timely and accurate data empower better decision-making, optimised resource allocation, and enhanced supply chain management. Maintenance and Monitoring Advancements, predictive maintenance, and advanced monitoring systems minimise downtime and ensure smoother operations, contributing to greater business continuity. Improved Product Quality: IoT-enabled quality control mechanisms and collaborative robots facilitate real-time monitoring and adjustments, resulting in higher-quality products. Promotion of Sustainability and Better Working Conditions: Industry 4.0 promotes sustainable practices through efficient resource usage and creates safer, more conducive working environments for employees. Personalisation and Consumer Trust: customised products and transparent, data-driven processes foster trust with consumers, driving brand loyalty and satisfaction. Challenges of Implementing Industry 4.0: Skills Gap, There is a pressing need to upskill the workforce to effectively operate and manage the sophisticated technologies integral to Industry 4.0.

Data Management and Privacy Concerns: Managing vast volumes of data while ensuring privacy and ownership rights poses significant challenges. Interoperability Issues: compatibility issues between different systems and technologies hinder seamless integration and data exchange. Cybersecurity Risks: The increased connectivity of Industry 4.0 systems introduces cybersecurity vulnerabilities, necessitating robust safety measures to safeguard confidential data and systems. Data Processing Complexity: Handling the influx of data in various formats and processing it efficiently using AI algorithms presents technical hurdles. Resistance to change, overcoming organisational resistance, and simulating a culture of innovation and adaptability are crucial for successful implementation. To overcome these challenges, manufacturers must adopt a comprehensive approach that includes technological investments, workforce development initiatives, regulatory conformance, and organisational restructuring. By addressing these hurdles effectively, companies can utilise the full power of Industry 4.0, gaining a fierce edge in the swiftly evolving manufacturing ecosystem.

The author [17], explained the development of Industry 4.0, indicated with the fusion of advanced technologies like AI, IoT, and robotics, is set to revolutionise business operations. Yet, this transformation also brings significant hurdles in terms of employment and organisational restructuring. One major concern is the need for workers to acquire new skills and adjust to the swiftly evolving technological landscape. The rapid pace of technological progress within Industry 4.0 demands that companies ensure their workforce remains capable of keeping up with these advancements. Another critical challenge is the cultural shift required by Industry 4.0. Traditional organisational structures and management practices may need reevaluation to accommodate new technologies and work methods. This shift can be challenging, requiring a substantial change in mindset and behaviour from employees and management alike. Additionally, implementing Industry 4.0 technologies often involves significant investments in new equipment, software, and training. This financial burden can be especially daunting for smaller companies, which may struggle to allocate the necessary resources. In response to these challenges, companies must develop strategies to support their employees through the transition to Industry 4.0. This might involve offering comprehensive training and upskilling programmes, as well as creating new job roles that leverage the capabilities of the latest technologies. Ultimately, the successful adoption of Industry 4.0 depends on companies' ability to navigate these challenges effectively and adapt to the new technological landscape. By doing so, they can unlock the benefits of improved efficiency, productivity, and innovation that Industry 4.0 offers. The authors Dong et al. [18] described this article as filling a significant void in supply risk management literature by tackling the complexities of correlated supply risks, a closer reflection of real-world dynamics.

By simulating suppliers' production processes with correlated proportional random yields, the study makes substantial strides in understanding how firms can effectively manage their supply base and leverage risk pooling through diversification. It defines the procurement problem for a monopoly firm that sources from many unreliable suppliers in a general n-supplier framework. It sheds light on relational sourcing decisions in a scenario involving two suppliers, illustrating the influence of yield relations on supply base selection and revealing that highly positively correlated suppliers might lead to sole sourcing from the retailer with a huge effective possession price and reliability, potentially reducing supply diversification and firm profit. The study extends these insights to the multiple-supplier scenario, assuming a multivariate normal distribution for yields. It underscores the necessity of considering yield relations, effective accession costs, and retailer reliability in concert for optimal supply base selection, warning against suboptimal outcomes if any factor is overlooked. The managerial implications emphasize the importance of adopting a holistic approach to choosing the ideal supply base, especially while grappling with correlated yield threats. Key terms such as supply diversification, correlated random yields, the newsvendor model, costing, and multivariate dependence encapsulate the central themes and contributions of the research.

The author [19], examined how firms make capital investment decisions amidst uncertain future output demand. The paper frames this challenge as a singular stochastic control problem, aiming to minimise the expected cost of capital investment while ensuring that capital stock remains within set constraints. Factors include Capital Stock: The firm operates within a finite capital stock that it can invest in a partially reversible manner. Output Demand: The paper models future output demand as a stochastic process, acknowledging its inherent uncertainty. Investment Cost: Capital investment incurs costs for the firm, which must be carefully weighed against potential returns. Ambiguity Aversion: The firm's manager may exhibit a heightened aversion to ambiguity, influencing their approach to investment decision-making. Methodology employed includes Formulating the firm's problem as a singular stochastic control problem, aimed at minimising expected capital investment costs while respecting capital stock constraints. Conducting comparative static analysis on various parameters, such as output demand volatility and the manager's level of ambiguity aversion. Findings encompass higher levels of output demand volatility, which tends to dissuade capital investment as the increased uncertainty prompts firms to postpone investment decisions until more clarity emerges. If the firm's manager is more averse to ambiguity, capital investment tends to be delayed as well, as they prefer to avoid making decisions under uncertain conditions. Presenting optimal investment strategies derived from variation inequalities, accompanied by numerical examples to illustrate these findings, the broadens implications for firms' investment decisions and the overall economy.

The authors [20] illustrated that Industry 4.0, the fourth industrial revolution, is reshaping the landscape of businesses and society through digitalization and connectivity. Key technologies like the Internet of Things (IoT), artificial intelligence (AI), and modern robotics are driving this transformation, particularly in supply chain management. These advancements are not only addressing ongoing megatrends such as globalization, demographic shifts, and sustainability concerns but also preparing businesses for future challenges. In the core of Industry 4.0 lies the integration of instantaneous data from various sources, including point-of-use data, supplier warehouse systems, and upstream processes. This integration facilitates a proactive approach to supply chain management, where AI plays a crucial role in predicting demand, optimizing production schedules, and streamlining procurement and logistics decisions. Moreover, collaboration across the supply chain is increasingly vital in the Industry 4.0 view. By breaking down data silos and embracing comprehensive data sharing, businesses can achieve lower inventory levels, shorter lead times, and improved responsiveness to disruptions. This collaborative mindset not only enhances operational efficiency but also enriches the overall customer experience. To succeed in the fourth industrial revolution, businesses must stay informed about current and anticipated megatrends. By understanding the evolving landscape and seizing opportunities for innovation, organizations can position themselves for competitive advantage in an ever-changing market environment. Authors explored their main purpose, which is to identify the main points affecting the similarities and make comparisons between the automotive, agricultural, and textile sectors. The Grey Relationship Analysis method identifies the best activities between business and business 4.0 technology. The findings highlighted that infrastructure is most important, followed by the organisation's strategy and capital investment. Interestingly, the automotive industry is the most successful industry in terms of tactical coordination of processes and technologies in production. Additionally, the problems and opportunities encountered in the use of these products in various production areas are presented. Additionally, cross-industry comparative analysis provides valuable guidance for companies looking to understand the benefits and barriers to implementing these technologies.

The authors throw light on Incorporating digital technologies into design and manufacturing processes is a cornerstone of Industry 4.0, representing a pivotal shift often dubbed the fourth industrial revolution. This evolution relies on leading-edge tools such as artificial intelligence (AI) and machine learning to transport through extensive information in manufacturing, aiming primarily to boost productivity, slash costs, and fortify workplace safety, thereby shaping the future landscape of manufacturing. The digitization of manufacturing involves migrating from paper-based systems to digital ones powered by smart technologies. This transition promises heightened product efficiency, increased worker output, and improved precision. It encompasses the digitization of various processes, notably design, now executed through 3D CAD files, and the deployment of digital twins to replicate objects under construction. Connectivity facilitated by the Internet of Things (IoT) ensures smooth data exchange, with designs seamlessly transmitted to tablets rather than traditional paper manuals. Digitizing manufacturing offers a host of advantages. It enables real-time data collection and analysis, resulting in significant productivity gains. Moreover, it streamlines the optimization of designs in response to real-world conditions, empowering engineers to focus on innovation. Furthermore, digitalization fosters agility and responsiveness among manufacturers, easing adaptation to evolving consumer preferences and new product launches. It also enables follow-the-sun manufacturing, ensuring continuous operations across different time zones. However, challenges accompany the digitization journey. Traditional companies may grapple with rigid systems, impeding the transition to agile methodologies. Moreover, a lack of familiarity with innovative technologies among employees can disrupt manufacturing processes. Despite these obstacles, the benefits of digitization in manufacturing are substantial. It drives up productivity, trims costs, enhances workplace safety, and bolsters market share and reputation. Companies that embrace this digital transformation stand to gain a competitive edge, while those hesitant to evolve risk lagging behind in the dynamic marketplace.

2.1 Summary of the Findings

Table 1 gives a brief description of the key findings, along with the methodologies and focus points of the references.

3 Industrial Progression: Traversing the Phases from 1.0 to 4.0

3.1 Industry 1.0: (1760–1820)

Industry 1.0, spanning from 1760 to 1820, represented a monumental shift from manual labour to mechanised processes fueled by steam and water power. This period witnessed the emergence of factories and marked the onset of industrialization. One of the foremost challenges during Industry 1.0 was the limited technological sophistication of early machinery, resulting in frequent breakdowns and inefficiencies in production processes. Workplace safety was a pressing concern as factories lacked regulations, leading to hazardous working conditions and frequent accidents. The introduction of machinery displaced many skilled artisans and craftsmen, contributing to social unrest and labour disputes. Establishing infrastructure for steam engines and water power posed significant challenges, requiring substantial investment and meticulous planning to support industrial growth.

3.2 Industry 2.0: (1870–1914)

Industry 2.0, spanning from 1870 to 1914, was characterised by the widespread adoption of electrical energy to power machines, revolutionising mass production and fostering economic growth. Innovations such as the assembly line became synonymous with this era. Key challenges during Industry 2.0 included the development of reliable electricity infrastructure, particularly in rural areas where access was limited. Despite technological advancements, labour conditions remained harsh in many factories, leading to labour strikes and worker protests demanding better treatment. Achieving standardisation in manufacturing processes and components was another challenge due to diverse regional practices and industrial norms. Industrialization exacerbated economic disparities between industrialised urban centres and agrarian regions, highlighting social inequalities and economic imbalances.

Table 1 Key findings and focus points of few references

References	Focus points	Methodology	Key findings	Limitations
[10]	Challenges and benefits of industry 4.0	Examination of challenges and benefits	Improved productivity and performance, creation of new value propositions and revenue streams, high costs and complexity, interoperability and data security concerns, organisational and cultural shifts, skill gaps, obstacles to digital transformation, development of new business models	High costs and complexity associated with implementing Industry 4.0 technologies
[11]	Challenges faced by SMEs in adopting Industry 4.0 technologies	Identification of challenges and gaps	Lack of infrastructure, skills, and financial resources, high investment costs, limited technical expertise, resistance to change among SMEs, potential solutions including government support initiatives collaborations with larger firms, and customized Industry 4.0 solutions tailored for SME's	SMEs in developing countries face barriers such as high investment costs and limited technical expertise
[12]	Challenges confronting industrial firms amidst global competition	Examination of challenges and implications	Intensified global competition, trade constraints, shrinking profit margins, need for resilience and adaptability, strategic implications and operational transformations, revolutionization of manufacturing practices, optimisation of efficiency and innovation, and fostering sustainable growth	The study focuses on challenges facing industrial firms amidst global competition and trade constraints, possibly overlooking other sectors

(continued)

Table 1 (continued)

References	Focus points	Methodology	Key findings	Limitations
[14]	Impact of the fourth industrial revolution, with a focus on AI	Examination of AI's role and historical evolution	Advancements and opportunities in AI, overview of historical technological evolution, importance of collaboration between natural and artificial intelligence, challenges posed by COVID-19 pandemic, AI's transformative role in driving innovation, reshaping industries, and shaping the future of business and society	It discusses the impact of AI on industries and society but may not delve deeply into the practical implications or implementation challenges
[15]	Significance of machine learning in manufacturing	Explanation of machine learning applications	Machine learning applications in production process optimisation, automation of tasks, quality control, predictive maintenance, demand forecasting, AI-powered customer support, transformative impact on manufacturing sector, and benefits including productivity gains, safety enhancements, and improved customer experiences	It covers various applications of machine learning in manufacturing but does not address specific industry challenges or scalability issues
[16]	Impacts and challenges of industry 4.0 in manufacturin g	Examination of positive impacts and challenges	Enhanced productivity and efficiency, real-time data visibility, maintenance and monitoring advancements, improved product quality, promotion of sustainability and better working conditions, challenges including skills gaps, data management concerns, and cybersecurity risks	While discussing the positive impacts of Industry 4.0, it could not provide detailed insights into overcoming implementation challenges

(continued)

Table 1 (continued)

References	Focus points	Methodology	Key findings	Limitations
[17]	Employment and organisational restructuring challenges posed by industry 4.0	Discussion on workforce and cultural shifts	Need for workforce upskilling and organisational restructuring, challenges in adapting to technological advancements, financial burden of technology adoption, strategies for supporting employees, successful adoption dependent on effective navigation of challenges	Challenges related to employment and organisational restructuring are highlighted but may lack comprehensive strategies for addressing these issues
[18]	Managing correlated supply risks within industry 4.0	Exploration of supply chain risk management	Utilisation of analytical hierarchy process (AHP) and grey relational analysis (GRA), identification of factors influencing supply risk management, strategic implications of correlated supply risks, importance of holistic approach in supply base selection, managerial implications for risk diversification and optimal supply base selection	The study focuses on supply risk management, which may not fully capture the broader implications of Industry 4.0 adoption for businesses
[19]	Capital investment decisions amidst uncertain future output demand	Examination of investment decisions	Formulation of stochastic control problem, consideration of capital stock constraints, output demand volatility, investment costs, and ambiguity a version, identification of optimal investment strategies, implications for firms' investment decisions	The study examines capital investment decisions but does not consider the specific challenges or opportunities related with Industry 4.0 adoption

(continued)

Table 1 (continued)

References	Focus points	Methodology	Key findings	Limitations
[20]	Transformation of supply chain management through industry 4.0	Discussion on supply chain integration	Combination of IoT, AI, and robotics in supply chain management, focus on real-time data integration, collaborative approach in supply chain, advantages including cost reduction and improved customer experience, importance of staying informed about megatrends and seizing innovation opportunities	It discusses reshaping potential of Industry 4.0 in supply chain management but practical implementation strategies are not present
	Alignment between operational excellence methodologiesand industry 4.0 technologies	Analysis of strategic alignment	Utilisation of analytical hierarchy process (AHP) and grey relational analysis (GRA), identification of critical factors influencing alignment, identification of the automotive industry as top performer in strategic alignment, benefits including cost reduction, inventory optimisation, and enhanced customer satisfaction	The research focuses on strategic distribution between operational quality and Industry 4.0 technologies but may not address sector-specific
	Digitising manufacturingprocesses within industry 4.0	Examination of digitization benefits and challenges	Focus on improved productivity, cost reduction, enhanced workplace safety, challenges including rigid systems and skills gaps, transformative potential of digitization, and the importance of embracing digital transformation for competitiveness in the manufacturing landscape	Highlighting the benefits of digitization in manufacturing, it may not offer detailed solutions to overcome challenges, particularly for traditional companies

3.3 3 3 Industry 3.0: (Late Twentieth Century)

Industry 3.0, emerging in the late 20th century, witnessed the integration of computer and communication technologies into manufacturing processes, leading to automation and digitalization. This era heralded a new age of efficiency and productivity in industrial production. Implementing Industry 3.0 technologies posed challenges such as high capital investment requirements and the need for technical expertise, mainly for small and medium-sized enterprises (SMEs). Workforce reskilling became imperative as workers needed to acquire new skills to operate and maintain automated systems effectively. Cybersecurity emerged as a critical concern with the increased reliance on digital systems, necessitating robust actions to safeguard sensitive data and frameworks. Integrating digital systems across supply chains presented challenges in interoperability and data exchange, requiring standardization and collaboration among stakeholders.

3.4 3 4 Industry 4.0: (Present, Twenty-First Century)

Industry 4.0, the current era, emphasizes the digitization and interconnectivity of production systems, leveraging technologies like IoT, AI, and data analytics for optimized operations and enhanced competitiveness. Industry 4.0 implementation entails substantial investment costs, including hardware, software, and workforce training, posing financial challenges for many organizations. Addressing the skills gap is crucial, with a shortage of professionals capable of managing and leveraging advanced technologies. Ensuring data security becomes paramount to protecting against cyber threats and maintaining the integrity of sensitive information in a connected environment. Interoperability among various digital systems and devices remains a challenge, necessitating standardization and protocols for seamless integration. Overcoming organisational resistance to change and cultural barriers is also essential for successful Industry 4.0 adoption, requiring effective change management strategies and leadership commitment.

Table 2 describes the industrial phase periods with their characteristics and major challenges.

Table 2 Industrial phase and its aspects

Industrial phase	Period	Key characteristics	Major challenges
Industry 1.0	1760–1820	Transition from manual labour to mechanised processes	• Limited technological sophistication of early machinery • Workplace safety concerns due to lack of regulations • Displacement of skilled artisans and craftsmen
Industry 2.0	1870–1914	Widespread adoption of electrical energy for mass production	• Development of reliable electricity infrastructure • Harsh labour conditions leading to protests • Standardisation challenges in manufacturing processes
Industry 3.0	Late twentieth century	Integration of computer and communication technologies	• High capital investment requirements • Workforce reskilling needs Cybersecurity concerns • Interoperability challenges in digital systems
Industry 4.0	Present	Emphasis on digitization and interconnectivity	• Substantial investment costs (hardware, software, andtheir training) • Addressing skills gap • Ensuring data security • Interoperability issues • Overcoming organisational resistance

4 Evolution of Industrial Revolutions: A Journey Through Time

4.1 Technological Innovations Driving Industrial Evolution

The progression of industrial revolutions can be traced back to the introduction of groundbreaking technological innovations that revolutionised manufacturing processes. Industry 1.0 saw the advent of steam engines and mechanized production, while Industry 2.0 brought electricity and industrial units. Industry 3.0 witnessed the integration of computers and automation, laying the foundation for digitalization. In Industry 4.0, technologies such as the Internet of Things (IoT), artificial intelligence (AI), and big data analytics are reshaping industrial landscapes, driving unprecedented levels of connectivity.

4.2 *Economic Implications of Industrial Revolutions*

Each industrial revolution has had profound implications for global economies, reshaping industries and driving economic growth. Industry 1.0 fueled the rise of factories and urbanization driving the transition from agrarian to industrial economies. Industry 2.0 accelerated mass production and consumption, leading to significant advancements in living standards and wealth distribution. Industry 3.0 facilitated globalization and the emergence of knowledge-based economies, driving innovation and specialization. Industry 4.0 is poised to further disrupt traditional economic models, fostering digital economies and reshaping global trade dynamics.

4.3 *Societal Transformations and Challenges*

The impact of industrial revolutions extends beyond economic realms, profoundly influencing societal structures and dynamics. Industry 1.0 spurred urbanization and population growth, leading to the emergence of factory towns and labor movements. Industry 2.0 brought about changes in living conditions and social hierarchies, fueling debates over worker rights and consumerism. Industry 3.0 saw the rise of the information age and digital divide, exacerbating social inequalities and cultural shifts. In Industry 4.0, societal challenges include concerns over job displacement, digital divide, and privacy rights amidst increasing reliance on technology.

4.4 *Environmental Considerations and Sustainability*

As industrial revolutions have advanced, so too have concerns over environmental sustainability and resource depletion. Industry 1.0 marked the beginning of widespread environmental degradation due to pollution and deforestation associated with industrialization. Industry 2.0 intensified environmental pressures with increased energy consumption and waste generation. Industry 3.0 introduced awareness of environmental issues and efforts towards sustainability through green technologies and regulations. Industry 4.0 presents opportunities for sustainable development through smart manufacturing, resource optimization, and renewable energy integration, yet challenges persist in balancing economic growth with environmental preservation.

4.5 Global Impacts and Collaborative Solutions

Industrial revolutions have transcended national boundaries, with global implications for trade, geopolitics, and cultural exchange. Industry 1.0 fueled colonial expansion and trade networks, shaping global power dynamics. Industry 2.0 accelerated industrialization worldwide, leading to economic competition and geopolitical tensions. Industry 3.0 facilitated globalization and interconnectedness, driving cross-border collaboration and technological exchange. In Industry 4.0, global cooperation is essential to label shared obstacles such as climate change, cybersecurity threats, and socioeconomic disparities, requiring collaborative solutions and collective action on a planetary scale, Table 3 gives insights into the revolutions of industries.

5 Challenges Across Industrial Eras

5.1 Industry 1.0: Establishing the Foundations

The transition to mechanized processes powered by steam and water during Industry 1.0 necessitated the development of robust technological infrastructure. Building factories equipped with steam engines and water power systems required significant capital investment and engineering expertise. However, the challenges extended beyond mere construction. Ensuring the reliability and efficiency of these early machinery posed considerable obstacles. Engineers faced frequent breakdowns and inefficiencies in production processes due to the limited technological sophistication of the era. Moreover, establishing the necessary infrastructure for steam engines and water power systems demanded meticulous planning and coordination, as well as addressing logistical challenges such as fuel supply and transportation networks. Table 4 mentions the industrial era and challenges.

5.2 Industry 2.0: Shifting Power Structures

The advent of electricity-powered machines in Industry 2.0 brought about profound changes in social and labor dynamics. While technological advancements promised increased productivity and efficiency, they also exacerbated existing labor rights issues. Harsh working conditions in factories led to labor strikes and worker protests demanding better treatment and fair wages. Moreover, achieving standardization in manufacturing processes and components proved challenging, as diverse regional practices and industrial norms had to be reconciled. Economic disparities between industrialized urban centers and agrarian regions widened, highlighting social inequalities and necessitating policy interventions to address them.

Table 3 About industrial revolutions

Industrial revolution	Period	Key technological innovations	Economic implications	Societal transformations	Environmental considerations and sustainability	Global impacts and collaborative solutions
Industry 1.0	1760–1820	Steam engines, mechanized production	Rise of factories and urbanization	Urbanization, labor movements, factory towns	Environmental degradation, pollution, deforestation	Colonial expansion, trade networks shaping global power dynamics
Industry 2.0	1870–1914	Electricity, assembly lines	Accelerated mass production, consumerism	Changes in living conditions, worker rights debates	Increased energy consumption, waste generation	Industrialization worldwide, economic competition, geopolitical tensions
Industry 3.0	Late twentieth century	Computers, automation	Globalization, knowledge-based economies	Digital economies, reshaping global trade dynamics, job displacement, digital divide	Opportunities for sustainable development, smart manufacturing, renewable energy integration	Awareness of environmental issues, social inequalities, privacy rights concerns, addressing shared challenges like climate change, cybersecurity threats, socioeconomic disparities
Industry 4.0	Present	IoT, cyber-physical systems, AI, additive manufacturing (3D printing), big data analytics, autonomous robots	Shift towards data-driven industries, creation of digital services, enhanced productivity, disruption of traditional	Rise of remote working, increased demand for digital literacy, changes in employment sectors, emphasis on lifelong learning skills	Improved resource efficiency through optimized processes, potential for decreased environmental footprints, challenges related to e-waste and energy consumption of data centers	Increased globalization, interconnectedness, emphasis on cross-border collaboration, technological exchanges to address global challenges like climate change and sustainable development

Table 4 Industrial era with their challenges

Industrial era	Challenges
Industry 1.0	• Developing robust technological infrastructure for steam and water power Systems • Ensuring reliability and efficiency of early machinery
Industry 2.0	• Addressing labor rights issues amidst harsh working conditions • Achieving standardization in manufacturing processes and components
Industry 3.0	• Affording high capital investment for adopting new technologies • Reskilling workforce to adapt to automation and digitalization
Industry 4.0	• Managing substantial costs associated with implementing advanced Technologies • Safeguarding against cybersecurity threats in interconnected systems
Overcoming challenges	• Prioritizing investments based on potential for long-term growth • Implementing robust cybersecurity measures and employee training

5.3 *Industry 3.0: Embracing Digitalization*

The result of Industry 3.0 is a major change in business and technology, the integration of computer systems and communications in production. While this era promised unprecedented efficiency and productivity gains, it also introduced new challenges. A major problem is the high investment required to use new technology. Small and medium-sized businesses (SMEs) often struggle to afford these resources, putting them at a competitive disadvantage. Additionally, the shift towards automation and digitalization necessitated workforce reskilling. Many workers lacked the requisite skills to operate and maintain automated systems effectively, highlighting the importance of comprehensive training and educational initiatives.

5.4 *Industry 4.0: Navigating the Digital Frontier*

In the Industry 4.0 era we are in, organisations face significant costs in the utilization of technologies such as cyber-physical systems, IoT and artificial intelligence. The upfront investment required for hardware, software, and workforce training can be prohibitive, particularly for smaller enterprises. Moreover, safeguarding against cyber threats in interconnected systems presents a critical challenge. With the proliferation of digital technologies, organisations are increasingly vulnerable to cybersecurity breaches and data breaches. Good security procedures and constant vigilance are essential to protect sensitive data and infrastructure. In addition, supporting employees to adapt to the digital environment is important in realizing the ability of Industry 4.0 technologies. organisations need to invest in coaching and training programs to groom their employees with the skills required to flourish in the digital age.

5.5 Overcoming Industry 4.0 Challenges

To overcome the challenges of Industry 4.0, organizations must adopt a strategic approach that encompasses careful planning, investment prioritization, and talent development. This includes identifying and addressing financial barriers to technology adoption, such as access to capital and affordability. Organisations should prioritize investments based on their potential to drive long-term growth and competitive advantage. Additionally, it is essential to implement powerful cybersecurity considerations to defend cyber threats and protect the integrity of sensitive information and infrastructure. This requires regular monitoring, regular updates and training of staff to follow safety procedures. In addition, it is essential to develop a culture of innovation and change to solve the complex problems of the digital environment. Organisations must encourage experimentation, collaboration, and knowledge sharing to drive continuous improvement and drive growth in the 4.0 era.

6 Industry 4.0: Navigating the Digital Frontier

6.1 Financial Considerations: Investing in Industry 4.0

Adopting Industry 4.0 requires organisations to make significant financial investments. This includes the procurement of advanced hardware and software systems, as well as the allocation of resources for comprehensive training programs. Increasing costs associated with Industry 4.0 implementation can be substantial, requiring careful budgeting and financial planning to ensure sustainable adoption. Moreover, organisations must consider the long-term return on investment (ROI) Industry 4.0 is starting, the benefits are being weighed terms of increased productivity, efficiency gains, and competitive advantage against the initial capital outlay.

6.2 Cybersecurity Challenges: Safeguarding Sensitive Data

The interconnected nature of cyber-physical systems introduces new cybersecurity risks for businesses. Protecting sensitive data and infrastructure from potential threats such as hacking, data breaches, and system vulnerabilities becomes paramount. Strong cybersecurity actions are necessary to reduce these threats and shield the integrity of digital operations. This may include using advanced encryption techniques, using multiple authentication methods, and performing regular security audits to identify and remediate vulnerabilities.

6.3 Workforce Development: Upskilling for the Digital Age

As technology continues to rapidly advance, the skills required to use and enable industry 4.0 technology also increase. Continuous employee promotion and retraining is essential to ensure employees continue to make good use of new technology. This will include providing education, training and career development to equip staff with the required digital skills. Additionally, organisations should stimulate a lifestyle of lifelong learning and transformation to motivate staff to adopt new technologies and change jobs and responsibilities.

6.4 Strategic Planning: Charting the Course for Digital Transformation

Navigating the complexities of Industry 4.0 requires strategic planning and foresight from organisational leaders. Developing a comprehensive strategy for digital transformation involves identifying business objectives, assessing technological requirements, and aligning resources accordingly. Effective strategic planning ensures that Industry 4.0 initiatives are aligned with broader organisational goals and objectives. Moreover, organisations must remain agile and adaptable in their strategic approach, anticipating changes in market dynamics, technological advancements, and competitive pressures.

6.5 Risk Management: Mitigating Potential Pitfalls

Industry 4.0 implementation introduces various risks and challenges that organisations must navigate effectively. From financial risks associated with investment decisions to cybersecurity risks threatening data security, effective risk management strategies are important. This requires recognizing potential threats, assessing their impacts, and taking steps to lessen or eliminate those threats. organisations also need to develop contingency plans to deal with unforeseen and disruptive issues, ensure business continuity, and be resilient in times of stress.

6.6 Collaboration and Partnerships: Leveraging Ecosystem Synergies

Collaboration and partnership play an important role in exploring the complexities of Industry 4.0. By forging alliances with technology providers, industry peers, and academic institutions, organisations can leverage synergies and expertise to drive successful digital transformation initiatives. Collaborative efforts enable organisations to share knowledge, resources, and best practices, accelerating the pace of innovation and adaptation. Moreover, partnerships with suppliers, customers, and other stakeholders can help organisations create integrated value chains and ecosystem-based business models that deliver enhanced value and customer experiences.

6.7 Continuous Improvement: Embracing a Culture of Innovation

Industry 4.0 is not a static place, but a journey of continuous development and innovation. To succeed in the digital age, organisations need to create a culture that embraces experimentation, creativity and change. A strategy that supports continuous learning and improvement allows employees to embrace change and foster innovation, positioning the organisation for long-term success in a fast-paced environment. By embracing new technologies, exploring new business models, and challenging the status quo, organisations can stay ahead and maintain a positive impact in the era of business 4.0.

6.8 Conclusion: Navigating the Digital Frontier with Confidence

In conclusion, navigating the complexities of Industry 4.0 requires a multifaceted pathway that addresses financial considerations, cybersecurity challenges, workforce development, strategic planning, risk management, collaboration, and continuous improvement. By recognizing and following these fundamental principles, organisations can confidently explore digital frontiers and set themselves up for success in the era of Industry 4.0. Through strategic investment, governing risk and commitment to innovation, companies can equip the transformative power of Industry 4.0 to increase productivity, competitiveness and sustainable development. By creating a vibe of collaboration, learning and innovation, organisations can adapt to the changing digital environment and seize diverse growth and innovation opportunities at Construction Four Corners.

This Table 5 gives us the key considerations that include financial planning, cybersecurity measures, workforce development, strategic planning, risk management, collaboration, and continuous improvement, to navigate Industry 4.0 successfully.

Table 5 Industry 4.0 key aspects and their strategic descriptions

Aspect	Description
Financial considerations	Significant financial investments are required for Industry 4.0 adoption, including procurement of hardware and software, and allocation for training programs. organisations need to carefully budget and plan for sustainable adoption, considering long-term ROI against initial capital outlay
Cybersecurity	It is critical for cyber-physical systems to protect sensitive data from hackers,
Challenges	Breaches, and vulnerabilities. Strong cybersecurity measures such as encryption protocols, authentication mechanisms, and timely security checks are necessary to protect digital operations
Workforce development	Continuous upskilling and reskilling are essential to keep pace with evolving technology. Training programs, educational opportunities, and fostering a culture of lifelong learning enable employees to acquire digital skills and adapt to changing job roles effectively
Strategic planning	Effective strategic planning involves aligning Industry 4.0 initiatives with organisational goals, anticipating market dynamics, and remaining adaptable to technological advancements. organisations must assess business objectives, technological requirements, and allocate resources accordingly to navigate digital transformation successfully
Risk management	Industry 4.0; It brings with it many risks, including financial, cybersecurity and operational complexity. An efficient governing of risk strategy involves pointing out potential risks, evaluating their impact, and advancing to reduce or eliminate those risks. Having an emergency plan ensures business continuity and protection in the event of a crisis
Collaboration & partnerships	Collaborative efforts with technology providers, industry peers, and academic institutions accelerate innovation and adaptation in Industry 4.0. Partnerships with suppliers, customers, and stakeholders create integrated value chains and ecosystem-based business models, delivering enhanced value and customer experiences
Continuous improvement	Embracing a culture of innovation, experimentation, and adaptability drives continuous improvement in Industry 4.0. organisations must encourage a mindset of learning and creativity among employees, enabling them to embrace change, challenge the status quo, and drive innovation to preserve a competing edge in the quickly evolving digital landscape
Conclusion	Navigating Industry 4.0 requires addressing financial, cybersecurity, workforce, strategic, risk management, collaboration, and continuous improvement aspects. By adopting a holistic approach and embracing these principles, organisations can navigate the digital frontier with confidence, driving productivity, competitiveness, and sustainable growth in the Fourth Industrial Revolution

7 Reshaping Workflows: Addressing Industry 4.0 Adoption Challenges

7.1 *Strategic Planning for Industry 4.0 Adoption Assessing Business Needs and Objectives*

Before embarking on the journey toward Industry 4.0 adoption, organisations must conduct a thorough assessment of their business needs and objectives. This involves identifying key areas for improvement, such as production efficiency, product quality, or supply chain management. By aligning technology implementation with strategic goals, organisations can ensure that investments yield tangible benefits and drive meaningful outcomes.Not all Industry 4.0 technologies will be equally relevant or beneficial to every organisation. Therefore, it's crucial to prioritize technology implementation based on business priorities and capabilities. This may involve identifying low-hanging fruit opportunities for quick wins, as well as longer-term strategic initiatives that require more extensive planning and investment. By focusing resources where they will have the greatest impact, organisations can maximize the return on investment from Industry 4.0 adoption.

7.2 *Cybersecurity and Data Protection Measures Establishing Robust Cyber Security Protocols*

As organisations transition to Industry 4.0, they become increasingly reliant on interconnected digital systems. This heightened connectivity introduces new cybersecurity risks, including data breaches, malware attacks, and system vulnerabilities. To mitigate these risks, organisations must establish robust cybersecurity protocols that encompass both preventive measures and incident response strategies. This may include implementing firewalls, encryption protocols, multi-factor authentication and regular security checks to verify and fix vulnerabilities.

7.3 *Ensuring Data Protection and Privacy Compliance*

In addition to cybersecurity measures, organisations must also prioritize data protection and privacy compliance in the era of Industry 4.0. This involves implementing mechanisms to safeguard sensitive data, such as customer information, intellectual property, and proprietary business data. organisations must comply with regulations and standards such as GDPR, HIPAA, or CCPA to ensure compliance and reduce the risk of fines or reputational damage commonly associated with data breaches or privacy breaches.

7.4 *Workforce Development Initiatives Offering Comprehensive Training Programs*

As Industry 4.0 technologies become increasingly integrated into organisational processes, the skillset required of the workforce evolves accordingly. To equip our employees with the skills required to flourish in the digital environment, organisations should prioritize workforce development initiatives. This may involve offering comprehensive training programs covering topics such as data analytics, artificial intelligence, cybersecurity, and advanced manufacturing technologies. By investing in employee training and productivity, organisations can help their employees adapt to technology and foster innovation in the organisation.

7.5 *Fostering a Culture of Innovation and Collaboration*

The success of Industry 4.0 goes beyond technology, it also requires change in organisational culture. For organisations to succeed in the digital age, they need to create a culture of innovation, collaboration and change. This includes creating an environment where employees feel empowered to try new ideas, collaborate across departments, and embrace change as a catalyst for growth. Strong leadership commitment and effective change management strategies are essential to foster cultural change and ensure employee engagement and participation in Industry 4.0 practices. Business and competition in the digital age. By being efficient and creative in adopting business 4.0, companies can battle obstacles and open the full power of digitalization. Through strategic planning, investments in cybersecurity and data protection, and early employee development initiatives, organisations can thrive in an increasingly digital world. Leveraging innovation, collaboration and change will be key to driving growth and remaining competitive in the era of Industry 4.0.

Table 6 gives us key strategies that include strategic planning, cybersecurity measures, and workforce development initiatives, to empower organisations in successfully embracing Industry 4.0 technologies and driving innovation in the digital era.

Table 6 Industry 4.0: key strategies

Aspect	Description
Strategic planning	Assessing business needs and objectives is crucial before embarking on Industry 4.0 adoption. Organisations must align technology implementation with strategic goals and prioritize technology based on business priorities and capabilities to maximize ROI
Cybersecurity and data protection	Establishing robust cybersecurity protocols is essential to mitigate risks associated with increased connectivity. This includes preventive measures such as firewalls, encryption protocols, and incident response strategies. Ensuring data protection and privacy compliance is also critical to safeguard sensitive data and adhere to relevant regulations and standards
Workforce development initiatives	Providing training programs to help employees adapt to Industry 4.0 technologies. Organisations must spend on advanced coaching programs that include topics such as data analytics, cybersecurity, and manufacturing technologies. Fostering a culture of innovation and collaboration is critical to fostering organisational change and maintaining employee engagement
Conclusion	Adopting Industry 4.0 requires support and advice. By investing in strategic planning, cybersecurity measures, employee development, and building a culture of innovation, organisations can overcome adoption challenges and thrive in the digital age

8 Empowering Innovation: Pioneering the Digitally Enabled Future

8.1 Embracing Digital Transformation Seizing Opportunities Amidst Challenges

The transition to Industry 4.0 introduces a plethora of challenges, ranging from financial constraints to cybersecurity risks. But within these challenges, there are many opportunities for organisations willing to embrace digital transformation. By using advanced technologies such as the Internet of Things, artificial intelligence and data analytics, companies to boost their operations, increase performance and profit effectively in the market.

8.2 Unlocking Productivity Gains

One of the key advantages of Industry 4.0 is the potential to unlock latest levels of productivity. With real-time monitoring through automation and physical integration, organisations can improve production processes, reduce downtime, and utilize capital at the highest level. This not only improves efficiency, but also allows businesses to better meet customer needs and accelerate growth.

8.3 Enhancing Agility and Flexibility

In the present day's rapid-paced business environment, agility and flexibility are critical to success. Industry 4.0 enables organisations to rapidly modify business and consumer preferences. Using data-driven insights and predictive analytics, businesses can make informed decisions in real time, allowing them to quickly respond to business trends, reduce risk and seize opportunities as they arise.

8.4 Revolutionizing Supply Chain Management

Supply chain management is the foundation of a successful business. Industry 4.0 is transforming supply chain management by creating a connected, data-driven ecosystem from suppliers to consumers. Thanks to IoT-enabled sensors, blockchain technology, and AI analytics, organisations can gain greater visibility, transparency, and traceability throughout their supply chains. This not only increases operational efficiency, but also improves coordination, shortens lead times and reduces disruptions.

8.5 Positioning for Long-Term Success

Embracing Industry 4.0 is not just about short-term gains; it's about positioning businesses for long-term success in an increasingly digitalized world. By investing in digital transformation initiatives today, organisations can future-proof their operations, adapt to evolving market dynamics, and outshine the competitiveness. Industry 4.0 offers the promise of sustained growth, innovation, and competitiveness, ensuring that businesses stay apt and resilient in the phase of technological disruption (Table 7).

This table offers key pointers for organisations to embrace digital transformation, unlock productivity gains, enhance agility, revolutionize supply chain management, and place themselves for everlasting triumph in the digital era, emphasizing the utilization of IoT, AI, information analytics, and cyber-physical systems for streamlined operations and sustained growth.

Table 7 Strategies for digital transformation success in industry

Aspect	Element
Embracing digital transformation	• Utilize IoT, AI, and data analytics for streamlined operations • Improve efficiency to gain a competitive edge • Address challenges through digitalization for long-term sustainability
Unlocking productivity gains	• Implement automation and real-time monitoring with cyber-physical systems • Enhance production to reduce downtime • Boost resource utilization for growth and scalability
Enhancing agility and flexibility	• Quickly adapt to changing business and customer preferences • Harness data-driven insights for informed Decision-making • Respond swiftly to market trends and mitigate risks
Revolutionizing supply chain management	• Create interconnected, data-driven ecosystems from suppliers to customers • Utilize IoT, blockchain, and AI for greater visibility and transparency • Improve collaboration and reduce lead times for enhanced efficiency and minimized disruptions
Positioning for long-term success	• Invest in digital transformation initiatives for future-proof operations • Adapt to evolving market dynamics and stay ahead of competition • Ensure sustained growth, innovation, and competitiveness in a digitalized world

9 Overcoming Barriers to Adoption

9.1 Addressing Financial Constraints

While the benefits of Industry 4.0 are clear, many organisations face significant financial constraints when it comes to implementation. The upfront costs of investing in hardware, software, and training might be prohibitive, mostly for small and medium-sized businesses (SMBs). To overcome this barrier, businesses must develop comprehensive investment strategies, prioritize technology investments based on ROI, and explore alternative financing options such as grants, loans, and partnerships.

Fig. 2 Cybersecurity Risk Matrix

9.2 Mitigating Cybersecurity Risks

As organisations become increasingly interconnected in the digital age, cybersecurity has emerged as a top priority. Industry 4.0 introduces new vulnerabilities and threats, ranging from data breaches to ransom ware attacks. To reduce cybersecurity risks, companies need to implement security measures including access, access control, and threat prevention and training efforts. Additionally, organisations must be aware of emerging threats and compliances to make-sure the integrity and confidentiality of their details and Fig. 2 indicates the risk matrix.

9.3 Bridging the Skills Gap

Another significant task in the adaptation of Industry 4.0 is the skills gap. Many organisations lack the expertise and talent needed to implement and leverage advanced technologies effectively. To bridge this gap, businesses must invest in workforce development initiatives, including training programs, certifications, and apprenticeships. By up skilling their employees and promoting lifelong learning, organisations build a talented workforce having the ability of driving digital transformation and change indicated in Fig. 3.

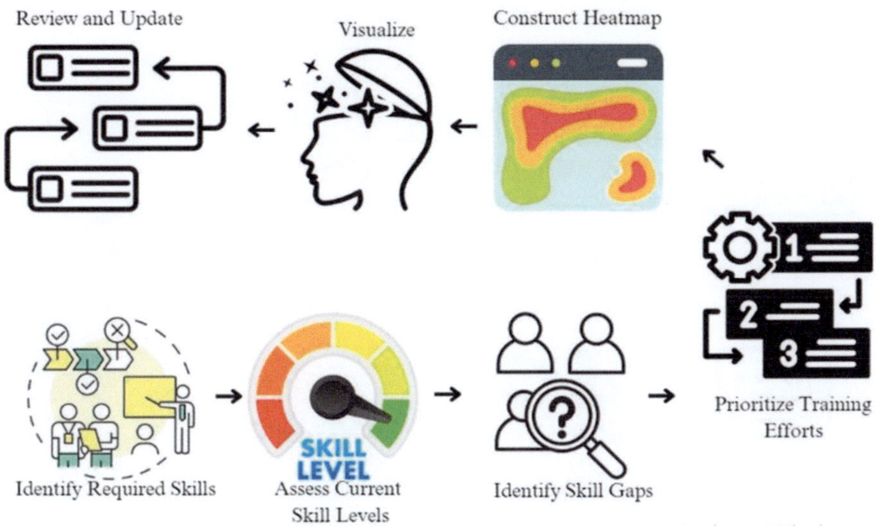

Fig. 3 Skills Gap Heatmap

9.4 Overcoming Organisational Resistance

Resistance to change is a common barrier to Industry 4.0 adoption, particularly among employees accustomed to traditional ways of working. To overcome organisational resistance, businesses must prioritize change management initiatives, communicate the benefits of digital transformation, and engage employees in the decision-making, process indicated in Fig. 4. By invoking innovation, collaboration, and continuous improvement, organisations can create an environment conducive to successful Industry 4.0 adoption.

This Table 8 outlines key strategies for addressing financial constraints, mitigating cybersecurity risks, bridging the skills gap, and overcoming organisational resistance to drive successful digital transformation initiatives. Prioritizing investment strategies, implementing robust security measures, investing in workforce development, and fostering a culture of innovation are highlighted as essential approaches for navigating the digital landscape effectively.

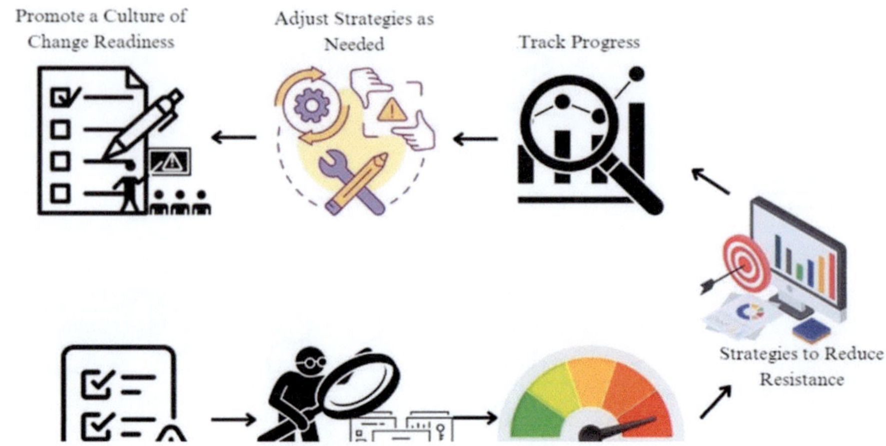

Fig. 4 Organisational resistance gauge

Table 8 Strategies for overcoming digital transformation challenges

Aspect	Key strategies
Addressing financial constraints	• Develop comprehensive investment strategies • Priorities technology investments based on ROI • Explore alternative financing options such as grants, loans, and partnerships
Mitigating cybersecurity risks	• Implement security measures, including access and access control • Stay abreast of emerging threats and compliance regulations • Conduct employee training programs
Bridging the skills gap	• Invest in workforce development initiatives including training programs and apprenticeships • Invoke lifelong learning and steady progress among the workforce
Overcoming organizational resistance	• Prioritize change management • Initiatives and impart the advantages of digital transformation • Involve employees in the decision-making process • Foster a culture of innovation

10 Navigating the Path to Success

10.1 Developing a Comprehensive Strategy

To encounter the issues of adopting Industry 4.0, organisations need to develop a strategy that aligns with their business goals. This strategy should encompass technology investments, talent development initiatives, cybersecurity measures, and change management strategies. By taking a holistic approach to digital transformation, businesses can maximize the benefits of Industry 4.0 while mitigating risks and challenges along the way.

10.2 Cultivating Strategic Partnerships

The key to victory in the digital era is collaboration. Companies must develop partnerships with technology vendors, industry organisations, research institutes, and government agencies to accelerate. Companies gain access to new technologies using external expertise, resources and partnerships to stay ahead and drive innovation across their businesses, driving excellence and business. Implements the concept.

10.3 Embracing a Culture of Innovation

Innovation is the lifeblood of Industry 4.0. Organisations should foster an innovative culture which promotes experimentation, creativity and steady learning. By empowering employees to think unique, take risks, and challenge adverse situations, companies can unlock new opportunities, foster innovation, and lead the competition. Additionally, organisations should encourage and reward innovation and create a supportive environment where ideas are encouraged, tested and implemented.

10.4 Embracing Industry 4.0: A Journey of Transformation

In conclusion, Industry 4.0 represents a journey of transformation for companies worldwide. By embracing digital transformation and overcoming barriers to adoption, businesses can unlock new opportunities, enhance productivity, and place themselves for eternal success in an increasingly digitalized world. Through strategic planning, investment in technology and talent, and a commitment to innovation, organisations can channelize the complications of Industry 4.0 and emerge as leaders in their respective industries. The future belongs to those who dare to innovate, adapt, and embrace change in the pursuit of excellence.

This Table 9 provides a comprehensive set of strategies for tackling financial constraints, mitigating cybersecurity risks, bridging the skills gap, overcoming organisational resistance, developing a comprehensive strategy, cultivating strategic partnerships, and embracing a culture of innovation to drive successful digital transformation initiatives. Aligning efforts with business objectives, fostering collaboration, and incentivizing innovation are highlighted as key principles for navigating the digital landscape effectively.

Table 9 Comprehensive strategies for successful digital transformation

Dimension	Facet
Addressing financial constraints	• Develop comprehensive investment strategies • Prioritize technology investments based on ROI • Explore alternative financing options such as grants, loans, and partnerships
Mitigating cybersecurity risks	• Implement security measures, including access and access control • Stay abreast of emerging threats and compliance regulations • Conduct employee training programs to enhance cybersecurity awareness
Bridging the skills gap	• Invest in workforce development initiatives including training programs and apprenticeships • Foster lifelong learning and improvement among the workforce
Developing a comprehensive strategy	• Align digital transformation efforts with business goals and Objectives • Incorporate technology investments, talent development, cybersecurity measures, and change management strategies • Take a holistic approach to digital transformation
Cultivating strategic partnerships	• Collaborate with technology vendors, industry associations, and research institutions • Leverage external expertise, resources, and networks to accelerate digital transformation initiatives
Embracing a culture of innovation	• Fostering encourages experimentation, creativity, and continual learning among the staff • Incentivize and reward innovation within the organisations • Create a supportive environment for testing and implementing new ideas

11 Conclusion and Future Scope: Embracing Industry 4.0

The change-over to Industry 4.0 is a transformative journey challenging traditional paradigms, reshaping how businesses operate globally. Embracing it is no longer optional but imperative for competitiveness and resilience. Strategic planning, technology, and talent investment, and a commitment to continuous improvement are keys to success. Industry 4.0 unlocks unprecedented opportunities for productivity, efficiency, and competitiveness through digital empowerment. It requires a culture of innovation, collaboration, and adaptability to drive meaningful change. By embracing it, organisations can lead in shaping the future of industry and commerce on a global scale. Investigation into emerging trends like quantum computing, AI, and cybersecurity is vital for responsible implementation. These areas drive sustainable practices, supply chain resilience, and workforce development in the Industry 4.0 landscape. Incorporating quantum computing capabilities can revolutionize data processing and optimization within Industry 4.0 systems, enabling more efficient resource allocation and complex problem-solving. Advanced AI algorithms empower predictive maintenance and quality control, enhancing operational efficiency while reducing downtime and waste. Robust cybersecurity measures are paramount in safeguarding confidential data and crucial framework from cyber risks, ensuring the integrity and reliability of Industry 4.0 systems. By prioritizing these areas, organisations can foster sustainable growth, fortify supply chains against disruptions, and cultivate a skilled workforce capable of navigating the evolving technological landscape with confidence and expertise.

References

1. Frank, A.G., Mendes, G.H.S., Ayala, N.F., Ghezzi, A.: Servitization and industry 4.0 convergence in the digital transformation of product firms: a business model innovation perspective. Technol. Forecast. Soc. Change Elsevier, (2019)
2. Dillinger, F., Bernhard, O., Kagerer, M., et al.: Industry 4.0 Implementation Sequence for Manufacturing Companies. Springer, Production Engineering Research and Development (2022)
3. Ortt, R., Stolwijk, C., Punter, M.: Implementing industry 4.0: assessing the current state. J. Manuf. Technol. Manag. (2020)
4. Lee, J., Bagheri, B., Kao, H.A.: A cyber-physical systems architecture for industry 4.0-based manufacturing systems. Manuf. Lett. Elsevier. (2015)
5. Lu, Y., Morris, K.C., Freese, M., Wang, X.: Industry 4.0: a Survey on Technologies, Applications, and Open Research Issues. Elsevier, Journaloff Industrial Information Integration (2017)
6. Schumacher, A., Erol, S., Sihn, W:. A maturity model for assessing industry 4.0 readiness and maturity of manufacturing enterprises. Proced. Cirp. Elsevier. (2016)
7. Stock, T., Seliger, G., Bauer, W.: Industrial internet of things and industry 4.0. IEEE Transactions on Industrial Informatics, (2018)
8. Xu, L.D., He, W., Li, S.: Internet of things in industries: a survey. IEEE Transactions on Industrial Informatics, (2014)

9. Kagermann, H., Wahlster, W., Helb J.J.: Recommendations for implementing the strategic initiative industries 4.0. acatech, Nat. Acad. Sci. Eng. (2013)
10. Gadre, M., Deoskar, A.: Industry 4.0—digital transformation, challenges and benefits, Int. J. Future Gener. Commun. Netw. (2020)
11. Atieh, A.M., Cooke, K.O., Osiyevskyy, O.: The role of intelligent manufacturing systems in the implementation of industry 4.0 by small and medium enterprises in developing countries, Wiley, (2022)
12. Friedli, T., Classen, M., Budde, L.: Industry 4.0: Navigating Pathways Toward Smart Manufacturing and Services. Springer, Connected Business (2021)
13. Milovanović, G., Milovanović, S., Popović, G.: The role of industry 4.0 in digitalization of production and supply chains, Ekonomika, (2022)
14. Imanova, Z., Mikayilzadeh, A., Dzhamalova, J.: 4th Industrial revolution and artificial intelligence, Equipment Technol. Mater. (2023)
15. Rahul, R., Tiwari, M., Ivanov, D., Dolgui, A.: Machine learning in manufacturing and industry 4.0 applications, Int. J. Prod. Res. (2021)
16. Barbosa Nunes, T.F., Zanini, R.R., Porto Rosa, A.F., Vergara, L.L.G.: Impacts and challenges of industry 4.0 in manufacturing: a systematic literature review, Sustainability, (2022)
17. Gajdzik, B.: Industry 4.0 as the challenge for employment change and for restructuring process, Acad. Human. (2020)
18. Dong, L., Geng, X., Xiao, G., Yang, N.: Procurement strategies with unreliable suppliers under correlated random yields, Manuf. Serv. Oper. Manag. (2021)
19. Tsujimura, M.: Capital investment under output demand and investment cost ambiguity, Econ. Bus. (2020)
20. Pessot, E., Zangiacomi, A., Marchiori, I., Fornasiero, R.: Empowering supply chains with industry 4.0 technologies to face megatrends, Innov. Supply Chain Manag. (2023)

Human–Robot Collaboration for Smart Manufacturing in Industry 4.0: A Review, Analysis, and Prospects

Nisha Banerjee

Abstract The onset of Industry 4.0 marks the beginning of a fresh phase in smart manufacturing, characterized by the adoption of emerging technologies such as artificial intelligence and the Internet of Things. Human–robot collaboration (HRC) is a crucial element in Industry 4.0's smart manufacturing, offering new opportunities to enhance efficiency, flexibility, and productivity that were not available before. By merging the capabilities of humans and robots, industry 4.0 enhances effectiveness and safety in the manufacturing process. People focus on decision-making, problem-solving, and creativity, while robots excel at tasks that involve heavy lifting and mechanical repetition. This partnership between individuals and artificial intelligence reduces mistakes, optimizes resource utilization, and elevates the overall quality of output. This study explores the basic concepts and presents practical uses of HRC in Industry 4.0, providing a straightforward overview of the field in the present era. The new concept of human–robot collaboration (HRC) is revolutionizing the industrial setting within Industry 4.0 by transforming smart production processes. This chapter thoroughly examines HRC's present situation in smart manufacturing, considering its benefits, limitations, and future potential. This study explores how human resource continuities (HRC) can transform production processes by analyzing background literature, case studies, and technological developments use cases to enhance flexibility, quality, and workplace security. This research delves into the socio-technical aspects of HRC, including concerns such as the necessity for improved AI training for employees, employee interactions, and ethical challenges. This paper offers detailed insights for researchers and policymakers interested in maximizing the potential of human–robot collaboration during Industry 4.0. This research also investigates the socio-technical aspects of HRC, addressing topics such as the necessity for improved AI training for employees, workforce dynamics, and moral quandaries. This article offers detailed insights for researchers and policymakers seeking to harness the potential of collaboration between humans and robots during the Industry 4.0 era. This is achieved by highlighting deficiencies in current studies and mapping out directions for future research.

N. Banerjee (✉)
Computing and Analytics, NSHM College of Management and Technology, Kolkata, India
e-mail: nishasonali2000@gmail.com

L. Krishnasamy et al. (eds.), *Distributed Deep Learning and Explainable AI (XAI) in Industry 4.0*, Information Systems Engineering and Management 55,
https://doi.org/10.1007/978-3-031-94637-0_6

Keywords Human robot collaboration · Artificial Intelligence · Industry 4.0 · Smart manufacturing

1 Introduction

Industry 4.0, also called the fourth business revolution, marks a giant shift in production. It's characterised through the deep integration of virtual technology into physical manufacturing techniques, blurring the strains among the bodily and virtual worlds. This integration creates a network of intelligent machines, systems, and people, basically reworking how we layout, produce, and manipulate merchandise. Cyber-Physical Systems (CPS) are sensible structures that combine bodily equipment with computational capabilities and networking. Sensors embedded in machines gather real-time information on performance, repute, and surrounding conditions. This records is then processed and analyzed via computer systems, taking into account self sufficient choice-making and manipulate of the bodily structures. Internet of Things (IoT) refers to the sizable network of bodily devices embedded with sensors and actuators that connect to the net. In Industry 4. Zero, machines, equipment, or even products themselves are ready with IoT sensors, permitting them to talk and change statistics with every other and with vital control systems. This creates a seamless go with the flow of statistics across the complete manufacturing chain. The tremendous quantity of facts generated via sensors and linked devices in Industry four. Zero falls under the umbrella of Big Data. Advanced analytics equipment are used to process and examine this data, extracting treasured insights into production techniques, gadget fitness, and capacity troubles. These insights permit for predictive protection, optimizing useful resource allocation, and enhancing usual efficiency. The area of robotics includes areas like object detection. Machine learning techniques, such as deep learning and neural networks, are crucial in this context, and a comparable system can be created using equivalent data [1]. Having a grasp of spatial relationships is crucial when constructing models for recognizing objects [2]. This is highly beneficial for tasks like self-driving robots, where utilizing AI for object recognition and related tasks can bridge the gap between AI and human intelligence, facilitating reliable decision-making, especially for novice users [3]. These factors play a role in the development of large-scale robotics and automation. Industry 4.0 anticipates a future where humans and robots will work together, with robots taking on tasks beyond simply replacing humans. In order to reduce accidents, robots should be able to anticipate hazardous scenarios and adjust to unforeseen incidents [4]. This necessitates that the robot possesses the capability to thoroughly analyze the surroundings and make effective decisions. Cyber Physical Systems (CPS) is a critical technology in achieving Industry 4.0. The 5C framework highlights that CPS deployment consists of five stages: smart connection, data and information conversion, network, information, and configuration. Smart connectivity emphasizes detecting the surroundings and communicating environmental data from any location. Data-to-data transformation involves machines utilizing environmental data to interpret their surroundings.

Cyber entails examining data from various systems to forecast future actions. Knowledge centers around converting statistical information into a format that is easier for professionals to comprehend. The focus is on the physics decision-making process involving information. AI and ML algorithms are crucial in facilitating human–machine interaction within Industry 4.0. They can analyze information to discover styles, are expecting equipment disasters, optimize manufacturing techniques, or even manage robots for complex obligations. Machine gaining knowledge of algorithms continuously research and enhance with new data, enabling a more dynamic and adaptable production surroundings. Cloud computing offers on-call for get entry to to computing resources like garage, servers, and databases. In Industry 4.0, production records can be saved and accessed securely on the cloud, facilitating real-time collaboration, data sharing, and far flung monitoring of manufacturing lines from everywhere. With Industry 4.0, robots are no longer isolated entities. HRC emphasizes robots operating along people, performing tasks applicable to their strengths. Robots can manage repetitive, risky, or physically disturbing responsibilities, while humans make contributions their hassle-fixing competencies, creativity, and dexterity for tasks requiring extra finesse. The combined impact of those center principles is the creation of "clever factories"—particularly automated, interconnected, and information-pushed manufacturing centers which could adapt to changing demands, optimize resource usage, and improve usual productivity. This paper aims to thoroughly evaluate Human–Robot Collaboration (HRC) within the framework of smart manufacturing in Industry 4.0. We will examine the innovative realm of HRC technology, investigating the different types of collaboration models and the technology that makes them possible. We will assess the vast potential of HRC, emphasizing its many benefits for productivity, security, and top-notch advancement. Additionally, we can severely look at the demanding situations related to HRC, which includes safety worries, skill gaps inside the group of workers, and the want for powerful human-robotic interaction layout. With a ahead-searching angle, we will discover the exciting opportunities of HRC within the destiny. This will contain discussing capability advancements in AI, device getting to know, and sensor era, their impact on human–robot communique and collaboration, and the broader implications for the future of manufacturing. By presenting an intensive evaluation and evaluation, this paper aims to shed mild on the modern panorama of HRC, its capacity benefits and challenges, and ultimately, its promising destiny within clever manufacturing for Industry 4.0.

2 Literature Review

From the perspective of this article, a good way to secure the intelligence of human–robot collaboration is to involve humans and robots in decision-making and planning processes, which should be based on discussion and consensus. With the principle of Goldberg's Consistency [5], namely. Creative and all-encompassing options: Diversity, meaning people collaborating with AI and robots for equilibrium. His vision shows the necessity for a new intelligent management that integrates machines and people into one cohesive system. According to Malone [6], the focus should move from having a "computer in the loop" to having a "computer in the farm" and should highlight the significance of optimizing machine development by combining human and computer abilities to generate smart solutions for convenience and effectiveness. This article will explain how humans, robots, and computers must collaborate as an intelligent community to achieve specific goals for successful and adaptable change. This article investigates the safe utilization of intelligent human–machine integration in product manufacturing. Instead of trying to remove all risks and imposing restrictions, humans should be seen as needing assistance, education, and training to navigate various levels of human–robot interaction (refer to [5, 7, 8]). In business and academia, the goal is to comprehend the purpose of interaction. Collaborative robots are designed to merge the benefits of both humans and robots, enhancing efficiency and productivity, as well as creating more streamlined and adaptable assembly lines [9]. [10] Franklin and colleagues highlighted that collaborative robots offer distinct economic advantages, including lowering maintenance expenses, enhancing building floor space, facilitating workers' faster return to work, and promoting utilization. Modern Process involves partial automation. Certain discoveries imply that they may be better off with increased independence, intelligence, and cognitive abilities. Yet, a prior research [11] revealed that security models for robots are consistently constrained in their ability to aid in intelligence and autonomy (to develop efficient and adaptable automation solutions) and overlook the incorporation of crucial human skills and knowledge. They fail to consider perceptual elements and fail to balance security with diverse business needs, resulting in ineffective and inadequately automated solutions [11]. [12] created a manual to assist designers in creating and assessing security, user-focused, and teamwork applications. Although safety and ergonomic standards serve as the foundation for body safety and body ergonomics, there are no established standards for safety and performance awareness in these areas, leading researchers to rely on translated findings from research studies. [13] discovered another flaw: a lack of a comprehensive framework for assessing HRC applications that takes into account all elements of human–robot interactions, such as human factors and abilities like autonomy, adaptability, and education. The writers offer various measurements for developers to think about and incorporate into their designs. While the majority of current literature on transformational automation focuses on machine collaboration, there are also numerous articles utilizing this idea to explore interactions between humans and machines. Miller and Parasuraman [14] propose that employees should have the ability to assign work tasks to automation and receive the same feedback as agents at the same location. People's organizations are efficient in various scenarios, at varied levels of specificity, and with diverse restrictions, regulations, circumstances, and other approaches. Miller and Parasuraman argue that adjusting the level of automation (LOA) according to the specific task can enhance efficiency. The idea of adaptive automation to enhance and streamline automation processes has been present for quite some time, such as [15, 16]. The concept here is that adjusting the level of automation is beneficial for various tasks in the production process. In contrast, the need for security precautions highlights the significance of security reform in addressing operational and security issues. Vision and forecasting algorithms can prevent collisions by altering robot paths or sequences [17], aiding humans with tools [18], or utilizing sophisticated sensors [19]. Security is frequently required to enable flexibility and facilitate connections and communication in order to develop a highly intelligent network. Sensors have the ability to monitor human movements and anticipate human intentions in specific scenarios, allowing robots to operate quicker than safety agents in those instances. Despite the significant potential of collaborative robots, their use in businesses has not met expectations due to safety worries, as stated by [20]. The authors stated that while ISO security standards provide guidelines for designing and assessing protection needs to minimize risk, it is crucial to consider the uniqueness of each application and various factors when developing new systems. Stringent regulations pose a significant challenge and are hard to meet given the demands of business operations. In order to address this issue, Saenz and colleagues suggest that an improved design process should consider and implement the model's regulations. [21] state in their study on interactions between humans and robotic systems that human–machine collaboration security involves physical and mental aspects. A comprehensive review was conducted on creating tactics and procedures for safety, categorizing them into four key areas: safety via management, planning, forecasting, and mental well-being. This study showcases the construction and execution of turn prediction systems (specifically nurse scrubs capable of interpreting doctors' multimodal communication and making predictions beforehand). The algorithm's performance was assessed using data from simulated surgical procedures, revealing that it can achieve human-level performance beyond its early stages. HRC is especially valuable in rehabilitation scenarios, as robots help individuals recover motor skills by aiding in movements and gestures. A robotic system was suggested by [22] that combined KUKKA LWR for carrying out conventional physical therapy. Based on this research, a robotic arm can mimic the visual guidance of a physical therapist with great precision and consistency. Alternatively, [23] conducted research on the willingness to use robotic assistance in healthcare. They analyzed users' feelings towards different types of remote-controlled robots by using three levels of remote control scenarios. This is the reason why managing navigation is the most challenging task. Although convenient, voice control lacks in security features [24] single. Smart manufacturing's flexibility allows for the creation of customizable products and adaptable production processes. Quick responses to situations can be enabled by smart manufacturing technologies [25]. This involves adapting to external changes and revising the product based on the decision, even in the absence of information. Analyzing data in smart manufacturing is essential for making rapid decisions and developing predictions and models. Humans play a vital role in smart manufacturing due to their intelligence, and collaborative robots work alongside them instead of replacing them. Collaborative robots offer numerous benefits compared to traditional robots, like the capability to operate alongside humans in the same area and handle a range of different tasks [26–30]. Using different types of robots or increasing the robot's space can raise expenses, potentially leading to traditional robots being more expensive than collaborative robots. Collaboration between humans and robots is vital and necessitates robots to be equipped for actual scenarios. Deep learning can be utilized for diverse tasks across various applications, depends on past knowledge or experience, and frequently demands extensive knowledge to be imparted. As a result, machines will not be able to make decisions like humans when faced with a new situation for the first time because of the absence of pre-training. This issue can be resolved through collaborative knowledge sharing (CSK) [31]. When machines are provided with CSK, they are able to mimic human behaviors and make instinctive choices that closely resemble human decision-making. An article that was released recently talks about the pros and cons of deep learning and CSK [32] and how they can complement each other. While delving into deep learning models, he also examines numerous intriguing projects within CSK; Thus, the extraction and compilation of CSK are crucial for advancing modern skills. Taking this into consideration, we will now explore relevant studies in this paper. Scientists in Germany at the University of Bremen are working on creating a robot with the ability to solve tasks using cognitive abilities [33]. The system of the robot saves the 3D coordinates of the present position; Consequently, the model system can retrieve the coordinates of any position at any given moment and display a 3D visualization of a complete tour. This data can help link visual patterns with spatial connections. An easy illustration involves linking 2D and 3D pictures with the dimensions and proximity of objects. You can make your job easier by merging these tasks and items. Research project demonstrates how artificial intelligence can result in improved efficiency in completing robotic tasks. Until now, humans and robots had to work separately because humans could interfere with robot tasks and robots could pose a risk to humans [34]. Robots are increasingly being made more secure and safeguarded. Physical Human–Robot Interaction (pHRI) involves the interaction between humans and robots and is connected to HRC. Nevertheless, these fixed assignments have a significant effect and may need certain enhancements in automation. Advancements in artificial intelligence and robotics now enable machines to work alongside humans across various fields, such as smart manufacturing. Robots have become smaller, more agile, and able to work side by side with humans safely, all due to the incorporation of machine learning algorithms. AI machines are created to collaborate with individuals, rather than to terminate their employment. Robots will rely on data and machine learning for solving basic tasks, while humans will utilize their common sense and intuition for solving more complex tasks. Progress in the collaboration between humans and robots. Tests and applications have become more extensive. In the days ahead, our integration of humans and machines, driven by information and requirements, may involve numerous essential elements. The use of common spatial senses in constructing object recognition models, as described in references [35, 36], which focus on object detection and automatic object recognition through sensory circuit adjustments, could be beneficial in this context. At times, laser sensors and vision may be needed for safety and controller adjustments. Hence, commercial cobots that do not involve additional hardware and installation expenses are a more optimal choice for companies. Collaborative robots, as proposed by [37] are ultimately created with distinct characteristics that set them apart from conventional robots [37]. In adherence to rules and ergonomic principles. It should also have extra functions that impact traditional robots, like power and energy monitoring, force limits, vision (camera), laser tech, collision avoidance, voice recognition, and/or teamwork with human operators. Deeds and deeds. To find out more, please see references [33, 37–41]. KUKA iiwa (2016–2018), ABB YuMi (FRIDA), and Universal Robots (UR), were all updated between 2014 and 2018. Robots are frequently selected for their efficiency and general usability; yet, the downside to this choice is the higher expenses and intricacy caused by the integration of numerous external sensors and limitations in HRC. The connection between the joint robot's motion and the task at hand is not clearly defined, as we think that various factors, like sensor presence on each axis, play a role in determining the appropriate robot for collaboration in this scenario. Nevertheless, it is important to mention that the assumption of kinematics, including the number of axes, is discussed in [42]. Future research will concentrate on confirming the results with kinematically redundant robots [43] or employing repetition for stronger manual guidance. Please paraphrase the text provided. Suggested [44]. Many review studies show how the methods they utilize will perform in the future, either by adding more complexity to the operator and/or the environmental model [45] or by employing alternative indicators to assess performance [46, 47] and choose the task. The text should be rephrased in the same language it was provided in. Some argue that the next move should involve broadening their research scope to encompass other fields of application. It is our belief that these objectives can be achieved without making alterations to current technologies or algorithms; To enhance safety, efficiency, and effectiveness, researchers should enhance individuals' ability to plan [48], comprehend their surroundings and tasks [49, 50], understand employees' needs, and design ergonomic workspaces [51]. Future collaborations aiming to enhance HRI systems will concentrate on enhancing the understanding of tasks and surroundings by both robots and workers, in a manner comprehensible to the operator, facilitated by product knowledge and knowledge exchange.

3 Background Technologies

3.1 Types of Human–Robot Collaboration (HRC)

HRC refers to an environment in which humans (human workers) and robots can communicate with each other to form a dynamic network to perform tasks [52–54]. To make the most of this cooperation, especially with the new innovations brought by change, to ensure good and effective cooperation. Fourth, robots are adopted with special techniques and are called collaborative robots (collaborative robots) [55, 56]. Robotics integration represents a natural evolution to solve complex problems faced by business that must be efficient while maintaining quality standards, including increasing concerns about work life and the quality of human workers involved in work [57]. [58] describe different types of HRC and specifically identify the following four categories: around but not interacting with each other. Figure 1 shows the Levels of Collaboration in HRC for Industry 4.0.

Synchronous: Human operator and robot work together to work around each other at different times.

Collaborative: Individual operators and colleagues They work together at approximately the same time, and each of the two organizations focuses on different tasks.

Collaboration: Human workers and cooperative workers work together, and the behavior of one organization affects the behavior of other organizations and directly affects it. Situation where operators work simultaneously in a shared workplace".

Within the realm of HRC, there exists a spectrum of interplay levels between people and robots. Here's a breakdown of the special styles of HRC, categorized by the level of interplay and proximity:

- **Segregated Collaboration**: This method continues a physical separation among human beings and robots. It's suitable for tasks concerning excessive dangers or risks for human beings.The 2 main Subcategories are Fixed Guarding wherein Robots operate within a fixed cage or barrier, absolutely break away the human workspace and Safety Interlocks in which Robots and humans percentage the identical area, however physical interplay is prevented by using safety interlocks that prevent the robotic if a human enters its detailed sector.
- **Cooperative Collaboration**: This kind includes a more in-depth proximity however with limited interplay between humans and robots. The 2 essential Subcategories are Independent Workstations where Humans and robots work at separate stations however on associated responsibilities. They can also communicate indirectly through shared information or control systems and Sequential

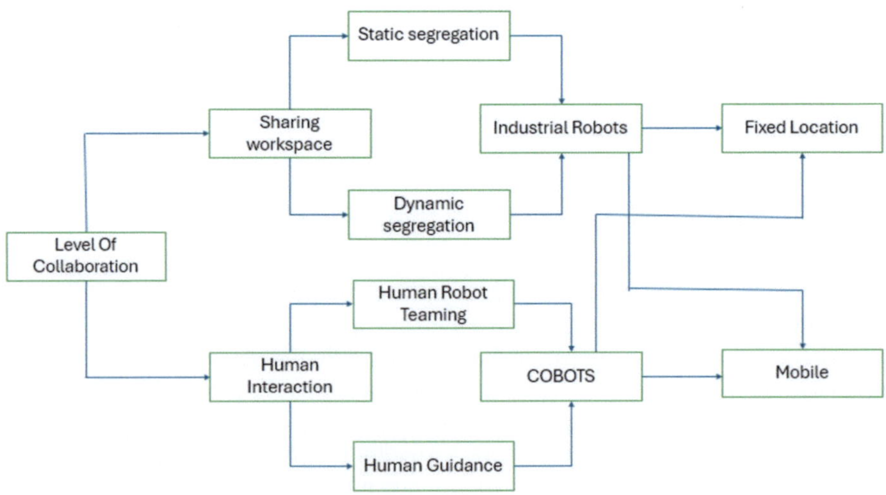

Fig. 1 Levels of collaboration in HRC for Industry 4.0

Work in which Humans and robots carry out extraordinary responsibilities at the same product or method, however in a sequential way. One assignment finishes earlier than the other begins.

- **Collaborative Collaboration**: This is the maximum advanced shape of HRC, wherein human beings and robots work in near proximity and interact immediately with each other. The 2 important Subcategories are Human–Machine Interface (HMI) Collaboration where Humans engage with robots thru a user interface, issuing instructions or receiving comments on robot actions, Physical Assistance Collaboration wherein Robots bodily assist human beings with tasks, such as handling heavy items or supplying assist in the course of assembly. This can contain cobots, in particular designed for safe interplay with humans and Adaptive Collaboration where Robots adjust their conduct primarily based on human movements or the surroundings via superior sensors and AI. This allows for a greater fluid and responsive collaboration.

Below in Fig. 2 the Taxonomy of Human–Robot Collaboration (HRC) in Industry 4.0 is explained well.

The selection of the maximum suitable HRC version relies upon on numerous factors, inclusive of The level of danger, complexity, and want for human intervention within the task, The capability dangers associated with the task and the want for bodily separation, The stage of automation and robotic intelligence to be had and The capabilities and education required for human workers to soundly and correctly collaborate with robots. As era advances, the lines between those types of collaboration will in all likelihood come to be an increasing number of blurred. We can expect more sophisticated robots capable of adaptive collaboration, blurring the traces between human and robotic workspaces.

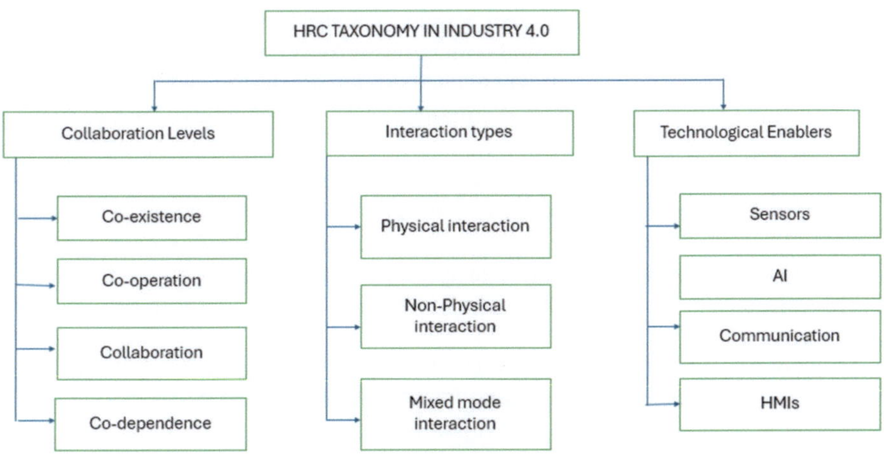

Fig. 2 Taxonomy of Human–Robot Collaboration (HRC) in Industry 4.0

3.2 *Enabling Technologies for Human–Robot Collaboration (HRC)*

Interaction between humans and computers relies on meaning and the organization of language. Robots must have the ability to understand complicated environments like production, storage, and mining to operate effectively. Derived semantics and learned semantics are the two concepts of semantics, with the former involving pre-programmed information in the robot before sending, while the latter involves the robot learning information before or during sending. Semantics holds promise in various fields such as sanitation, SEO, intelligent transit, and online communities. The growth of HRC relies on essential enabling technologies. This technology not only enhances the success of robots but also closes the gap between human and robotic capabilities, promoting secure and efficient teamwork.

- **Sensors**: Cameras and depth sensors permit robots to "see" their surroundings, hit upon mans and items, and navigate thoroughly in shared workspaces.Force Sensors measure the quantity of force exerted via a robot, allowing secure interplay with people and stopping accidents.Proximity Sensors stumble on the presence of a human in close proximity, allowing robots to regulate their moves or forestall absolutely to keep away from collisions.
- **Collaborative Robots (Cobots)**: These robots are mainly designed for secure interaction with people. They are normally lightweight, have rounded edges, and function with restrained strength and pace. This lets in them to work along people without the need for considerable safety cages.Cobots can be programmed to carry out numerous obligations and are frequently person-pleasant, taking into account easy variation to converting production needs. Cobots, short for collaborative robots, are seen as a leader in the field of industrial robotics and are thought to have the potential to enhance the economy by enhancing simplicity, efficiency, and productivity. Both developers and collaborative robot experts agree that collaborative robots have the capacity to enhance product flexibility and provide protection against rapid supply chain disruptions, market fluctuations, and escalating production demands [59, 60].
- **Artificial Intelligence (AI) and Machine Learning (ML)**: AI algorithms enable robots to examine and adapt to their environment. They can examine sensor statistics to recognize human actions and expect human wishes.Machine gaining knowledge of permits robots to continuously improve their performance via revel in. This can empower them to collaborate with human beings extra efficiently through the years.
- **Human–Machine Interface (HMI)**: HMIs provide a person-friendly interface for people to interact with robots. These interfaces may be voice-activated, contact-primarily based, or use other intuitive methods to allow people to manipulate robots, issue commands, or acquire feedback.
- **Communication Protocols**: Reliable and steady verbal exchange protocols are vital for clean collaboration. These protocols make certain seamless records

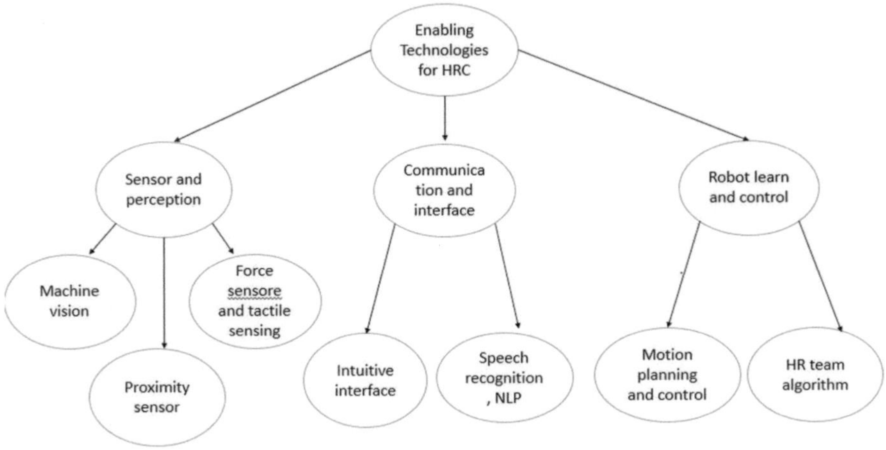

Fig. 3 Enabling Technologies for Human–Robot Collaboration (HRC)

trade among robots, control systems, and human operators, enabling actual-time coordination and assignment control.

- **Safety Standards and Regulations**: As robot abilities develop, so too do safety concerns. Standardized protection protocols and policies are essential for ensuring safe human–robot interaction. These requirements cope with aspects like robot design, risk assessment, and schooling for human employees in safe interplay practices with robots.

As shown below, Figure 3 explains about the Enabling Technologies for Human-Robot Collaboration (HRC)

The synergistic mixture of those technologies lays the muse for a future of secure, efficient, and productive human–robot collaboration. As technology keeps to conform, we will count on even more superior abilities that further beautify the ability of HRC.

4 The Rise of Human–Robot Collaboration (HRC) in Industry 4.0

While Industry 4.Zero brings a wave of automation with robots and clever machines, it also ushers in a brand new generation of collaboration—Human–Robot Collaboration (HRC). This partnership between human beings and robots is rapidly turning into a important factor for success in smart production inside Industry 4.0. Here's why HRC holds such developing significance: Humans excel at responsibilities requiring creativity, essential questioning, hassle-solving, and complicated selection-making. Their dexterity and flexibility lead them to valuable for tasks requiring quality motor abilties and coping with sensitive items. Robots convey unequalled strength, speed, and precision to repetitive, strenuous, or risky responsibilities. They can operate tirelessly in harsh environments and are programmed to perform responsibilities with excessive accuracy and consistency. However there are huge Synergistic Benefits. HRC lets in people and robots to work together, leveraging their specific strengths. Robots deal with the heavy lifting and repetitive duties, liberating up human beings to cognizance on higher-cost activities. This optimized workflow ends in significant gains in universal production output. Robots' precision and consistency limit mistakes in manufacturing approaches, main to a higher high-quality of completed merchandise. With robots looking after standardized tasks, human employees can effortlessly adapt to changing manufacturing necessities. This flexibility lets in manufacturers to reply quick to market demands and convey a greater diversity of merchandise. Robots can be programmed to deal with dangerous obligations, reducing the danger of administrative center injuries and injuries for human workers. HRC necessitates a shift in group of workers capabilities. Workers broaden new capabilities related to operating, programming, and keeping robots, creating a extra skilled and adaptable group of workers. However, the current market requires the use of simple and versatile assembly machines, as they require a short period of time and major adjustments [61]. Small and medium-sized businesses (SMEs) rely heavily on these particular needs. Cobots, also known as collaborative robots [62], are a groundbreaking innovation that has the potential to address issues in production and assembly work. They enable interaction with humans in shared workspaces, operate effectively without the need for experts, and can be quickly reprogrammed for various tasks within changing workflows. Working together with robots is seen as an effective method to boost productivity and cut down on production expenses, allowing individuals to make decisions and communicate effectively [57]. Industry 4.0 promotes a higher degree of collaboration that goes beyond traditional robot automation. Collaborative robots, known as cobots, are specifically created to work safely alongside humans in shared work environments. Progress in sensor technology and artificial intelligence allows robots to sense their environment, respond to human actions, and even learn from their interactions. As Industry 4.0 keeps to evolve, so too will HRC. We can anticipate even greater seamless and complicated collaboration between people and robots. Imagine robots that may assume human actions, research new tasks on the fly, and adapt to changing environments. This level of collaboration holds immense potential for in addition productiveness gains, improved product satisfactory, and a more dynamic and adaptable manufacturing panorama. In conclusion, HRC isn't just a trend but a middle pillar of clever production in Industry four.0. By leveraging the complementary strengths of people and robots, producers can liberate a new technology of efficiency, flexibility, and innovation. As shown in Table 1, the Comparison of Traditional and HRC Automation is discussed.

HRC refers to a piece surroundings where people and robots paintings collectively in a shared space to reap a not unusual goal. This collaboration leverages the precise strengths of both. In essence, HRC represents a shift from human replacement to human-robotic teamwork, fostering a greater collaborative and efficient destiny of labor.

Table 1 Comparison of traditional and HRC automation

Feature	Human–Robot Collaboration (HRC)	Traditional automation
Work environment	Shared workspace	Isolated workspace
Focus	Collaboration, leveraging strengths of both	Replacement of human labor
Task allocation	Complementary tasks	Standardized tasks
Adaptability	High adaptability to changing situations (real time interaction)	Limited adaptability
Safety	Prioritized through design	Primarily relies on engineering safeguards

a. Case Studies: successful Implementations of HRC in Various Industries

Here are some case studies showcasing successful implementations of Human–Robot Collaboration (HRC) across special industries:

4.1 Ergonomics

When discussing safety and performance, it is essential to bring up ergonomics. The primary distinction in job performance between an electric machine and a human worker is the impact of ergonomic factors on the latter [58]. For instance, assembly tasks demonstrate how cobots can offer valuable assistance to workers, utilizing specific communication and shared workspace. Nevertheless, it is important to note that, as previously stated, the robot must have the ability to adjust and modify its actions to function ergonomically and securely without impacting human productivity. Numerous simulation tools with human models can be used to assess the efficiency of the HRC environment by replicating actual working conditions in a simulated setting [63, 64]. Ergonomics is crucial in agriculture as many tasks require awkward postures and heavy lifting for extended periods. [65] performed trials in agricultural HRC. A joint project involved humans and robots making decisions together for strawberry picking plans. Different sensors and navigation systems designed for autonomous vehicles to enable efficient and safe driving are showcased in [66]. The primary action of the vehicle involves tracking a row of trees, recognizing the conclusion of the row, deviating from it, and subsequently pivoting and entering the subsequent row. [67] focused on incorporating gesture recognition to facilitate communication between humans and robots. In order to achieve this goal, six different machine learning classifiers were assessed and the Robot Operating System (ROS) software was employed to convert the robot's commands into actions.

4.2 Industry

The two main jobs of this structure are assembly and welding. [68] studied digital twins of mobile devices that collaborate with robots to perform assembly tasks alongside humans in a device state-driven simulation. A business case study with a power outlet is used to prove the case study. In order to automate manual or integrated lines, [69] a decision allowing worker division has been proposed. The preparation process has been tested in scientific literature (especially for the automotive industry) and helps speed up the dispensing and preparation of mixed products and work more efficiently. [70] examined the HRC assembly process to compare the decision-making of humans and robots. The algorithm was utilized in a height-adjustable door installation case study to showcase its efficiency in operation planning. [71] suggested a collaborative welding system in virtual reality where humans and robots can collaborate on welding tasks. The evaluation of a GTAW welding machine was utilized for research purposes. Battery production at BMW has transformed into responsibly managing and assembling large, weighty lithium-ion battery modules for electric vehicles. The Solution turned into found that BMW carried out a crew of collaborative robots along human employees. The cobots help with responsibilities like lifting and maneuvering battery additives, at the same time as human beings awareness on precision meeting and quality manage. The primary blessings are Reduced risk of musculoskeletal accidents for workers, stepped forward manufacturing speed, and more suitable safety in battery coping with. The principal undertaking became Efficiently drill and rivet big plane wing additives with excessive precision and repeatability. The Solution changed into discovered that Airbus makes use of robotic palms ready with vision systems for automated drilling and riveting responsibilities. Human workers collaborate with the robots, tracking the procedure, conducting pleasant assessments, and performing complicated meeting steps requiring dexterity and judgment. The major advantages are extended manufacturing pace and accuracy, advanced consistency in wing assembly, and reduced chance of mistakes. DHL Warehouse Automation: The main undertaking found was Efficiently manage and type a excessive extent of applications in a massive warehouse environment. The solution found was DHL employs a mixture of self reliant cell robots (AMRs) and collaborative robots to automate diverse tasks of their warehouses. AMRs handle tasks like transporting pallets, at the same time as cobots help human workers with picking and packing person items. The fundamental advantages are Increased order achievement speed and accuracy, reduced reliance on manual labor, and advanced common warehouse performance.

4.3 Agricultural Industry

Robots are used in agriculture, as demonstrated by robotic assistance in fruit tree planting [72]. There are three types of fruit planting robot: mule type, tempo type and scaffold type. In mule mode, the robot follows a group of workers, helping with tasks such as harvesting. The speed model involves a robot operating in a specific area. Scaffolding mode allows robots to walk while acting as a scaffolding for humans to stand on. What makes robots useful is that workers can cut wood twice as fast in scaffolding mode than using a ladder. It also makes people's jobs easier. This example demonstrates how robots in HRC can positively impact human workers by improving efficiency and comfort, enhancing overall productivity. Sustainable production brings about additional advantages for society, leading to improved working conditions and an increase in job opportunities. Disassembly is the primary production method in sustainable production for remanufacturing to conserve resources, energy, and decrease emissions. While robots are capable of performing certain monotonous and messy dismantling tasks, there are still some duties that must be carried out by humans. PCDEE-Circle is an HRCD system that employs "multimodal perception, multi-objective cognition, decision making, and knowledge creation and evolution."

4.4 Food Industry

The European economy is significantly impacted by the food industry, where certain stores are capable of producing as many as 10,000 meals daily [73]. The process consists of three primary phases: cultivation, manufacturing, and ultimately delivering the finished goods to the market. Food industry executives are aiming to update their marketing tactics in order to meet the growing demand. This is particularly crucial in the initial phases of the COVID-19 outbreak, as it hinders the recovery of supply chains [48]. Hence, the food industry requires the quick adoption of digitalization and Business 4.0 technology to implement changes that will enhance the business's sustainability. Agricultural robots enhance understanding of farming, soil, and crop cultivation. Utilizing sensors in smart packaging enhances system reliability by providing accurate information on product expiration dates [49]. Utilizing a robot equipped with a gripper camera for the selection and examination of fruits [50]. In Nestlé Chocolate Packaging The main task became Safely and hygienically select and vicinity delicate chocolate bars into packaging machines at high speeds. The solution discovered was Nestlé makes use of cobots ready with specialized grippers for managing chocolate bars. The cobots work seamlessly with human employees who oversee the packing system and make sure first-rate manage. The most important blessings are Improved hygiene requirements, reduced danger of product damage, and multiplied production throughput for chocolate packaging.

4.5 Automobile Industry

The automotive industry is the biggest economic sector globally. In the UK alone, the private car sector provides employment for 3.7 million individuals and adds around $26 billion to the UK economy [51]. Assembly cells are crucial in the automotive sector; assembly tasks are carried out in 83% of production cells [52]. Nevertheless, certain manuals require additional flexibility and strength in order to excel; therefore, depending only on robots to carry out these duties might not be a feasible answer since human abilities are invaluable. Hence, the emphasis is on uniting the capabilities of people and robots to cooperate effectively while prioritizing safety and accident prevention in the workplace [53]. Ford VOME (Vehicle Operation Manufacturing Engineering): The primary mission changed into Safely and correctly alter fog lighting fixtures at some stage in automobile assembly. The Solution became found that Ford partnered with Kuka to enforce a collaborative robotic (cobot) for fog light adjustment. The cobot's touchy sensors and lightweight design permit it to work correctly along human workers. The predominant advantages are Improved ergonomics for employees, decreased hazard of injuries, and elevated performance within the meeting procedure. These are only some examples demonstrating the ability of HRC across numerous industries. As generation advances, we will expect even greater progressive applications of HRC, further remodeling the way human beings and robots work collectively within the future.

5 Methodology: Proposed Architecture

The use of assembly machines in businesses can be customized, but it may result in decreased product quality and increased labor costs [54]. Comparing human workers with electronic machines highlights how ergonomic factors significantly impact manual assembly performance by restricting the weight of the product and the precision of the human operator [55]. As a result, these constraints restrict the operator's capacity to manage and store bulky/large items. These devices can be used with transport machines like jib cranes: they can be viewed as significant workstations in assisting robots with moving heavy items [56]. Nevertheless, based on the authors' understanding, there is currently no market available that enables these machines to carry out intricate tasks like assembly or direct recording due to their usual constraints of efficiency and accuracy [57]. Conventional robots [58] are linked to the surrounding area, offering workers high-tech tools (such as the FANUC M-2000 series capable of carrying 2.3 tons [74]) and continue to demonstrate strong performance. Nevertheless, the cost or feasibility of accomplishing complex assembly tasks with traditional robotic systems may be too high or unattainable [75]. This gap can be bridged by incorporating the strengths of conventional robots with the adaptability and efficiency of a human operator. The collaborative robot is perfect for tasks in assembly, particularly when workers are present. In Fig. 4, the steps in HRC process are shown as used in Industry 4.0.

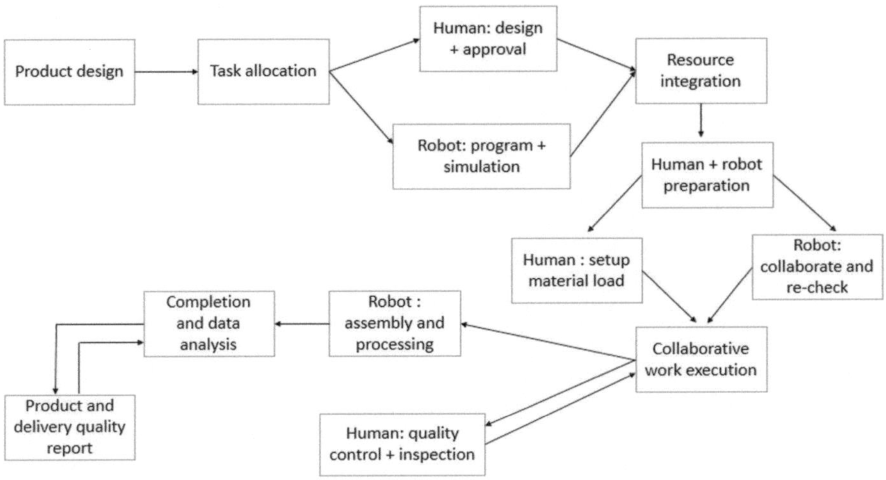

Fig. 4 Steps in HRC process in Industry 4.0

5.1 Smart Manufacturing

During Industry 3.0, automation focuses on optimizing manufacturing procedures and overseeing various parts with sensors and actuators. This enables workers to oversee the manufacturing process and make necessary adjustments to the work environment. The objective is to enhance effectiveness and precision by still depending on human attention. Nevertheless, this method is not without its constraints, and with the shift to Industry 4.0, numerous collaborative strategies have been suggested [59]. In this new digital era, numerous businesses are now seeking innovative ways to utilize technology for better productivity and success. One method that is often used is smart manufacturing, which utilizes technologies like artificial intelligence and robotics to carry out intricate tasks with greater accuracy and precision. Smart manufacturing allows for enhanced regulation, increased safety protocols, and heightened versatility during the production phase. In contrast to Industry 3.0, automation solutions prioritize developing production processes and depend on human workers to oversee and make changes in the workplace, while Industry 4.0 emphasizes digitalization and technology enhancements for enhancing the production process. Through this approach, businesses can enhance their efficiency and output while cutting down on energy usage and expenses, resulting in a sustainable operation and improved collaboration. This scenery is characterized by four primary strategies for efficient manufacturing: Internet of Things (IoT), cloud computing, big data, and analytics [60]. Hence, implementing Industry 4.0 can unify the complete manufacturing procedure. In the age of Industry 4.0, it is crucial for companies to stay competitive by enhancing productivity, efficiency, and cost savings using technologies like AI, robotics, and the Internet of Things. Nonetheless, achieving a prosperous digital transformation journey necessitates more than simply adopting new technologies. A strong leadership commitment and a clear future vision are needed for a major change in thinking to occur throughout the entire organization [61]. Leaders need to grasp the possibilities of digital technology and convey its advantages to the organization in an effective manner. Encouraging innovation, adaptability, and ongoing learning is essential for driving digital transformation. This involves promoting trying new things, incentivizing taking chances, and establishing a space where employees are motivated to embrace new technologies. Involving employees in the early stages of change, collecting their feedback, and addressing their worries are crucial actions. Training programs and advanced courses are essential in order to provide employees with the necessary skills to utilize new technologies. Beginning with a pilot program allows companies to try out new technologies in a controlled setting first, showcasing the success and productivity of digital transformation [62]. Partnering with technology providers, research centers, and industry professionals fosters knowledge exchange and ensures the company remains current on emerging trends and top practices. Moreover, intelligent manufacturing merges supply chain management, customer service, and manufacturing business processes for enhanced collaboration and information management. This results in enhanced productivity, improved product quality, and increased customer satisfaction [57]. Utilizing HRC in manufacturing provides a hopeful option aside from automation, simplifying processes and enhancing teamwork between humans and robots. Interactive thinking is promoted by interfaces that are easy for users to navigate [58]. HRC machines have the potential to enhance the effectiveness, output, and adaptability of the manufacturing process, paving the way for the future of manufacturing. Nevertheless, the current application of collaborative robotics is restricted to uncomplicated production tasks due to its complexity, affecting user trust and decision-making in crucial scenarios. Design and evaluate human responsibility is crucial, along with upholding security and accessibility [74]. Industry 4.0 heavily relies on Human–Robot Collaboration (HRC) to achieve increased efficiency and personalization. The HRC system can be divided into various important steps.

1. **Data Collection and Analysis**: This preliminary step involves amassing records approximately the responsibilities which might be being taken into consideration for HRC. This statistics can be used to decide the feasibility of collaboration, as well as to identify any potential safety dangers.
2. **Task Planning and Programming**: Once the facts has been analyzed, the specific responsibilities with a purpose to be executed by the robotic and the human employee(s) can be described. This consists of programming the robotic's moves and designing the human-robotic interface.
3. **Risk Assessment and Safety System Design**: A important step is to perceive and investigate any capability dangers related to the HRC technique. This will tell the layout of protection systems and techniques to mitigate those dangers. Safety requirements outlined in ISO suggestions are important for commercial settings to create a secure paintings surroundings.
4. **System Integration and Validation**: The robotic, the human-robotic interface, and any protection systems are all incorporated right into a single gadget. This

system is then very well tested to make sure that it capabilities effectively and successfully.

5. **Deployment, Operation, and Maintenance**: After the machine is verified, it may be placed in the manufacturing setting. However, the work is not carried out—the HRC machine will want to be monitored and maintained on an ongoing foundation. This may contain accumulating new statistics, updating packages, and revising protection approaches as wanted. By following those steps, businesses can put into effect HRC in a manner this is both secure and powerful.

6 Results and Discussion

The decision to integrate humans and robots is frequently influenced by motivation for work, workplace satisfaction (ergonomics and human factors), and optimal utilization of workspace. One more benefit is that the robot can easily handle the required task [76]. Furthermore, the most popular method of learning is through demonstration [77]. Furthermore, cooperation is easier and faster due to its simplicity, as distributing items to other areas of the factory voluntarily is more straightforward without the need for tight security measures for shared units; orders need frequent modifications [78]. Nevertheless, similar to other traditional systems, there is a need to restrict high-risk applications, leading to reduced flexibility. This assessment of current research presents a convincing perspective on how HRC could transform smart manufacturing in Industry 4.0. Figure 5 shown below shows the importance of HRC for Industry 4.0.

Here are the key takeaways. HRC gives huge blessings, along with extended productiveness and efficiency, advanced product first-class, and more suitable safety in production environments. Overcoming safety issues, addressing talent gaps in

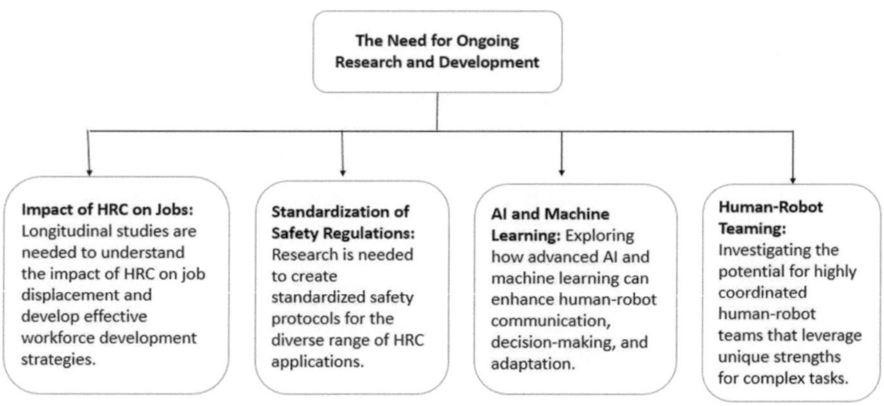

Fig. 5 The importance of HRC for Industry 4.0

the group of workers, and making sure powerful human–robot interplay layout are essential for successful HRC implementation. Strategies for upskilling and reskilling the team of workers are crucial to prepare workers for the changing demands of participating with robots. Robust safety requirements and policies are paramount to mitigate risks and ensure worker protection throughout collaboration. Designing person-friendly interfaces and ergonomically sound workstations are important for powerful human-robotic interplay and employee properly-being. HRC is not only a fashion; it's a center pillar of clever manufacturing in Industry 4.0. Manufacturers can develop a new technology by combining the strengths of both people and robots. Collaborative robots can deal with repetitive obligations, releasing human people to attention on better-fee sports, leading to big production gains. Robots' precision and consistency reduce mistakes, main to better fine products with fewer defects. With robots looking after standardized tasks, human people can simply adapt to converting manufacturing needs. By prioritizing ongoing research and development, we will ensure that HRC reaches its full capacity, shaping a future where people and robots collaborate seamlessly to force innovation and progress across numerous industries.

6.1 Research Gaps

Based on the reviewed literature, here are some critical research gaps and areas for further investigation in Human–Robot Collaboration (HRC):

1. **Impact of HRC on Job Displacement and Workforce Development**: While a few research improve worries about job losses, a more complete know-how calls for longitudinal studies to track the real effect of HRC on employment across one-of-a-kind sectors and talent tiers. Research is needed to discover the evolving nature of work in the face of HRC. This ought to contain investigating new activity roles that emerge because of automation and collaboration, and the particular skillsets required for the ones roles. Research is wanted to develop and compare powerful training programs for upskilling and reskilling the staff to evolve to the demands of running with robots.
2. **Safety Standards and Regulations for HRC**: Existing safety requirements might not cope with the whole spectrum of HRC packages. Research is needed to increase standardized safety protocols for a much wider range of duties and operating environments.Clear recommendations and processes for robotic certification and compliance with safety standards are vital.With elevated robotic autonomy and connectivity, studies is needed to deal with capacity cybersecurity threats and make certain steady verbal exchange within HRC structures.
3. **Human Factors and Ergonomics in HRC Design**: Research is wanted to recognize and degree the cognitive load on human employees in HRC situations. This can tell the layout of robots and HRI structures to decrease mental pressure and optimize human performance.The psychological impact of operating with robots

desires similarly research. Studies may want to discover worker attractiveness, consider, and potential feelings of isolation in collaborative work environments. There's a want for standardized metrics to assess the effectiveness of HRI structures. This ought to involve measuring factors like ease of use, communication performance, and usual collaboration overall performance.

Further research is needed to discover how advanced AI and system mastering can beautify human-robotic communication, choice-making, and adaptation inside collaborative responsibilities. As robots become more state-of-the-art, moral considerations surrounding activity displacement, worker privateness, and the capability for self reliant selection-making by means of robots want to be addressed. Long-term research with a forward-searching angle is wanted to discover the capacity future of HRC, along with the opportunity of more seamless human–robot teamwork or even human-robotic symbiosis. By investigating those essential research gaps, we are able to pave the manner for a destiny where HRC benefits each businesses and people, fostering a secure, green, and adaptable production panorama.

6.2 Impact of HRC on Job Displacement and Workforce Development

The reviewed literature highlights the potential benefits of HRC for efficiency and productivity gains.

- **Potential for Job Displacement**: Some studies, like the ones noted in the barriers phase, increase concerns approximately process displacement because of automation. As robots take over repetitive tasks, a few jobs might grow to be obsolete. This raises worries about Jobs closely reliant on repetitive tasks like meeting or material dealing with is probably most inclined. Workers whose jobs are automated may additionally need retraining or upskilling to adapt to new roles in the changing work environment.
- **Workforce Development Strategies**: However, different studies endorse possibilities for team of workers improvement along automation. The want for human people will in all likelihood shift toward tasks requiring higher-order capabilities like hassle-fixing, important wondering, creativity, and complicated selection-making. Manufacturers want to invest in education packages to equip their workforce with the competencies needed to collaborate successfully with robots. This could encompass schooling in running, programming, and maintaining robots. The destiny of work may contain a collaborative model where people and robots paintings together, leveraging each different's strengths. Workers will want abilties in conversation, coordination, and teamwork to characteristic effectively in those collaborative environments.
- **Importance of Safety Standards and Regulations for HRC**: The reviewed literature, specially studies emphasizing the importance of safety standards and guidelines for successful HRC implementation. Robots are effective machines,

and there's constantly a capability for accidents if right safety measures are not in region. Standardized protocols can assist reduce these risks. Regulations can mandate safety features in robots, inclusive of sensors and emergency forestall buttons, to protect workers from injuries throughout collaboration. Standardized risk assessment strategies ought to be applied to identify capability hazards related to specific HRC tasks and increase mitigation techniques.

- **Role of Human Factors and Ergonomics in Designing Effective HRC Systems**: Effective HRC hinges on well-designed human-robotic interplay (HRI) systems that consider human factors and ergonomics.HRI systems ought to be intuitive and consumer-pleasant for human employees to engage with robots without difficulty. This consists of clean visible and auditory cues, and interfaces that accommodate exceptional skill stages.Workstations designed for collaboration have to do not forget the bodily limitations and capabilities of human people to prevent fatigue, discomfort, or musculoskeletal accidents.HRC structures have to be designed to decrease the cognitive load on human workers. This ought to involve obligations requiring much less mental effort or imparting robots with a few stage of decision-making capability to unburden human people.

The overview of present literature showcases the colossal capacity of HRC in smart manufacturing. However, it's crucial to acknowledge the demanding situations related to activity displacement and ensure right group of workers improvement strategies are in area. Additionally, prioritizing safety via strong standards and regulations, and specializing in human factors and ergonomics in HRI design, are essential for a a success and sustainable future of human-robotic collaboration. Figure 6 shows the impact of HRC in smart manufacturing.

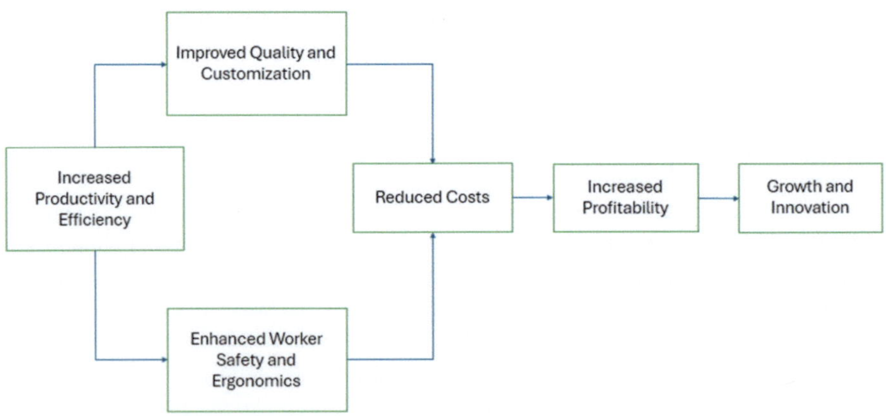

Fig. 6 Impact of Human–Robot Collaboration (HRC) in smart manufacturing

6.3 Benefits of HRC as a Collaborative Approach

The utilization of HRC enhances production methods, resulting in significant improvements in performance, adaptability, and ultimately, increased productivity. The key to HRC's success is combining the unique capabilities of both humans and robots. The analysis emphasizes HRC's transformative impact on efficiency, adaptability, and productivity within Industry 4.0 manufacturing. Figure 7 shows the benefit of HRC.

- **Efficiency**: Robots tirelessly take care of repetitive tasks like welding, assembly, or cloth managing, considerably reducing the time it takes to finish a product. Humans can cognizance on fee-brought activities like high-quality checks or changes.Robots are much less susceptible to fatigue or breaks compared to human beings. This reduces downtime and continues manufacturing strains jogging smoother. Collaborative robots, or cobots, are designed to function properly along human beings. This allows for a extra dynamic workflow in which robots take care of the heavy lifting or hazardous duties, even as humans manipulate the more nuanced steps, leading to a extra green normal manner.
- **Flexibility**: Cobots are typically smaller, lighter, and easier to program compared to traditional industrial robots. This allows them to be flexible in response to evolving manufacturing needs. Production lines can be quickly reconfigured to support new products or models with minimal interruption.HRC enables producers to scale manufacturing up or down unexpectedly. During periods of high demand, robots can take on extra responsibilities, at the same time as human people can recognition on areas needing extra attention. Conversely, throughout gradual durations, robots may be redeployed to distinct tasks, maximizing aid usage.

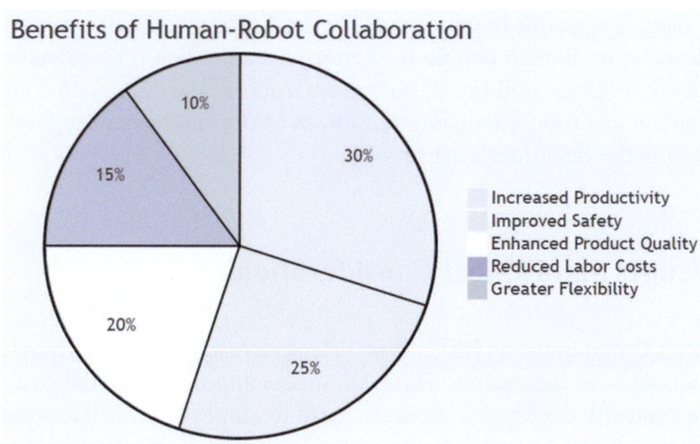

Fig. 7 Benefits of human robot collaboration

Table 2 HRC as a collaborative approach

Area	Humans	Robots
Decision-Making	Strategic thinking, ethics, complex situations	Data analysis, simulations, objective suggestions
Problem-Solving	Improvisation, critical thinking, adaptation	Repetitive procedures, data-driven solutions, tireless testing
Physical Tasks	Dexterity, hand–eye coordination, adaptability	Strength, endurance, hazardous environments, precision

- **Productivity**: By streamlining workflows and reducing downtime, HRC at once interprets to improved manufacturing output.Robots perform duties with excessive precision and repeatability, minimizing mistakes that may lead to rework or scrap. Humans can awareness on obligations in which their judgment and selection-making are essential, similarly enhancing typical high-quality and decreasing waste.Repetitive or physically disturbing duties can be tiring and result in injuries. By delegating those responsibilities to robots, HRC facilitates reduce worker fatigue and strain, potentially main to stepped forward morale and higher productiveness.

As shown below, Table 2 depicts HRC as a Collaborative Approach.

By combining these strengths, HRC creates a powerful synergy. Humans provide the judgment and adaptableness, while robots offer tireless execution, data-driven insights, and the ability to deal with physically demanding or hazardous tasks. This permits for a stronger and efficient approach to choice-making, problem-fixing, and physical duties within the manufacturing surroundings. HRC minimizes errors, optimizes aid use, and enhances common output quality in several methods. Robots excel at repetitive obligations with high precision, reducing mistakes caused by fatigue or human oversight. Humans can consciousness on exceptional control responsibilities where their judgement is important. By automating repetitive responsibilities, robots unfastened up human people for better-cost activities. This ensures the most promising use of both the abilities of human and robots. The combined cognizance of human judgment and robot precision minimizes errors and ensures regular excellent at some stage in the manufacturing process.

7 Challenges and Ethical Considerations

Human-Robot Collaboration (HRC) is an effective resource in intelligent manufacturing, but it does have constraints. This is of utmost importance. In the past, industrial robots were typically enclosed to prevent harm to employees. Collaborative robots, also known as cobots, are specifically created to operate closely with humans, yet challenges may still arise. Essential safeguards such as speed and force limits, along with improved human-robot communication, are necessary. Robots must be more intelligent for effective collaboration. Figure 8 depicts the challenges of HRC.

In the regulatory framework of the EU, businesses, institutions, and individuals must guarantee that the utilization of robotics complies with the necessary health and safety standards set forth in the Machinery Directive (2006/42/EC) [79]. Provide feedback or follow the Consensus approach. The key and relevant standards are ISO 10218:1 and ISO 10218:2 (category c) as well as ISO/TS 15066 for Robots and robotic products—specification for integrated Robots. Abiding by the standards is not required, however, it is compulsory to conduct risk assessment for all robotic applications in compliance with the Machinery Directive [80, 81]. This approach better state of affairs attention, being able to are expecting human moves, and having greater bendy decision-making abilities. Advancements in synthetic intelligence (AI) are key right here. Some people might also fear being changed by way of robots. Training and clean verbal exchange approximately how HRC can increase human abilties is critical. As HRC becomes greater huge, there needs to be clean standards and regulations to make certain safety and worker properly-being. Implementing HRC structures can contain huge upfront prices for brand new technology and workspace redesign. The long-time period blessings want to justify the investment (Table 3).

Researchers think that by acknowledging these constraints, HRC could completely change the way smart manufacturing operates, enhancing productivity, flexibility, and efficiency. It is evident that safety is crucial in the HRC setting and all scientific data indicates that research is conducted safely and steadily evolving. Safety regulations can be met with the use of various sensors, whether they are already integrated into modern machines or temporarily added to older machines. There is a growing amount of research and development focused on industrial-grade sensors. The categorization of sensor types in the HRC environment is determined by the separation of contact sensors and non-contact sensors. Additional details are available in references [80, 81, 86–89]. The primary objective of the former is to identify accidents in real-time and minimize their impact.

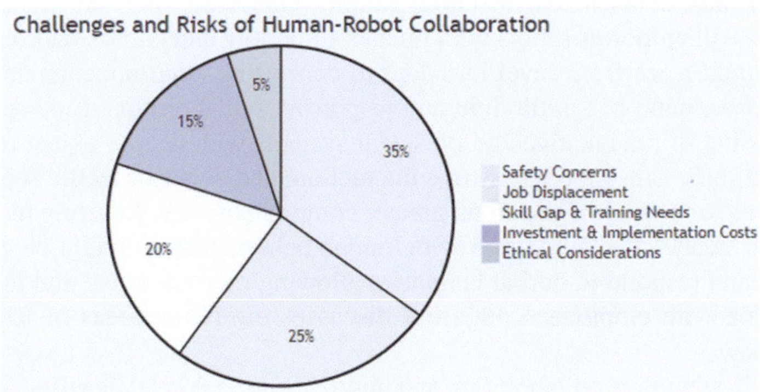

Fig. 8 Challenges of HRC

Table 3 Challenges of implementing HRC in smart manufacturing

Challenge	Description	Example References
Safety Concerns	Ensuring safe interaction between humans and robots	[82]
Skill Gaps in Workforce	Upskilling workers to operate, program, and maintain robots	[83]
Cost of Implementation	High initial investment for acquiring and integrating robots	[84]
Design of Human–Robot Interaction (HRI)	Importance of user-friendly interfaces for effective communication	[85]

8 Future Directions

A major challenge ahead is enabling the incorporation of social robots in society that can effectively interact and communicate with humans, beyond just being used for business reasons. Recently, there has been an increase in the research and development of social robots in the field of education [90]. [91] explored the possibility of employing robotic trainers to aid in collaborative projects. They compare tablets with robots in this manner. Hence, robots are beneficial for supporting educators by addressing various student issues. Students mentioned that interacting with a robot was preferable to interacting with a human guide in certain aspects because they felt less scrutinized by the robot. [92] suggested a way of teaching robotics to computer science students through a case-based approach in a human–computer interaction setting. Multimodal collaborative robots are utilized to replicate genuine human–robot collaboration situations within an educational setting, like in a classroom. The findings and assessments received indicate that students acknowledge and approve of the robotic teaching method necessary for this study. The destiny of HRC in clever manufacturing is brimming with thrilling possibilities fueled by way of advancements in AI, machine gaining knowledge of, and sensor era. AI algorithms will empower robots with greater autonomy and choice-making abilties. Robots should learn from revel in, adapt to converting environments, and assume human moves, main to a extra fluid and responsive collaboration. Enhanced sensor era, consisting of imaginative and prescient systems with deeper object reputation, progressed force sensors for sensitive interaction, and superior tactile sensors, will allow robots to "perceive" the arena greater comprehensively, fostering more secure and greater nuanced collaboration with human beings. Robots might be capable of recognize and respond to herbal language, allowing more intuitive and human-like conversation with employees. Figure 9 discusses the future areas of Research as shown below.

This will streamline collaboration and improve undertaking delegation. Advanced AI should lead to the improvement of shared mental fashions, where human beings and robots can anticipate each other's movements and intentions, main to a greater

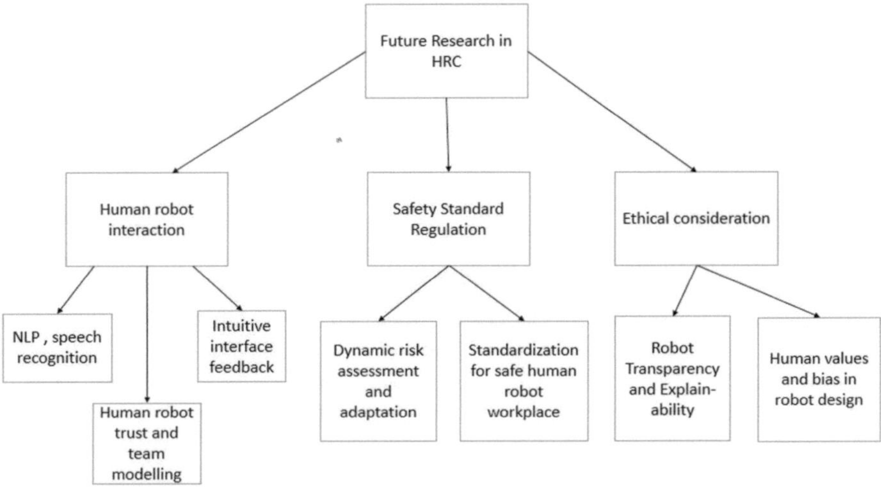

Fig. 9 Future areas of research

seamless and efficient collaboration. Robots and people could interact in collaborative getting to know, where robots examine from human understanding and people learn from robot capabilities, growing a at the same time useful feedback loop. The capacity for HRC extends past the area of clever manufacturing. Robots ought to collaborate with surgeons in running rooms, assisting with sensitive methods or supplying actual-time information analysis. Robots could assist caregivers in presenting companionship and aid for the aged, while humans consciousness on emotional connection and complicated care wishes. Robots could work alongside human rescue teams in dangerous environments, acting duties like debris elimination or locating survivors. The future of HRC may not simply be approximately collaboration however also approximately teaming. Humans and robots could form noticeably coordinated groups, leveraging their precise strengths to obtain complicated dreams. This could involve robots dealing with the bulk of records processing and evaluation, while people consciousness on strategic choice-making and innovative hassle-fixing. While these improvements hold large promise, demanding situations stay. As robots come to be extra state-of-the-art, moral troubles surrounding task displacement, bias in AI algorithms, and the potential for independent selection-making by using robots will want careful attention. As capabilities evolve, so too need to safety requirements. Regulations will want to evolve to cope with the complexities of human–robot teaming in diverse environments. Building accept as true with and setting up clean roles and duties may be essential for effective human–robot teaming. The future of HRC is brimming with possibilities. By addressing the challenges and harnessing the electricity of technological improvements, we will create a destiny in which human–robot groups work together seamlessly, unlocking new ranges of efficiency, innovation, and development throughout various industries.

9 Conclusion

The review of HRC for smart manufacturing in Industry 4.0 recognized several key findings. HRC has the capacity to boost productiveness and performance at the same time as improving protection in factories. This collaboration leverages the strengths of both human beings and robots: robots cope with repetitive or risky tasks, whilst human beings cognizance on creativity, problem-solving, and complex work. The evaluation reinforces the significance of HRC for clever manufacturing. As Industry four.0 integrates automation and data into manufacturing, HRC guarantees a place for human ingenuity alongside advanced technologies. This collaborative method is crucial for navigating the complexities of modern-day production. However, the evaluation also highlights the need for ongoing research and development in HRC. Future studies must cope with areas like protection protocols, human–robot communique, and powerful training applications. By continuing to broaden HRC technologies, we can free up the whole ability of human-robotic collaboration in clever manufacturing for Industry 4.0.

References

1. Pandey, A., Puri, M., Varde, A.: Object detection with neural models, deep learning and common sense to aid smart mobility. In: Proceedings of the IEEE 30th International Conference on Tools with Artificial Intelligence, pp. 859–863 (2018). https://doi.org/10.1109/ICTAI.2018.00134
2. Garg, A., Tandon, N., Varde, A.S.: I am guessing you can't recognize this: generating adversarial images for object detection using spatial commonsense (student abstract). In: (2020) Proceedings of the Association for the Advancement of Artificial Intelligence Conference, pp. 13789–13790 (2020)
3. Persaud, P., Varde, A.S., Robila, S.: Enhancing autonomous vehicles with commonsense: Smart mobility in smart cities. In: Proceedings of the IEEE 29th International Conference on Tools with Artificial Intelligence (ICTAI), pp. 1008–1012 (2017). https://doi.org/10.1109/ICTAI.2017.00155
4. Garcia, M., Rauch, E., Vidoni, R., Matt, D.: AI and ML for human–robot cooperation in intelligent and flexible manufacturing. In: Implementing Industry 4.0 in SMEs, pp. 95–127. Palgrave Macmillan (2021)
5. Hollnagel, E., Wears, R.L., Braithwaite, J.: From Safety-I to Safety-II: A White Paper. Technical Report, University of Southern Denmark; University of Florida; Macquarie University (2015)
6. Goldberg, K.: Robots and the return to collaborative intelligence. Nature Mach. Intell. **1**(1), 2–4 (2019). https://doi.org/10.1038/s42256-018-0008-x
7. Malone, T.W.: Superminds: The surprising power of people and computers thinking together. Little, Brown (2018). https://books.google.se/books?id=Qe0zDwAAQBAJ
8. Alenljung, B., Lindblom, J.: User experience of socially interactive robots: its role and relevance. In: Vallverdú, J. (Ed.), Handbook of Research on Synthesizing Human Emotion in Intelligent Systems and Robotics. IGI Global (2015)
9. Wang, L., Gao, R., Váncza, J., Krüger, J., Wang, X.V., Makris, S., Chryssolouris, G.: Symbiotic human-robot collaborative assembly. CIRP Ann. **68**(2), 701–726 (2019)
10. Krüger, J., Lien, T.K., Verl, A.: Cooperation of human and machines in assembly lines. CIRP Ann. **58**(2), 628–646 (2009)
11. Franklin, C.S., Dominguez, E.G., Fryman, J.D., Lewandowski, M.L.: Collaborative robotics: New era of human-robot cooperation in the workplace (n.d.)

12. Hanna, A.: Towards Intelligent and Collaborative Automation of Automotive Final Assembly (Mälardalen University) (2021)
13. Gualtieri, L., Rauch, E., Vidoni, R., Matt, D.T.: Safety, ergonomics, and efficiency in human-robot collaborative assembly: Design guidelines and requirements. Proc. CIRP **91**, 367–372 (2020)
14. Gervasi, R., Mastrogiacomo, L., Franceschini, F.: A conceptual framework to evaluate human-robot collaboration. Int. J. Adv. Manuf. Technol. **108**, 841–865 (2020)
15. Miller, C.A., Parasuraman, R.: Designing for flexible interaction between humans and automation: delegation interfaces for supervisory control. Hum. Factors **49**(1), 57–75 (2007)
16. Rouse, W.B.: Adaptive aiding for human/computer control. Hum. Factors **30**(4), 431–443 (1988)
17. Scerbo, M.W.: Theoretical perspectives on adaptive automation. In: Automation and Human Performance: Theory and Applications, pp. 37–63. CRC Press (2018)
18. Pulikottil, T.B., Pellegrinelli, S., Pedrocchi, N.: A software tool for human-robot shared-workspace collaboration with task precedence constraints. Robotica Computer-Integr. Manuf. **67**, 102051 (2021). https://doi.org/10.1016/j.rcim.2020.102051
19. Michalos, G., Kousi, N., Karagiannis, P., Gkournelos, C., Dimoulas, K., Koukas, S., Mparis, K., Papavasileiou, A.: Seamless human-robot collaborative assembly—an automotive case study. Mechatronics **55**, 194–211 (2018). https://doi.org/10.1016/j.mechatronics.2018.08.006
20. Dean-Leon, E., Ramirez-Amaro, K., Bergner, F., Dianov, I., Cheng, G.: Integration of robotic technologies for rapidly deployable robots. IEEE Trans. Industr. Inf. **14**(4), 1691–1700 (2018). https://doi.org/10.1109/TII.2017.2766096
21. Saenz, J., Elkmann, N., Gibaru, O., Neto, P.: Survey of methods for design of collaborative robotics applications—why safety is a barrier to more widespread robotics uptake. In: ACM International Conference Proceedings Series (Part F137690), pp. 95–101 (2018). https://doi.org/10.1145/3191477.3191507
22. Zhou, T., Wachs, J.P.: Early prediction for physical human-robot collaboration in the operating room. In: Autonomous Robots (Special Issue on Learning for Human-Robot Collaboration) (n.d.)
23. Antunes, R.G.C.: A robotic system for musculoskeletal rehabilitation of the shoulder. In: Biomedical Engineering (2016)
24. Lee, W.H., Park, J., Park, C.H.: Acceptability of tele-assistive robotic nurse for human-robot collaboration in medical environment. In: Proceedings of HRI'18 Companion: Conference on ACM/IEEE International Conference on Human-Robot Interaction, Chicago, IL (2018)
25. Thoben, K.D., Wiesner, S., Wuest, T.: "Industrie 4.0" and smart manufacturing—a review of research issues and application examples. Int. J. Autom. Technol. **11**(1), 4–16 (2017)
26. Wang, W., Li, R., Chen, Y., Sun, Y., Jia, Y.: Predicting human intentions in human-robot hand-over tasks through multimodal learning. IEEE Trans. Autom. Sci. Eng. (2021). https://doi.org/10.1109/TASE.2021.3074873
27. Suchan, J., Bhatt, M.: Commonsense scene semantics for cognitive robotics: towards grounding embodied visuo-locomotive interactions. In: Proceedings of the IEEE International Conference on Computer Vision Workshops (ICCVW), pp. 742–750 (2017). https://doi.org/10.1109/ICCVW.2017.93
28. Chen, Y., Wang, W., Abdollahi, Z., Wang, Z., Schulte, J., Krovi, V., Jia, Y.: A robotic lift assister: a smart companion for heavy payload transport and manipulation in automotive assembly. IEEE Robot. Autom. Mag. **25**(2), 107–119 (2018)
29. Whitney, D.E., Lozinski, C.A., Rourke, J.M.: Industrial robot forward calibration method and results. J. Dyn. Syst. Meas. Contr. **108**(1), 1–8 (1986). https://doi.org/10.1115/1.3143737
30. Pettersen, T., Pretlove, J., Skourup, C., Engedal, T., Lokstad, T.: Augmented reality for programming industrial robots. In: Proceedings of the 2nd IEEE/ACM International Symposium on Mixed and Augmented Reality, pp. 319–320 (2003)
31. Wang, W., Li, R., Diekel, Z.M., Chen, Y., Zhang, Z., Jia, Y.: Controlling object hand-over in human–robot collaboration via natural wearable sensing. IEEE Trans. Human-Mach. Syst. **49**(1), 59–71 (2019). https://doi.org/10.1109/THMS.2018.2841052

32. Tandon, N., Varde, A.S., de Melo, G.: Commonsense knowledge in machine intelligence. ACM SIGMOD Rec. **46**(4), 49–52 (2018). https://doi.org/10.1145/3186549.3186562
33. Razniewski, S., Tandon, N., Varde, A.: Information to wisdom: commonsense knowledge extraction and compilation. In: Proceedings of the ACM WSDM, pp. 1143–1146 (2021)
34. Beetz, M., Bessler, D., Haidu, A., Pomarlan, M., Bozcuoglu, A.K., Bartels, G.: KnowRob 2.0—a 2nd generation knowledge processing framework for cognition-enabled robotic agents. In: Proceedings of the IEEE International Conference on Robotics and Automation (ICRA), pp. 512–519 (2018). https://doi.org/10.1109/ICRA.2018.8460964
35. Garg, A., Tandon, N., Varde, A.S.: I am guessing you can't recognize this: generating adversarial images for object detection using spatial commonsense (student abstract). In: Proceedings of the Association for the Advancement of Artificial Intelligence Conference, pp. 13789–13790 (2020)
36. Xu, F.F., Lin, B.Y., Zhu, K.Q.: Automatic extraction of commonsense LocatedNear knowledge. In: Proceedings of the 56th Annual Meeting of the Association for Computational Linguistics, pp. 96–101 (2018). https://doi.org/10.18653/v1/p18-2016
37. Gaskill, S., Went, S.: Safety issues in modern applications of robots. Reliab. Eng. Syst. Saf. **53**(3), 301–307 (1996). https://doi.org/10.1016/0951-8320(96)00016-6
38. Wang, L., Gao, R., Váncza, J., Krüger, J., Wang, X.V., Makris, S., Chryssolouris, G.: Symbiotic human-robot collaborative assembly. CIRP Ann. **68**(2), 701–726 (2019). https://doi.org/10.1016/j.cirp.2019.05.002
39. Müller, R., Vette, M., Mailahn, O.: Process-oriented task assignment for assembly processes with human-robot interaction. Proc. CIRP **44**, 210–215 (2016). https://doi.org/10.1016/j.procir.2016.02.043
40. Wang, X.V., Kemény, Z., Váncza, J., Wang, L.: Human–robot collaborative assembly in cyber-physical production: classification framework and implementation. CIRP Ann. **66**(1), 5–8 (2017). https://doi.org/10.1016/j.cirp.2017.04.002
41. Krüger, J., Lien, T.K., Verl, A.: Cooperation of human and machines in assembly lines. CIRP Ann. **58**(2), 628–646 (2009). https://doi.org/10.1016/j.cirp.2009.09.003
42. Michalos, G., Makris, S., Tsarouchi, P., Guasch, T., Kontovrakis, D., Chryssolouris, G.: Design considerations for safe human-robot collaborative workplaces. Proc. CIRP **37**, 248–253 (2015). https://doi.org/10.1016/j.procir.2015.08.042
43. Whitsell, B., Artemiadis, P.: Physical human–robot interaction (pHRI) in 6 DOF with asymmetric cooperation. IEEE Access **5**, 10834–10845 (2017). https://doi.org/10.1109/ACCESS.2017.2670783
44. Safeea, M., Bearee, R., Neto, P.: End-effector precise hand-guiding for collaborative robots. In: Iberian Robotics Conference, pp. 595–605. Springer (2017)
45. Mendes, N., Safeea, M., Neto, P.: Flexible programming and orchestration of collaborative robotic manufacturing systems. In: Proceedings of the 2018 IEEE 16th International Conference on Industrial Informatics (INDIN), pp. 913–918 (2018)
46. Rozo, L., Calinon, S., Caldwell, D.G., Jimenez, P., Torras, C.: Learning physical collaborative robot behaviors from human demonstrations. IEEE Trans. Rob. **32**(3), 513–527 (2016). https://doi.org/10.1109/TRO.2016.2532689
47. Rahman, S. M., Liao, Z., Jiang, L., Wang, Y.: A regret-based autonomy allocation scheme for human-robot shared vision systems in collaborative assembly in manufacturing. In: Proceedings of the 2016 IEEE International Conference on Automation Science and Engineering (CASE), pp. 897–902 (2016)
48. Koch, P.J., van Amstel, M.K., Debska, P., Thormann, M.A., Tetzlaff, A.J., Bøgh, S., Chrysostomou, D.: A skill-based robot co-worker for industrial maintenance tasks. Proc. Manuf. **11**, 83–90 (2017)
49. Unhelkar, V.V., Lasota, P.A., Tyroller, Q., Buhai, R.D., Marceau, L., Deml, B., Shah, J.A.: Human-aware robotic assistant for collaborative assembly: Integrating human motion prediction with planning in time. IEEE Robot. Autom. Lett. **3**, 2394–2401 (2018). https://doi.org/10.1109/LRA.2018.2847732

50. Hamabe, T., Goto, H., Miura, J.: A programming by demonstration system for human-robot collaborative assembly tasks. In: Proceedings of the IEEE International Conference on Robotics and Biomimetics (ROBIO), pp. 1195–1201. IEEE (2015). https://doi.org/10.1109/ROBIO.2015.7419267
51. Tlach, V., Kuric, I., Zajacko, I., Kumicáková, D., Rengevic, A.: The design of method intended for implementation of collaborative assembly tasks. Adv. Sci. Technol. Res. J. **12**(3), 244–250 (2018). https://doi.org/10.12913/22998624/96384
52. El Makrini, I., Merckaert, K., Lefeber, D., Vanderborght, B.: Design of a collaborative architecture for human-robot assembly tasks. In: Proceedings of the IEEE/RSJ International Conference on Intelligent Robots and Systems (IROS), pp. 1624–1629. IEEE (2017). https://doi.org/10.1109/IROS.2017.8206212
53. Bauer, A., Wollherr, D., Buss, M.: Human-robot collaboration: a survey. Int. J. Humanoid Rob. **5**(1), 47–66 (2008). https://doi.org/10.1142/S0219843608001349
54. Krüger, J., Lien, T.K., Verl, A.: Cooperation of human and machines in assembly lines. CIRP Ann. Manuf. Technol. **58**(2), 628–646 (2009). https://doi.org/10.1016/j.cirp.2009.09.008
55. Ajoudani, A., Zacchettin, A.M., Ivaldi, S., Albu-Schäffer, A., Kosuge, K., Khatib, O.: Progress and prospects of the human-robot collaboration. Auton. Robot. **42**, 957–975 (2018). https://doi.org/10.1007/s10514-018-9774-8
56. Matheson, E., Minto, R., Zampieri, E.G.G., Faccio, M., Rosati, G.: Human-robot collaboration in manufacturing applications: a review. Robotics **8**(4), 100 (2019). https://doi.org/10.3390/robotics8040100
57. Colgate, J.E., Edward, J., Peshkin, M.A., Wannasuphoprasit, W.: Cobots: robots for collaboration with human operators. In: Proceedings of the ASME International Mechanical Engineering Congress and Exposition, pp. 433–439. American Society of Mechanical Engineers (1996)
58. Guerin, K.R., Lea, C., Paxton, C., Hager, G.D.: A framework for end-user instruction of a robot assistant for manufacturing. In: Proceedings of the IEEE International Conference on Robotics and Automation (ICRA), pp. 6167–6174. IEEE (2015). https://doi.org/10.1109/ICRA.2015.7227533
59. Villani, V., Pini, F., Leali, F., Secchi, C.: Survey on human-robot collaboration in industrial settings: safety, intuitive interfaces, and applications. Mechatronics **54**, 115–130 (2018). https://doi.org/10.1016/j.mechatronics.2018.02.009
60. Ericsson: Creative Machines: How Artificial Intelligence will Impact the Future Labor Market (2021). https://www.ericsson.com/en/reports-and-papers/consumerlab/reports/creative-machines
61. Sadik, A.R., Urban, B.: CPROSA-holarchy: an enhanced PROSA model to enable worker-cobot agile manufacturing. Int. J. Mech. Eng. Robot. Res. **7**(3), 296–304 (2018). https://doi.org/10.18178/ijmerr.7.3.296-304
62. Barbazza, L., Faccio, M., Oscari, F., Rosati, G.: Agility in assembly systems: a comparison model. Assem. Autom. **37**(4), 411–421 (2017). https://doi.org/10.1108/AA-03-2017-067
63. Tsarouchi, P., Matthaiakis, A.-S., Makris, S., Chryssolouris, G.: On human-robot collaboration in an assembly cell. Int. J. Comput. Integr. Manuf. **30**(6), 580–589 (2017). https://doi.org/10.1080/0951192X.2016.1187295
64. Rosenstrauch, M.J., Krüger, J.: Safe human-robot collaboration: Introduction and experiment using ISO/TS 15066. In: Proceedings of the 3rd International Conference on Control, Automation and Robotics (ICCAR), Nagoya, Japan, April 24–26, pp. 740–744 (2017). https://doi.org/10.1109/ICCAR.2017.7942742
65. Edmondson, N., Redford, A.: Generic flexible assembly system design. Assem. Autom. **22**(2), 139–152 (2002). https://doi.org/10.1108/01445150210420894
66. Battini, D., Faccio, M., Persona, A., Sgarbossa, F.: New methodological framework to improve productivity and ergonomics in assembly system design. Int. J. Ind. Ergon. **41**(1), 30–42 (2011). https://doi.org/10.1016/j.ergon.2010.11.002
67. Sawodny, O., Aschemann, H., Lahres, S.: An automated gantry crane as a large workspace robot. Control. Eng. Pract. **10**(12), 1323–1338 (2002). https://doi.org/10.1016/S0967-0661(02)00094-4

68. Krüger, J., Bernhardt, R., Surdilovic, D., Spur, G.: Intelligent assist systems for flexible assembly. CIRP Ann. **55**(1), 29–32 (2006). https://doi.org/10.1016/S0007-8506(07)60306-8
69. Rosati, G., Faccio, M., Carli, A., Rossi, A.: Fully flexible assembly systems (F-FAS): a new concept in flexible automation. Assem. Autom. **33**(1), 8–21 (2013). https://doi.org/10.1108/01445151311294694
70. FANUC Italia S.r.l.: M-2000—The Strongest Heavy-Duty Industrial Robot in the Market (2019). Retrieved from https://www.fanuc.eu/it/en/robots/robot-filter-page/m-2000-series
71. Hägele, M., Schaaf, W., Helms, E.: Robot assistants at manual workplaces: effective cooperation and safety aspects. In: Proceedings of the 33rd ISR (International Symposium on Robotics), Stockholm, Sweden, October 7–11 (2002)
72. Vinitha, K., Ambrose Prabhu, R., Bhaskar, R., Hariharan, R.: Review on industrial mathematics and materials at Industry 1.0 to Industry 4.0. Mater. Today Proc. **33**, 3956–3960 (2020). https://doi.org/10.1016/j.matpr.2020.02.788
73. Fakhruldeen, H., Maheshwari, P., Lenz, A., Dailami, F., Pipe, A.G.: Human robot cooperation planner using plans embedded in objects. IFAC-PapersOnLine **49**(1), 668–674 (2016). https://doi.org/10.1016/j.ifacol.2016.10.106
74. Müller, R., Vette, M., Geenen, A.: Skill-based dynamic task allocation in human-robot cooperation with the example of welding application. Proc. Manuf. **11**, 13–21 (2017). https://doi.org/10.1016/j.promfg.2017.07.053
75. ISO: ISO 10218-1: Robots and Robotic Devices—Safety Requirements for Industrial Robots—Part 1: Robots. International Organization for Standardization (2011)
76. Korostynska, O.: Sensors for smart packaging in healthcare and food industry. In: Proceedings of the 2021 IEEE Sensors, Virtual, October 31–November 3, p. 1 (2021)
77. Grobbelaar, W., Verma, A., Shukla, V.K.: Analyzing human robotic interaction in the food industry. J. Phys: Conf. Ser. **1714**, 012032 (2021). https://doi.org/10.1088/1742-6596/1714/1/012032
78. Bakalis, S., Gerogiorgis, D., Argyropoulos, D., Emmanoulidis, C.: Food Industry 4.0: Opportunities for a digital future. In: Juliano, P., Buckow, R., Nguyen, M.H., Knoerzer, K., Sellahewa, J. (Eds.), Food Engineering Innovations Across the Food Supply Chain, pp. 357–368. Academic Press (2022)
79. Segura, P., Lobato-Calleros, O., Ramírez-Serrano, A., Soria, I.: Human-robot collaborative systems: structural components for current manufacturing applications. Adv. Ind. Manuf. Eng. **3**, 100060 (2021). https://doi.org/10.1016/j.aime.2021.100060
80. ISO: ISO 10218-1:2011—Robots and Robotic Devices—Safety Requirements for Industrial Robots—Part 1: Robots (2011a). Retrieved from https://www.iso.org
81. ISO: ISO 10218-2:2011—Robots and Robotic Devices—Safety Requirements for Industrial Robots—Part 2: Robot System and Integration (2011b). Retrieved from https://www.iso.org
82. Wang, W., Coutras, R., Zhu, M.: Empowering computing students with proficiency in robotics via situated learning. In: Smart Learning Environments, vol. 8, issue 24 (2021). https://doi.org/10.1186/s40561-021-00177-4
83. Marinelli, M.: Human–robot collaboration and lean waste elimination: conceptual analogies and practical synergies in industrialized construction. Buildings **12**(12), 2057 (2022). https://doi.org/10.3390/buildings12122057
84. Onnasch, L., Hildebrandt, C.L.: Impact of anthropomorphic robot design on trust and attention in industrial human-robot interaction. ACM Trans. Human-Robot Interaction (THRI) **11**(1), 1–24 (2021). https://doi.org/10.1145/3460574
85. Dakulagi, V., Yeap, K.H., Nisar, H., Dakulagi, R., Basavaraj, G.N., Galindo, M.V.: An overview of techniques and best practices to create intuitive and user-friendly human-machine interfaces. In: Artificial Intelligence and Multimodal Signal Processing in Human-Machine Interaction, pp. 63–77 (2025). https://doi.org/10.1016/B978-0-12-820430-2.00005-4
86. Asokan, A., Pothen, A.J., Vijayaraj, R.K.: ARMatron: a wearable gesture recognition glove for control of robotic devices in disaster management and human rehabilitation. In: Proceedings of the 2016 Robotics and Automation for Humanitarian Applications (RAHA), IEEE, 1–5 (2016). https://doi.org/10.1109/RAHA.2016.7883947

87. Cha, Y., Seo, J., Kim, J.-S., Park, J.-M.: Human–computer interface glove using flexible piezoelectric sensors. Smart Mater. Struct. **26**(5), 57002 (2017). https://doi.org/10.1088/1361-665X/aa676f
88. Dong, W., Xiao, L., Hu, W., Zhu, C., Huang, Y., Yin, Z.: Wearable human–machine interface based on PVDF piezoelectric sensor. Trans. Inst. Meas. Control. **39**(4), 398–403 (2017). https://doi.org/10.1177/0142331216655527
89. Hong, S., Lee, H., Lee, J., Kwon, J., Han, S., Suh, Y.D., Cho, H., Shin, J., Yeo, J., Ko, S.H.: Highly stretchable and transparent metal nanowire heater for wearable electronics applications. Adv. Mater. **27**(32), 4744–4751 (2015). https://doi.org/10.1002/adma.201501371
90. Park, J.J., Hyun, W.J., Mun, S.C., Park, Y.T., Park, O.O.: Highly stretchable and wearable graphene strain sensors with controllable sensitivity for human motion monitoring. ACS Appl. Mater. Interfaces. **7**(11), 6317–6324 (2015). https://doi.org/10.1021/acsami.5b00295
91. Liu, S., Wu, X., Zhang, D., Guo, C., Wang, P., Hu, W., Li, X., Zhou, X., Xu, H., Luo, C.: Ultrafast dynamic pressure sensors based on graphene hybrid structure. ACS Appl. Mater. Interfaces. **9**(28), 24148–24154 (2017). https://doi.org/10.1021/acsami.7b06796
92. Haddadin, S.: Towards Safe Robots: Approaching Asimov's First Law. Springer (2014)

The Impact of Digital Twin in Industry 4.0 Using Graph Neural Network: An Approach to Explainability in the Manufacturing Industry

P. Jayadharshini, S. Santhiya, C. Vasuki, T. Vanaja, S. Archanaa, and K. Samyuktha

Abstract The advent of Industry 4.0 has transformed the manufacturing landscape, enabling unprecedented levels of automation and data-driven decision-making. Central to this transformation are Digital Twins (DT) and Graph Neural Networks (GNNs), which together offer innovative solutions for optimizing manufacturing processes. This paper delves into the integration of GNNs with digital twins and establishes their relevance in tapping into real-time data, whereby industries can enhance processes such as maintenance forecasting, improve supply chain efficiency, assure quality management, and promote sustainable complex relational modeling. GNNs also enable better insights and more explainability in manufacturing systems. Yet, challenges exist, such as data quality and scalability. Unlocking these advancements hinges on regulatory compliance. The conclusion ends with a thorough discussion of future directions, highlighting significant aspects such as data management, model

P. Jayadharshini (✉) · S. Santhiya
Assistant Professor, Department of Artificial Intelligence, Kongu Engineering College, Perundurai, Erode 638060, Tamil Nadu, India
e-mail: jayadharshini.ai@kongu.edu

S. Santhiya
e-mail: santhiya.cse@kongu.edu

C. Vasuki
Assistant Professor, Department of Information Technology, Nandha Engineering College, Perundurai, Tamil Nadu, India
e-mail: Vasuki.chinnappan@nandhaengg.org

T. Vanaja
Assistant Professor, Department of CSE, Kongu Engineering College, Perundurai, Erode 638060, Tamil Nadu, India

S. Archanaa · K. Samyuktha
Student, Department of Artificial Intelligence, Kongu Engineering College, Perundurai, Erode 638060, Tamil Nadu, India
e-mail: archanaas.21aim@kongu.edu

K. Samyuktha
e-mail: samyukthak.21aim@kongu.edu

L. Krishnasamy et al. (eds.), *Distributed Deep Learning and Explainable AI (XAI) in Industry 4.0*, Information Systems Engineering and Management 55,
https://doi.org/10.1007/978-3-031-94637-0_7

interpretability, and training the workforce toward fruitful benefits in GNNs and the implementation of digital twin technology in the manufacturing industry.

Keywords Digital Twin · Industry 4.0 · Graph Neural Networks · Smart Manufacturing · Predictive Maintenance · Supply Chain Optimization · Quality Control · Data Quality · Scalability · Explainability

1 Introduction

1.1 Industry 4.0 and Its Evolution

Industry 4.0, that is the Fourth Industrial Era, is a totally new approach toward manufacturing. This era uses industrial innovations that bring changes in traditional ways of producing things. Industry 4.0 sticks to the developed stages of the previous industrial era but ensures the unification of cyber-physical systems, IoT, cloud infrastructure, and analysis of data to evolve intelligent and interconnected frameworks [1]. This has resulted in the increased development of ever more efficient, data-centric operational methodologies within self-governing artificial intelligence/machine learning-driven manufacturing systems.

Some of the underlying forces behind Industry 4.0 include the requirement to increase productivity, continuous monitoring, a more effective decision-making process, and the need for more flexible production systems. Integration of advanced technologies enables manufacturers to achieve maximum operational effectiveness to overcome the changing market conditions.

1.2 The Role of Digital Twin in Industry 4.0

A digital twin is thereby defined as a virtual copy of a real, tangible entity, system, or process that replicates the actual performance through continuous exchanges of data. Industry 4.0 digital twins allow for an always-on, dynamic form of manufacturing process, providing predictive insights and enabling real-time modifications. The advanced concept of the digital twin allows for comprehensive simulations, evaluations, and optimization processes without needing any actual prototype, thereby reducing downtime and showing much enhancement in efficiencies.

The applications of digital twins in manufacturing can range from prediction of maintenance and quality assurance to supply chain efficiency and process simulation. The constant flow of information between the virtual and real worlds enhances better decision making while allowing for improved product lifecycle management, as well as adaptive and autonomous manufacturing systems.

1.3 Motivation for Explainability in AI-Driven Manufacturing

Algorithmic systems form an integral part of the Industry 4.0 framework, through which companies can develop an ability to automate and predict better and in simple ways run their operational processes. However, as their complexity advances, they have increasingly become part of the "black box" models following deep learning [2]. The lack of transparency offers a multitude of manufacturing difficulties where the reason behind AI-generated decisions needs to be understood for trust, accountability, and safety.

Explanatory AI attempts to explain the complex nature of these nontransparent models by providing an insight into the justification of particular predictions made by a model. In the industrial domain, the need for interpretable models is strongly required to instill confidence in operators of AI systems, maintain regulatory compliance, and further enable quick system malfunction or error detection. XAI promotes transparency and reliability within industrial activity while offering understandable explanations for predictions made by AI.

2 Overview of Digital Twin Technology

2.1 Definition and Architecture of Digital Twins

A digital twin essentially is a virtual representation or imitation of an actual system, object, or process, thereby establishing a correspondence between the virtual environment and the real-world context [3]. Generally, a digital twin consists of three components: a tangible entity, such as a machine or system and its corresponding virtual model, with a continuous data stream between the two. This enables real-time data flow, supporting the "digital twin" in mimicking, analyzing, and predicting the current state of the physical counterpart.

Digital twins are dynamic entities because they operate and change along with their physical counterparts, as they are continuously updated to mirror all changes in the system. This inherent dynamic nature makes them of tremendous value in simulating different scenarios, optimizing operation processes, and enhancing making decision capacities without actually altering the physical system itself.

2.2 Applications of Digital Twins in Smart Manufacturing

Digital twins add new innovations and solutions to age-old industrial problems. Predictive maintenance is one of the major applications in which the digital twin can predict possible failure or degradation of equipment based on constant analysis

Table 1 Application and description of digital twins in industry

Application	Description
Predictive Maintenance	Predicts equipment failures and schedules maintenance to prevent downtime
Process Optimization	Identifies inefficiencies in manufacturing processes and suggests improvements
Product Lifecycle Management	Tracks products through their lifecycle from design to disposal, ensuring optimal management
Supply Chain Optimization	Enhances supply chain efficiency by simulating different supply chain scenarios

of machine data, thus preventing costly downtimes and smooth production. The application of digital twins can be inferred from Table 1.

Besides, they use digital twins for the process optimization; they help to smoothen down the manufacturing workflows and identify the weak points and that on the basis of the improvements come out. Which in turn increases the production, reduces waste, and enhances productivity in general. Beyond that, product lifecycle management stands as a key application where, in tracking a full lifeline of a product, from design to development and then to the operation and disposal, the digital twin is applied.

2.3 Integration of Digital Twins with AI

When the trend of digital twins assimilates with developments in artificial intelligence, the very foundation of intelligent manufacturing can get redefined. Artificial intelligence may be added to digital twins when there is an enablement of the latter to extract meaningful insights into huge amounts of data they deal with. It would then imply implementation of machine learning algorithms within digital twins to refine predictive capabilities and optimize complex systems and empower decision-making processes.

For instance, coupling digital twins with artificial intelligence enables the building of self-optimizing systems that can adapt the production parameters to automatically optimize their performance. This places it significantly above current industry standards, and it allows companies to align their processes with the fast-changing requirements of today's manufacturing.

3 Graph Neural Networks: A Primer

3.1 Fundamentals of Graph Neural Networks (GNNs)

Graph Neural Networks, or GNNs for short, are specifically designed neural models that have been established to deal with graph-based data structures. Unlike traditional neural networks which treat fixed-size inputs such as images or text, GNNs are able to process information based on the relation of the different elements as important as the elements themselves. Nodes in a graph represent entities, and edges indicate the relationship or connections between those nodes [4]. This is why GNNs work really well in manufacturing industries where data is typically structured as a network of inter-related components.

The main idea of GNNs is the transmission of information between connected nodes of the graph. For each node, it aggregates information from the neighboring nodes and allows the model to learn both local and global patterns of data. GNNs construct deep knowledge of the entire graph by successive layers of information transmission and, therefore, are suitable for application in manufacturing settings-anomaly detection, fault prediction, and optimization tasks.

3.2 Use Cases of GNNs in Industrial Contexts

GNNs have a number of practical applications in industrial settings. For example, given the analysis of sensor data from machines arranged in a graph structure, a GNN can predict that equipment is probably going to fail. GNNs can therefore determine the interrelationship between various parts of a system and come up with the root cause of failures by providing early warnings.

Another major application is anomaly detection; strange patterns within a complex network of interdependent components can be identified using GNNs. The potential, therefore, is significant in identifying defects or inefficiencies in a production line even before they become major issues. The ability of GNNs to model interdependencies in manufacturing industries with heavy interdependence in large-scale processes improves efficiency in operation and reduces downtime.

3.3 Key Differences Between GNNs and Other Deep Learning Models

Those traditional deep learning models, such as CNNs, perform very well on structure data, especially images, whereas GNN does pretty well in handling unstructured or semi-structured data whose interconnectivity of data points are pretty informative. Unlike CNNs having a rigid grid-based structure, GNN works with graphs and changes it dynamically with more data added. Another big difference among them relates to the flow of information of GNN. While traditional neural networks rely on a fixed-size of input, GNNs continually update their perceptions through cross-communications between nodes in the graph. It increases the strength of GNNs in dynamic environments where conditions and the process in manufacturing are constantly changing. Figure 1 depicts the architecture of GNN.

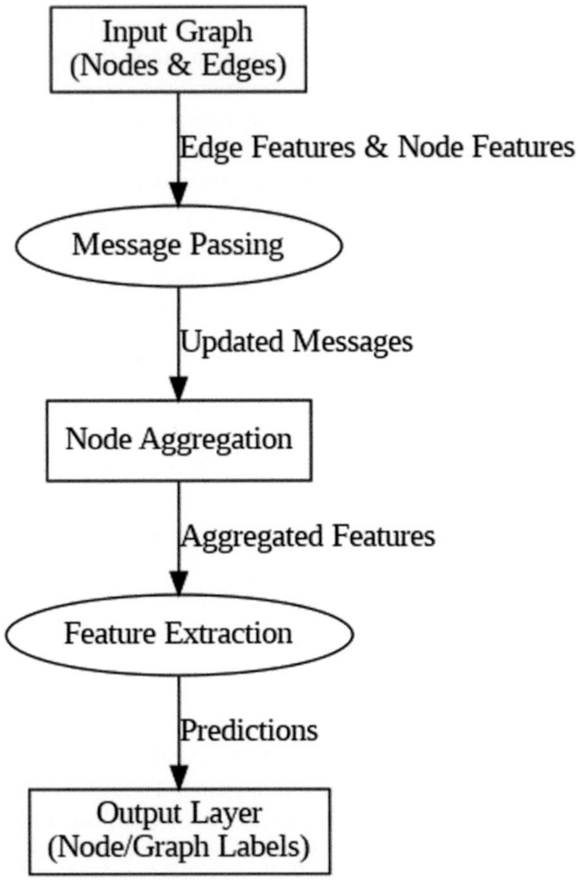

Fig. 1 Architecture of GNN

4 Explaining AI Models in Manufacturing: A Necessity

4.1 What is Explainability in AI?

Explainability in artificial intelligence refers to the ability to describe or explain the decisions of machine learning models, or more generally, what the model has learned [5]. In practice, determining exactly why intricate AI systems are arriving at the conclusions they do with deep learning or some other advanced methodology is a challenge. This lack of transparency raises issues related to trust, accountability, and regulatory compliance, especially in domains that depend on manufacturing.

This objective of XAI is to provide, in a transparently and interpretable manner, insights into why a given AI model produced a particular outcome. In manufacturing, where AI tasks nowadays have been applied to predictive maintenance, quality control, and process optimization, there also are good reasons that the outputs of models must be understood by those decision-makers, operators, and engineers. The inability of AI models to explain how they arrive at their output will be branded only a "black box." It prevents adoption and also decreases effectiveness.

4.2 Digital Twins Explainable AI (XAI) Approaches

Explainability is distinctly vital wherever digital twins engage with the physical world, because those systems are by necessity intertwined with industrial processes. For instance, in feature attribution, XAI techniques can be used to provide transparency models by indicating what factors are most pivotal in altering the model's prediction so that engineers can develop an understanding of what is most important within their data streams.

Another one is model distillation. This distills an AI system by a simpler, more interpretable model. Such a simple model may yield insights into how the original AI works at the cost of slight loss in predictiveness. Visualization tools form many XAI techniques that graphically present model behavior toward improving users' intuition of complex decision-making processes.

4.3 Role of Graph Neural Networks in Explainable AI

The interpretability aspect becomes an important challenge of Graph Neural Networks, as their decision-making will no longer depend on individual data points but also upon the relationships that exist among them. However, with the recent advancement in XAI methods, specifically for GNNs, it does become very feasible to visualize and interpret the reasoning behind a model's output. Techniques for node importance scoring can facilitate an engineer's discovery of those nodes and edges in the graph that contributed the most toward one particular prediction [6]. Graph attention mechanisms help engineers to focus on the most important nodes of the graph as the process of training becomes more understandable and the ability of explanation of the decisions made typically is improved. Therefore, using such techniques in digital twin systems allows manufacturers to better understand the underlying mechanisms of the AI model, thus eliminating the problem of the crucial decisions being opaquely operated by the model.

5 Graph Neural Networks for Digital Twin-Based Systems

5.1 Modeling Industrial Data as Graphs

The data concerning industrial processes in manufacturing is complex and connected. Therefore, the conceptual framework of GNNs sounds like a perfect match to this kind of data representation because it enables someone to describe relations among a complex set of entities in the form of graphs. Graphs are basically mathematical structures described by nodes-that is, individual elements like machines or production stages-and edges that describe the connections or dependencies between the different elements. Take, for example, a manufacturing plant. Here, nodes are the machines, and their interconnected bonds are represented by edges, such as material flow, communication channels, and interdependent tasks.

Representation of industrial processes as graphs helps GNNs to learn behavior both from individual components and from how they interact with one another. For example, the efficiency of processing in a multiple-stage manufacturing line could depend directly on one machine's performance compared to that of subsequent steps in the processing line. In modeling the data in a graphical form, GNNs are able to propagate information better among the nodes to capture local dependencies and global relationships thus having a better understanding of the overall manufacturing system as a whole.

It is very useful in optimizing such industrial complexity processes where every piece dynamically connects with all others. GNNs can analyze such graphs to help identify hidden patterns, optimize workflows, predict potential bottlenecks or failures before they happen, and provide actionable insights that drive improvements in efficiency for manufacturers.

5.2 GNNs in Digital Twin Architectures

GNNs may be integrated into digital twin architectures to provide a more holistic approach toward enhancing the predictability and decision-making in manufacturing operations. The procedure starts with the creation of a mapping between the physical system and its digital twin. It converts real-time sensor data, operational logs, and historical performance data into graph structures that the GNN processes for predictions, optimizing procedures, or simulating possible outcomes.

This means GNNs have the huge advantage of handling any kind of data from structured to unstructured. For example, for a production line, where each machine functions differently, GNNs can adapt to the changing dynamics among the different machines and capture real-time relationships with them. This makes the GNN continuously update its learning so that predictions and optimizations are always made with the latest operational data the digital twin evolves with.

GNNs can also model digital twin systems [7]: because a complex production process can be thought of as a subgraph in a large graph, the systems can be analyzed with their interactions not only through the systems themselves. With GNNs, manufacturers may optimize their processes to have maximum efficiency with minimum unexpected interruption by simulating the production line as a graph and continually training a GNN on new data.

5.3 Case in Point: Predictive Maintenance and Process Optimization

One of the very interesting applications of GNNs with digital twins is for predictive maintenance. Traditional maintenance times are usually either time-based or reactive; that is, they are scheduled or done at a time when failure has already happened. GNNs can help digital twins create predictive insights from the live data to indicate possible problems to the maintenance crew before they turn catastrophic.

For instance, sensors fitted within the machinery at a manufacturing facility can send data regarding temperature, vibration, loads, and other similar parameters of operation in real time. GNNs can then parse this data as a graph where every node represents a sensor, and the edges are associated with that sensor with the other sensors based on the flow of operation data. Analysis of these graphs will reveal some of the patterns by which GNNs can predict anomalies or exactly when a machine is likely to break down, thus enabling proactive maintenance measures. Although it reduces the downtime, this approach extends the life of such critical equipment.

This is beyond predictive maintenance and truly a capability for process optimization. An end-to-end graph model of the production line can be analyzed by GNNs concerning interdependencies across different machines and their respective processes, identification of inefficiencies or bottlenecks, and so on; thus it enables manufacturers to make informed adjustments in order to optimize throughput, energy consumption, and streamline operations to improve overall productivity and costs.

6 Contributing to Explainability of Digital Twins with Graph Neural Networks

Graph Neural Networks offer great potential when combined with digital twin structures, especially for improving decision-making processes within complex manufacturing environments in terms of explainability and interpretability. Explainability within artificial intelligence systems is increasingly becoming a priority, particularly within industrial applications, as this is critical to transparency and safety but also to all regulatory compliance and standards, of course. Given their use in monitoring, optimization, and prediction for Industry 4.0, such industrial systems require insights that are not only clear but also understandable to engineers, operators, and even decision-making authorities. GNNs stand uniquely poised given the capacity of modeling relations in connected data, hence an excellent tool to decipher complicated manufacturing systems.

6.1 Graph Neural Networks and Their Specific Capabilities

The idea fits well within the center of GNNs, that one might represent data as a graph in which entities such as machines, components, and sensors are represented as nodes, while interaction with each other in terms of material flow, data exchange, or dependency constitutes edges. GNNs capture relationships and dependencies between nodes very differently than traditional neural networks, which primarily deal with feature-based data. In manufacturing, complex inter-relations exist among different systems of another.

Maybe in a real manufacturing setup, there are various stages where different machines and raw materials interact with each other under varying conditions. The graph Neural Networks can encode such complicated interactions through graph structures, which enables the digital twin to account for the whole production network rather than taking a machine-based approach for each one separately [8]. Modeling such systems and learning from them may provide much more accurate details, but they greatly improve the interpretability of outputs returned as a pair.

These are especially suitable for use cases where the system state depends on a set of interconnected elements, such as predictive maintenance, supply chain optimization, or production planning. Such interactions in classical machine learning models are either simplified or completely ignored, limiting the robustness of predictions by those models. But GNN excels in such settings and enables, additionally, that the inspection of the system is holistic.

6.2 *Improving Interpretability with Node-Level and Graph-Level Explanations*

One of the major concerns associated with complex AI models such as GNNs is that they are inherently overcomplicated and seem to evoke a "black box" effect, whereby the process by which decisions are made is not easily understood. GNNs, however, allow totally unique capabilities for generating outputs to be interpretably seen at both node and graph levels. It would permeate the manufacture even beyond the length and breadth of a singular machine or process but would include the entire production network.

At the node level, GNNs could then determine the most critical or important components or machines in a network. For instance, in predictive maintenance settings, GNN might predict an impending failure in a particular machine. Then by node-level explanation, the GNN can reveal which aspects of the machine's operational data, or its interaction with other machines contributed the most in determining that failure. Undoubtedly, this type of explanation is really helpful for operators themselves to take focused action about certain issues, whether it involves an increase in temperature or degradations of a critical component.

The overall structure and flow of operations at the graph level can be analyzed to provide deeper insights into an entire production process using GNNs. For example, in supply chain optimization, GNNs may represent the flow of materials across several suppliers and production lines in terms of bottlenecks or inefficiencies. These explanations at the graph level furnish decision-makers with an understanding of interactions among various components of a system, thereby adding to overall performance and showing an exact route for optimization.

6.3 *Attention Mechanisms in GNNs for Better Explainability*

One of the most appealing methods to enhance the interpretability of GNNs is through the application of attention mechanisms. For the last few years, the attention mechanism has been the strongest tool in providing applications for enhancing focus on the important features in data for many neural network architectures. For instance, when the focus is on GNNs, the mechanisms of the attention focus enable the model to pay attention to which nodes and edges of the graph are relevant in its decision-making process.

For instance, certain machines or sensors contributing most to specific outcomes within a system may be identifiable in a graph neural network with attention applied to applications on digital twins—such as delays in production forecasts or a drop in operational efficiency. Visualization of such contributions helps operators to have a clearer picture of the system's behavior. More pertinent in high-stakes manufacturing environments, where such decisions must clearly and explicitly be traceable for particular factors, is enhanced transparency.

The attention mechanisms also support the adaptive learning task, so that GNNs can adaptively shift their focus with every addition of new data. Adaptability in fast-changing industrial settings with the rapid shifting of processes and conditions ensures that a GNN is always relevant and interpretable even with the change of the system [9]. This enhances not only explainability but overall robustness of the digital twin.

6.4 Explainability for Compliance and Decision-Making

In safety and compliance-related industries, explainability is a both a legal requirement as well as a business necessity. The aerospace, automotive, and pharmaceutical sectors are the ones subjected to the highest levels of regulation and, therefore, must understand the impact their decisions have on safety or quality. For example, such as in the forecast of failure or non-conforming production standards, to understand what drives the digital twin powered by GNNs to make a prediction of something in particular or recommendation, it needs the operators too.

GNNs allow the process to make decisions aligned with regulations, by enabling node- and graph-level explanations. For instance, a GNN-based system predicts a defect in a few products, whereas those defects may later be interpreted to belong to specific machines or process steps, along with some additional data from sensors or operation logs. It gives such detail that the root causes can finally be clearly defined, if corrective actions are to be maintained when an audit or regulatory investigation is initiated.

The factor of explainability also incites trust among the human operators and decision-makers toward artificial intelligence systems. Users are more likely to place trust in the insights as provided by the system when they get a clear explanation of the rationale behind specific recommendations made by that system; thereby, it becomes easier to incorporate AI into existing workflows. This is specifically very important in scenarios where human oversight remains critical, even as automation and artificial intelligence are increasingly implemented.

6.5 Tools and Techniques for Explainable GNNs in Industry

Some of the emerging tools and techniques developed for increasing explainability specifically in GNN industrial settings. Such platforms can visualize graph structure which an operator needs to understand relationships that exist between different entities. The operators understand a digital twin's decision-making process with help from graph visualization and flow of information between nodes.

Of its many approaches, it uses saliency maps and feature attribution methods that focus on the most critical nodes, edges, or features that give rise to a particular outcome. Besides increasing transparency, these tools enhance the system's performance in the sense that they help identify weaknesses or bottlenecks within the models' structure.

Therefore, incorporating these explainability tools into the framework of a digital twin guarantees that AI-driven systems [10] put to use by the manufacturing firms remain transparent, interpretable, and trustable even as their operational complexities escalate.

7 Challenges in Implementing Graph Neural Networks and Digital Twins in Manufacturing

Digital twin using GNN has tremendous potential to change the manufacturing paradigm under Industry 4.0. Thereby, it shall revolutionize the manufacturing as well as factor production processes. Although the realistic factory setup, with the technical, operational, and organisational challenges on the back of manufacturers in real-world settings, is high. Thus, this section will offer a deeper insight into such challenges enabling a successful transition towards smarter, more efficient factories.

7.1 Data Collection and Management Issues

Data-based GNNs and digital twins suffer with problems that arise because of their large volume, diversity, and speed in this current world of industry. The manufacturing systems continuously generate data from sensors, control systems, and even manual entries. These sources have formats, frequencies, and accuracies varied, which poses distinct challenges in collecting and managing the data.

Data Integrity and Accuracy

Generally, noise or aberrant data causes flawed models producing unreliable insights which, in extreme cases, may lead to an entire breakdown of the production cycle. For example, a temperature sensor connected to a machine sometimes gives noisy values for data readings due to wear and tear, thereby contributing noise to the dataset. In a GNN configuration, incorrect data may provide wrong structures or relationships between nodes and the system may predict something that proves flawed.

Manufacturing systems, besides, often lead to real-time streaming data. This further complicates the gathering of data. Managing real-time data calls for high-processing power solutions, like edge computing or distributed computing architectures, so that data gets processed and analyzed as close to the moment it becomes available as possible. Not processing such data in real time translates into a delay in decisions taken-in the case of predictive maintenance or in real-time optimization, valueless.

Data Volume and Storage

The amount of data generated in a factory setup is extremely huge. In the smart factory model, thousands of sensors produce gigabytes to terabytes of data every day, which captures metrics for machinery, including temperatures, pressure, speed, and vibration. Such a large volume of data is further bolstered with much more extensive infrastructure in terms of storage and processing, using solutions such as high-capacity storage and adapted data processing frameworks.

Data curation is concerned with establishing the value of data for analysis, a process manufacturers oftentimes require to be performed. With unneeded data sitting idle over long periods of time, extreme consumption of storage is created, and the possible analyses and decision-making capabilities are limited by an overflow of information. In contrast, a clearly defined approach to data governance would preserve only the most valuable data and allow for the erasure or archiving of less valuable information.

Data Standardization and Integration, is the issue of raw data from these disparate systems is also in the data standardization and integration. Because a manufacturing facility probably comprises old and new machines having proprietary formats, the resulting data is bound to vary appreciably. For example, the output from an old lathe is likely to be just basic status signals, whereas the output from a new CNC machine might resemble a high frequency time series.

This implies that the information put together needs to be represented in a single, coherent format.

A production line can consist of several stages with a combination of software systems, including, but not limited to, enterprise resource planning systems and supervisory control and data acquisition systems [11]. In this scenario, integrating these heterogeneous systems into a single framework is an arduous challenge that has to be supported by significant investments in information technology infrastructure. Manufacturers have to implement middleware solutions or alternatively adapt new technologies that facilitate the integrated integration of different systems to maintain the running process of real-time data in the manufacturing site.

7.2 Computational Complexity and Scalability

When graph neural networks are used in the digital twin system, the set of computational challenges is posed. One primary attribute of GNN is its ability to model complex relationships. However, a large graph would heavily increase its computational complexity and this really tests in large-scale industrial environments where hundreds or thousands of machines and processes interplay with each other.

Graph Complexity

In the case of a manufacturing facility, all of the equipment, sensors, and systems can be thought of as nodes in a graph, and the relationship between them—material flow, communication pathways, or dependencies—represent edges. As the line adds more machines to the production line, the graph becomes more complex, placing additional computational stress on the Graph Neural Network (GNN).

The size of the graph can explode exponentially, especially in a case when interconnections of the machines are rather large, as in highly automated plants, where systems are constantly exchanging data. Such complexity demands substantial computational power, often necessitating the use of supercomputing clusters or specialized hardware like Graphical Processing Units (GPUs) or Tensor Processing Units (TPUs). Without such resources, the GNN could be severely challenged in properly processing the graph, which may lead to delays or errors in the insights produced by the digital twin [12].

Scalability Challenges

Scaling is another major challenge. For GNNs to be practical for the high volume of manufacturing environments, they must be scalable to large growing datasets. An expansion of factory lines or new machinery will result in larger frameworks when the GNN framework does scale to accommodate the added complexity. However, scaling GNNs comes with various challenges, including data partitioning and distributed computing.

Many times, it is not possible to process the graph entirely on a single machine due to constraints with respect to memory and computation. Hence, manufacturers would have to add distributed GNN algorithms such that the graph gets split into smaller subgraphs that can be subsequently processed in parallel across multiple machines. While this improves the scalability of the computation, communication efficiency between different machines within the distributed system has to be ensured along with graph consistency across partitions.

7.3 Integrating Legacy Systems

The major challenge of implementation of modern digital twin and GNN solutions relates to legacy systems. Most manufacturing plants, including the older ones, still use outdated machines and control systems that are not based on the point of view of an era of digital transformation. Sensors for such machinery often are not installed or data communication capabilities, which could enable their integration with the digital twin architecture, are missing.

Retrofitting Legacy Equipment

One means of integrating legacy systems is retrofitting. In this, IoT sensors, actuators, and other devices are added onto older machines that collect data from the machine. For example, the existing stand-alone legacy lathe machine can be retrofitted with sensors that will measure the speed of the spindle, temperature, and tool wear. All these will be inputs for creating a digital version of the lathe, thereby integrating it into a GNN-generated system.

Retrofitting is expensive, particularly for those installations that keep a great deal of old equipment. Additionally, retrofitted systems are sometimes less precise in terms of data as compared to modern equipment. This limits the extent of insights that such systems can make. Thus, manufacturers need to balance the convenience of retrofitting against total replacement of legacy equipment.

Interoperability and Communication Protocols

Another challenge lies in interoperability between legacy machines and modern IT infrastructure. Legacy machines often communicate using proprietary protocols that are incompatible with modern systems. For example, a legacy machine may run under serial protocols like RS-232, but modern digital twin platforms use internet-based protocols like MQTT or HTTP [13]. So, in some cases, interoperability would necessitate a middleware solution to translate one protocol into another so that data flows between systems without much interference. The producers must also consider the security implications arising from connecting older systems to today's digital infrastructures. Many older computers were designed without security in mind, making them vulnerable to hacking or data breaches. Arming these machines with IoT devices provides even more avenues of attack that must be combated through strong cybersecurity, such as firewalls, encryption, and intrusion detection.

7.4 Ensuring Explainability and Transparency

Explainability of an AI system is crucial when such decisions, based on GNN insights, have severe operation or financial consequences in an industry. Transparency of the GNN-based model becomes hard to achieve due to complexity in relationships. Unlike traditional machine learning models, where the reasoning behind a prediction can be traced back to individual features, GNNs rely on the intricate interactions between nodes and edges, making it harder to explain their decision-making process.

Interpretability Challenges in GNNs

For a GNN, neighbouring nodes influence the representation of every node by propagating information across the entire graph, but with some uncertainty with respect to which particular node or connection is actually responsible for some specific prediction or recommendation.

For example, if a GNN makes a prediction that a specific machine in the production line will break down, it is challenging to tell whether it did so based on the operational data of the machine or due to its relationship with the upstream processes, or any combination of both.

This may be quite difficult in those industries that are characterized more generally by high standards for safety and regulatory compliance, where it becomes a matter of essential importance for operators to understand the rationale behind every decision. Just imagine the automotive or aerospace industry, in which safety should be paramount, relying upon a "black box" for critical decisions about maintenance.

Explainable AI Techniques for GNNs

In a bid to address some of these challenges, producers are increasingly embracing customized Explainable AI (XAI) methods for Graph Neural Networks (GNNs). The goal of the methods is to enhance GNN explainability by focusing the most influential nodes, edges, or subgraphs relevant to the prediction in question. For instance, applying the node importance score allows operators to identify which machines or processes had the maximum influence on a given failure prediction, which may help them understand the root cause of the problem.

Visualization tools are another important part of XAI for GNNs [14]. Graph creation and emphasis on the most relevant relationships will, in many ways, make clearer how different parts of the production line impact specific outcomes to the operators. Graph attention mechanisms will also allow focusing attention on the most relevant parts of the graph responsible for the decision-making.

8 Real-World Applications and Case Studies of Digital Twins and GNNs in Manufacturing

Integration of DT and GNNs has revolutionized manufacturing in several industries, including automotive and aerospace, energy and pharmaceutical. These technologies bring new solutions to otherwise complex manufacturing-related problems, by increasing the efficiency of the operation, by reducing downtime, providing predictive maintenance, as well as optimization. The next section will have real applications and case studies in which digital twins, powered by GNNs, have bought such revolutions in manufacturing.

8.1 Predictive Maintenance in the Automotive Industry

For the automotive industry, one of the significant applications of digital twins and GNNs is in the maintenance prediction feature. In large-scale automobile plants, machinery breakdowns often result in terrible delays and increase the costs a great deal. The real-time status of the machines on the production line, right from the robotic arms to conveyor belts [15], can be monitored through digital twins. Manufacturers can forecast a time when a machine is likely to fail by continuously collecting sensor data and feeding the same into a GNN, thereby facilitating proactive maintenance before any disruption occurs.

Case Study: Ford Motor Company

Ford successfully implemented a GNN-based digital twin solution across all production lines to reduce the likelihood of uncontrolled shutdown. Temperature and vibration sensors are installed on each machine and continually send data about temperature, vibrations, and usage rates. These models calculate this information and identify how one equipment is related to another, and if any machine is predicted to fail based on its current state and dependency with other equipment nearby.

Ford's system herein alerts maintenance teams to serve the machines even before break down occurs. This thus saves millions of repair bills and delays in the vehicle's production. The digital twin also threw more insight regarding the points of production bottlenecks from the interdependency between machines. GNN performed well in providing suggestions on workflow adjustments, which improved the production and, subsequently, reduced the assembly time for the vehicles to about 15%.

8.2 Supply Chain Optimization in the Electronics Industry

Major electronics manufacturers have significant problems optimizing supply chains covering thousands of suppliers, manufacturers, and distributors, and often spanning entire continents. This makes it now possible to consider taking a graph view of the entire supply chain [16]. Companies can now utilize GNNs combined with digital twins to capture complex relationships between suppliers, factories, routes of transport, and customers. Based on this foundation, firms will be able to detect bottlenecks, optimize flow, and minimize risks of disruption.

Case Study: Samsung Electronics

Samsung migrated its global semiconductor supply chain with a digital twin, aiming at optimizing the global manufacturing supply chain. GNN-based digital twin mapped all supply nodes from raw material suppliers through the manufacturing plants and distributors. By monitoring real-time data on supply levels, production rates, and transportation times, Samsung's system could predict possible delays and disruptions.

For example, if it rains and there are transportation delayors with a supplier of silicon wafers, the GNN may recommend switching orders to a different supplier to avoid an extra day in queue at the manufacturer.

The system also proved insightful in indicating which areas of the supply chain were not operating properly. The entire supply chain was visualized as a graph, and through this, Samsung noted that one particular supplier was consistently delaying orders due to inefficiencies in their internal manufacturing processes. Working with this supplier to improve the procedures resulted in a 10% lead time reduction across the entire supply chain.

8.3 Quality Control and Defect Detection in Aerospace Manufacturing

Aerospace manufacturing is not a field, which can afford even a minor error. High standards of safety are required to produce aircraft. A defect at any point may lead to disastrous failure such as recalling or grounding the aircraft and sometimes complete safety failures [17]. Digital twins associated with GNNs have found applications in aerospace manufacturing; errors are detected at early stages and correction actions are taken in real time.

Case Study: Boeing

Boeing uses digital twins for monitoring the whole aircraft assembly process from component manufacture to final assembly, using GNNs to model relationships between various components such as wings, fuselage, and engines, thus enabling them to analyze real time data off automated tools on material quality, assembly tolerance, and sensor readings to be able to identify possible defects in parts and deviations in assembly processes.

For instance, Boeing's system highlighted a minor anomaly in the thickness of a composite material in an aircraft wing, which human eyes failed to notice when inspecting. The GNN indicated this variation as a risk by linking its properties with the piece of the aircraft as a whole. Engineers were able to correct the situation early so that rework didn't cost millions of dollars later on.

More importantly, the digital twin has given Boeing a chance to understand how changes in the assembly process-reither temperature or pressure-affected the quality of the product. Here, changing assembly parameters based on GNN's recommendation has brought about a 20% reduction in material defects and stayed within the strict regulatory bounds.

8.4 Energy Efficiency and Sustainability in Manufacturing

As industries reach for a greener, low-carbon approach, energy efficiency and sustainability have come to the forefront [18]. Digital twins and GNNs can assist companies in their quest to monitor and optimize energy usage throughout an enterprise level of production lines and facilities, allowing manufacturers to identify inefficiencies and implement strategies to minimize energy consumption and waste.

Case Study: Siemens

Energy consumption was optimized in one of the turbine manufacturing plants by establishing a GNN-powered digital twin from Siemens. It was monitoring the entire process of energy consumption across cutting metals, welding, and heat treatment. The GNN model identified the processes consuming more energy than usual based on real-time data collected and recommended optimization in consumption.

The system therefore showed that an energy-intensive heat treatment process was running on 10% more energy than that originally required due to worn-out furnace parts. Siemens was able to replace the faulty equipment and reduce energy consumption as well as extend the usable life of the furnace. In addition, GNN recommended rescheduling certain high-energy-consuming processes to off-peak periods when energy prices were lower, thereby achieving a saving of 12% in the total energy spent.

It further assisted Siemens in gaining its goal towards sustainability through identifying disparate areas where usage of raw materials can be optimized. Through the GNN model analyzing the relation between material inputs, energy consumption, and outputting products, it helped reduce material waste in Siemens processes by 15%, effectively achieving the environmental objectives set by the company.

8.5 Smart Factory Implementation in the Pharmaceutical Industry

In the pharmaceutical industry, the concept of digital twin is being implemented to make drug manufacturing more efficient as per the challenges posed by tough regulatory mandates [19]. Implementing GNNs enables companies to work on optimized production procedures, quality control processes, and regulatory compliance, all while making the whole production lifecycle transparent and traceable.

Case Study: Pfizer

Pfizer used a digital twin platform developed with GNNs in order to optimize the production of its vaccines and biologics. It monitors each phase of the production of drugs, from raw material inputs to final packaging. The GNN model analyzes how different production steps are related and thus allows Pfizer to find areas for improvement, predict possible quality deficits, and ensure that the products meet regulatory standards.

In one instance, Pfizer used the digital twin to optimize a complex drug formulation process that required precise temperature control during mixing. The GNN model identified that minor fluctuations in temperature were affecting the consistency of the final product. By adjusting the temperature control systems based on the model's recommendations, Pfizer improved product consistency and reduced batch failures by 8%.

The digital twin provided information related to the state of production machinery; Pfizer could plan their maintenance operations more efficiently. Time spent idling equipment came down, and more production lines came up and running, for example, when manufacturing COVID-19 vaccines.

9 Challenges and Future Directions of Integrating GNNs with Digital Twins in Manufacturing

Even though GNNs and Digital Twins have tremendous potential in revolutionizing manufacturing into better practice, they are yet to be achieved because they are surrounded by several challenges that need to be overcome if their true potential is to be realized. This chapter identifies some of the key challenges at integration and evaluates potential futures that may enable GNNs in combination with digital twins to be more effective and used more widely.

9.1 Data Quality and Availability

The main challenge in integrating GNNs with digital twins is data quality and availability. To make GNN models work, it is essential to have accessible high-quality real-time data coming from a wide range of sources-sensors, machines, production logs-and much more [20]. However, in various manufacturing environments, data may be inconsistent, incomplete, or noisy. As a result, predictions might be relatively less accurate, harming the performance of the model.

One of the common challenges manufacturers face is data silos-in which data is held in different systems and formats that are not commonly aggregate in a way to provide good analysis. Better challenges of such a phenomenon entail manufacturers investing appropriately in robust data management systems to allow integration of data across different platforms. Improved standardized data protocols and cleaning procedures also enable better data quality and GNN model training.

A crucial requirement, in this case, is data consistency across all stages involved in the manufacturing process. Misaligned sensor readings, for instance, may result from error calibration or environmental factors and will naturally affect GNN's performance. Manufacturers have to adhere strictly to data validation protocols that ensure differences are detected and corrected before feeding into the GNN models.

9.2 Scalability of GNNs in Large-Scale Manufacturing Systems

As manufacturing systems scale, scalability issues arise in GNNs. Although they represent the relationships in data so effectively, with increasing graph sizes, GNNs can get costly computational resources. In expansive manufacturing systems where there are several nodes (machines, sensors, and components) as well as edges that describe the interaction, computation processes have to be efficient.

Researchers find different strategies to counter scalability problems. Graph sampling techniques reduce the size of the graph and keep the important information that GNN models will essentially require. Techniques like mini-batch training or distributed computing can handle large datasets and make real-time processing of vast data possible [21].

Moreover, applications in edge computing will further enhance the scalability of GNNs in manufacturing. Manufacturers will reduce the stream of data to centralized servers by processing data closer to their source-the factory floor-and enable them to respond sooner and decrease the computational load on GNN models. This is particularly beneficial in real-time decision-making environments.

9.3 Model Interpretability and Trust

Although GNNs have offered quite a few advantages in explainability, full interpretability is still challenging to achieve. Complex graphs work on nodes and edges; hence, very difficult to understand for which reasons the GNN made a prediction, due to which operators and decision-makers are skeptical.

To gain the trust of people in GNN-driven systems, manufacturers need to invest in tools that improve model interpretability, such as visualization techniques, which help developers understand how GNNs arrived at particular predictions. Clear visualizations of relationships between nodes and their contributions to the outputs of the model would improve transparency in the system and help develop confidence in it.

There is a necessity for developers and validators to involve end-users in developing the GNN models. By making room for the voices of operators and domain experts, manufacturers can prove that the models fit real practice and thus serve practical purposes. A great collaborative approach will foster more confidence and uptake of GNN-based solutions.

9.4 Regulatory Compliance and Data Privacy Concerns

The manufacturing industry is a space where rules and regulations are very strictly adhered to, as in pharmaceuticals, for instance, and aerospace [22]. Combining GNNs with digital twins requires ensuring adherence to such requirements set by the industry, especially concerning data management, storage, and security.

In addition, GNNs rely much more heavily on data, hence requiring manufacturers to address issues regarding data privacy, especially if it involves sensitive information. Indeed, the adoption of robust data governance structures that stipulate how data is being collected, processed, and shared in a firm is pivotal in maintaining compliance with the requirements of the GDPR.

To overcome these challenges, manufacturers should use privacy-preserving techniques, like differential privacy, which enables organizations to analyze data without the loss of individual privacy. They should also develop clear data policies that convey how data is used and protected to build trust with its stakeholders and customers.

9.5 Future Directions and Opportunities

Despite the difficulties, GNNs and digital twins have a bright future in manufacturing. As technology changes, several trends and developments might be realized to further improve the efficiencies of such systems:

Integration with AI and Machine Learning: High-level integration techniques of AI with GNNs may result in far more adaptive and intelligent manufacturing systems [23]. Hybrid models, for example, may self-adjust in real-time feedback loops for further operation optimization.

Industry 5.0 and Human–Machine Collaboration: Industry 5.0 further unfolds into the realm of human–machine collaboration. GNNs integration in digital twins may open up real-time insights to aid in decision-making and enable workers. Human–machine interfaces that are enhanced by integration would further lead to intuitive interaction with increased productivity.

Sustainability Initiatives: GNNs can greatly contribute in optimizing resource use, reducing waste generation, and minimizing energy consumption in manufactured processes. Complex interactions between resources and production outputs may be modeled with the GNN, thus serving to help an organization in regulatory compliance and sustainability goals.

Interoperability and Standardization: Interoperability between the variety of digital twin and GNN solutions available will improve the sharing and collaboration of data across diverse manufacturing ecosystems. Industry standards and frameworks can help integrate more smoothly, thus allowing manufacturers to use GNNs with their digital twins across different systems.

There is a growing need for professionals with skills in developing, implementing, and managing these systems with the proliferation of GNNs and digital twins. Hence, investment in education and training programs for AI, machine learning, and data science would be required to prepare the necessary workforce.

By capitalizing on these future directions as said in Fig. 2, while overcoming the current challenges, manufacturers will be able to use the true potential of GNNs and digital twins in driving transformational changes in efficiency, productivity, and innovation.

10 Conclusion

Industry 4.0 technologies pointing toward automation and data interchange coupled with advanced analytics are significantly changing the manufacturing sector. Among the widely recognized technologies associated with Industry 4.0, Digital Twins and

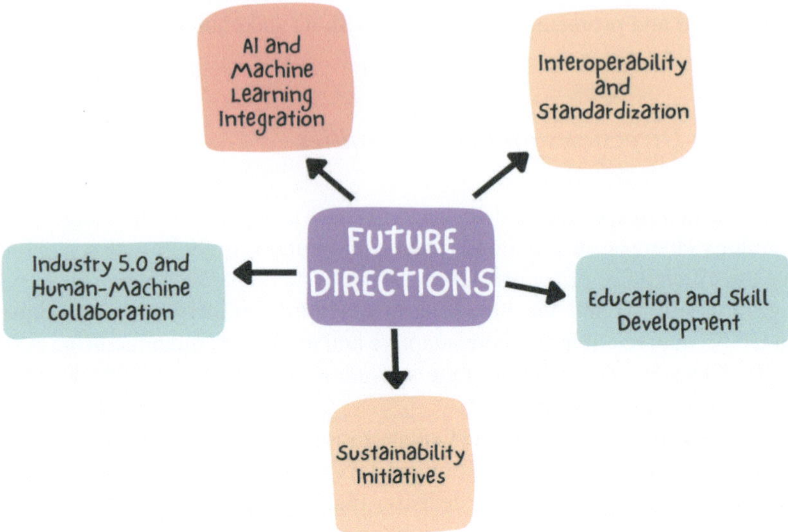

Fig. 2 Future directions and opportunities

Graph Neural Networks are recognized as key instruments for engineering transformations that could bring a sea change to traditional manufacturing processes. This chapter provides a comprehensive study on the convergence of GNN and DT [24], diversified applications and challenges faced, and possible future directions.

10.1 Summary of Key Insights

The integration of digital twins with GNNs will benefit manufacturers in a significant manner. Manufacture can make optimal usage of resources, make an optimum system for increased productivity, and significantly reduce downtime [25]. The connected devices and virtual replicas of physical systems will easily allow manufacturers to get greater operational efficiency in real-time. GNN will enable organizations form highly complex relationships inside data, thus enabling predictive maintenance, optimization of supply chains, quality control, and sustainability initiatives.

Various case studies reported from diverse sectors such as automotive and electronics, aerospace, and pharmaceuticals about real-world implementation of GNNs and digital twins. Ford's predictive maintenance reduced unexpected downtime, and supply chain optimization by Samsung improved agility in a global marketplace. Some examples include Boeing quality control and Pfizer's drug manufacturing procedures to show that the technology improves product quality and maintains the set regulatory standards.

10.2 Challenges and Future Directions

This would certainly help both the integration of GNNs and digital twins but some of the challenges include: data quality, scalability, model interpretability, regulation compliance, and data privacy. These challenges would warrant investment in robust systems for managing data, developing interpretable models, and strict adherence to regulatory frameworks.

The prospects are bright for GNNs and digital twins in manufacturing, with further potential integration into AI and machine learning to enhance human and machine collaboration toward more sustainability, interoperability, and standardization. The increased demand for skilled professionals puts an added emphasis on education and training to help qualify workers for this new future.

10.3 Final Thoughts

Concluding Reflections Accepting Digital Twins and Graph Neural Networks is, therefore, a strategic turning point in manufacturers' operation processes, placing them better to gain the upper hand of having an increasingly fluid global market through the power of these technologies [26]. The continuous pursuit of change and development by industries will certainly force the interaction between digital twins and GNNs to influence the future of manufacturing, guiding the industry toward efficiency, sustainability, and resilience. Conclusion: This is still at the introductory point as far as the full exploitation of GNNs and Digital Twins is concerned, hence, industrials would have to take this challenge head-on and capitalise in the opportunities that are set to pave a way for transformative advancements which shall characterise the next era of industrial innovations.

References

1. Boschert, S., Rosen, R.: Digital twin—the simulation of reality. In: Advances in Engineering Design, pp. 59–74. Springer, Heidelberg (2016)
2. Kritzinger, W., Karner, M., Hermann, C., Dallasega, P.: Digital twin in manufacturing: a review. In: Advances in Production Management Systems, pp. 115–122. Springer, Heidelberg (2018)
3. Lee, J., Lapira, E., Bagheri, B., Kao, H.A.: Smart manufacturing. Ann. Rev. Control Robot. Autonomous Syst. **2**, 27–51 (2013)
4. Tao, F., Zhang, M., Liu, Y.: Digital twin and cyber-physical systems: a new paradigm of smart manufacturing. IEEE Trans. Industr. Inf. **15**(3), 1895–1904 (2018)

5. Barata, J., Oliveira, A., Ferreira, J.: Graph-based modelling for digital twins in manufacturing. Proc. Manuf. **13**, 1311–1318 (2017)
6. Rajesh Kumar, D., Rajkumar, K., Lalitha, K., Dhanakoti, V.: Bigdata in the Management of Diabetes Mellitus Treatment. In: Chakraborty, C., Banerjee, A., Kolekar, M., Garg, L., Chakraborty, B. (eds.), Internet of Things for Healthcare Technologies. Studies in Big Data, vol. 73. Springer, Singapore (2021). https://doi.org/10.1007/978-981-15-4112-4_14
7. Wu, Z., Zhang, X., Yu, Z.: A comprehensive survey on community detection with deep learning. IEEE Trans. Knowl. Data Eng. **33**(1), 42–62 (2020)
8. Abirami, T., Mapari, S., Jayadharshini, P., Krishnasamy, L., Ragavendra Vigneshwaran, R.: Streamlined deployment and monitoring of cloud-native applications on AWS using kubernetes, keda, argocd, prometheus and grafana, ICAICCIT-23, IEEE Delhi section, Nov 2023
9. Hinton, G.E., Salakhutdinov, R.R.: Reducing the dimensionality of data with neural networks. Science **313**(5786), 504–507 (2006)
10. Zhang, H., Liu, S., Xu, Y.: Modeling and control of cyber-physical systems in smart manufacturing. IEEE Trans. Syst. Man Cybern. Syst. **50**(6), 2228–2239 (2018)
11. Cheng, J., Liu, X., Zhang, Y.: Graph neural networks: a survey of methods and applications. IEEE Trans. Neural Netw. Learn. Syst. **32**(1), 4–24 (2020)
12. Jiang, L., Li, S., Wang, H.: A survey on the state of the art of graph neural networks. J. Comput. Sci. Technol. **34**(4), 787–803 (2019)
13. Ge, L., Yang, Z., Zhang, H.: Digital twin: a new concept for smart manufacturing. IEEE Access **8**, 203285–203295 (2020)
14. Rojas, J.A., Gonzalez, R.: Digital twin in the aerospace industry. J. Aerosp. Eng. **231**(1), 25–35 (2018)
15. Liu, Y., Wang, Y.: The role of digital twins in smart manufacturing: a systematic literature review. Int. J. Adv. Manuf. Technol. **110**(9–12), 2959–2974 (2020)
16. Tolk, A., Diallo, S., Turnitsa, C.D.: Modeling and Simulation Support for System of Systems Engineering Applications. John Wiley & Sons (2013)
17. Becker, J., Wills, M.: Industrial Internet of Things: Digital Twin Technology in Manufacturing. McKinsey & Company (2018)
18. Gao, T., Wang, L., Zhang, S.: Digital twin and cyber-physical systems for sustainable manufacturing. J. Clean. Prod. **271**, 122743 (2020)
19. Marzouk, M., Abdel-Wahab, M.: Digital twin technology: opportunities and challenges. Int. J. Eng. Res. Technol. **9**(1), 45–55 (2020)
20. Wang, K., Chen, Y., Liu, Z.: Digital twin-driven smart manufacturing: a comprehensive review. IEEE Access **7**, 74318–74334 (2019)
21. Rosen, R., Schauer, T., Kritzinger, W.: About the importance of digital twins in smart manufacturing. IEEE Trans. Autom. Sci. Eng. **13**(3), 909–920 (2015)
22. Zhou, J., Li, L., Zhang, Z.: Graph neural networks: new directions for the study of learning on graphs. ACM SIGKDD Explorations Newsl. **20**(2), 1–20 (2018)
23. Jin, W., Zhao, Z., Liu, J.: Graph neural networks for real-time manufacturing system performance prediction. Int. J. Prod. Res. **58**(20), 6110–6126 (2020)
24. Sinha, A., Singh, A.: Applications of GNNs in industrial internet of things. J. Ind. Inf. Integr. **22**, 100207 (2021)
25. Schmidt, L., Wang, H., Zhang, M.: Artificial intelligence in manufacturing: a review. CIRP Ann. **70**(1), 64–87 (2021)
26. Duflou, J.R., Gauthier, C., Liu, D.: Sustainable manufacturing: a comprehensive approach. J. Clean. Prod. **21**(1), 1–3 (2012)

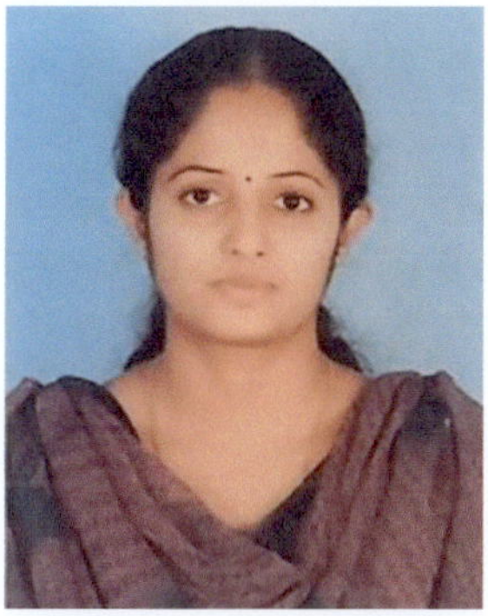

P. Jayadharshini

S. Santhiya

C. Vasuki

T. Vanaja

S. Archanaa

K. Samyuktha

Synergies of Human–Robot for Smart Manufacturing in Industry 4.0

C. N. Vanitha, P. Anusuya, and Rajesh Kumar Dhanaraj

Abstract Industry 4.0 has brought about a new era of manufacturing that is defined by automation, interconnection, and data-driven decision-making. The fusion of human laborers and robotic systems, which ushers in the era of Human–Robot Collaboration (HRC), is one of the key features of this industrial revolution. This paper overviews HRC's guiding principles, difficulties, and developments in Industry 4.0's smart product manufacturing. In production environments, HRC has many advantages, such as increased flexibility, safety, and productivity. However, to fully utilize HRC, several organizational, sociocultural, and technical issues must be resolved. These problems include everything from resolving issues with job displacement and workforce training to guaranteeing smooth human–robot interaction and interoperability. Developments in artificial intelligence, robotics, and sensor technologies have made it easier to create collaborative robots, or co-bots, that can operate alongside humans. To enable secure and effective cooperation in shared workspaces, these co-bots are outfitted with sophisticated sensing capabilities, adaptive control algorithms, and safety features. Furthermore, real-time monitoring, predictive maintenance, and HRC system optimization are now possible thanks to the development of digital twins, cloud computing, and Internet of Things (IoT) technologies. Manufacturers may make better decisions, allocate tasks more effectively, and increase system performance by utilizing data analytics and machine learning approaches. The accomplishment of HRC in smart manufacturing is contingent upon cultivating a collaborative culture, advocating for human-centric design principles, and offering sufficient training and assistance to human laborers. Regulations must also change to consider the ethical and legal aspects of HRC, including liability risks, privacy concerns, and safety requirements.

C. N. Vanitha · P. Anusuya (✉)
Department of Information Technology, Karpagam College of Engineering, Coimbatore, India
e-mail: anusuyamathan.ece@gmail.com

C. N. Vanitha
e-mail: drcnvanitha@gmail.com

R. K. Dhanaraj
Symbiosis International (Deemed University), Pune, India
e-mail: sangeraje@gmail.com

L. Krishnasamy et al. (eds.), *Distributed Deep Learning and Explainable AI (XAI) in Industry 4.0*, Information Systems Engineering and Management 55,
https://doi.org/10.1007/978-3-031-94637-0_8

Keywords Robotics · Automation · Internet of things · Cyber-physical systems · Human–robot interaction

1 Introduction

Industry 4.0 is a revolutionized technological evolution in which the existing industrial units are incorporated with advanced robotics and artificial intelligence. It is sometimes referred to as the Fourth Industrial Revolution. It appears to be a revolution in business industries with advanced technology that makes the disparity between the virtual world and the real- world invisible. Unlike its predecessors, which were focused on mechanization, mass production, and the use of computers and electronics to automate processes, the fourth revolution of the industry is caused by a combination of cyber and physical systems, the IoT, networks and big data, and AI and cloud computing. Such clustering allows the development of "smart factories" enabling everything, including machines, systems, and products to be "networked" with each other and even be associated with one another.

Technology revolutionized the industry by creating production systems. Labelled as the Second Industrial Revolution, the changes applied also involved computing to automate processes across other noted levels such as connectivity and data-directed thinking and strategizing. In this reformation, there emerges collaboration between man and machines in the form of robots and robotic systems. This creates an idea people nowadays call Human–Robot Interaction. This is a recent shift from the country's established manufacturing policies on working with robots. HRC is where the synergic effect of human labor and core technological advances of robots will create the most effect in my current manufacturing industry highlighting the current trends towards productivity and safety in production. HRC remains a radical departure from ways of thinking about, and operating manufacturing systems in practice [1], technology is not taken for Armageddon risk. So, it is a culture change, in this case, an organizational culture that is practical and calls for many things. Benefits from this perspective of HRC vary in many aspects. Robots accomplish to perform hazardous and monotonous tasks with higher efficiencies and strength than human use of labor while flexibility, creativity, and problem-solving come from human labour [2]. Overall, they form a manufacturing system that is likely to be more flexible and adaptable.

Nevertheless, the successful incorporation of HRC comes with a host of challenges. The companies need to resolve potential concerns on job redundancy and skills upgradation within the collaboration. Another challenge is to cope with the evolving expectations and responsibilities of human associates in a joint work environment. The technical challenge regarding the barrier to effective communication is to implement methods that will allow for very high levels of interaction between men and robots. Current developments in artificial intelligence, robotics, and sensor technology have made practical the manufacture of structures known as collaborative robots or co-bots. These robots are integrated with sensors, control strategies, and safety structures designed to allow human–robot interaction without any threat. The possibilities of incorporating digital twin, cloud computing, and IoT make HRC better by providing management support through remote surveillance, maintenance on a condition basis, and predictive engineering as well as optimization through analytic and algorithm techniques.

Put differently, the enthusiastic extent to which HRC implements smart manufacturing depends on the adoption of collective practices, human factors in design engineering methods, and adequate education and training of employees. Also, there will be a need to consider the legal and ethical aspects of the HRC that will be evolving regulations including liability, privacy, and HRC safety concerns. By answering these questions and considering technological enhancements, it will be possible to understand economic advantage overcoming the compliance Hurdles associated with industry 4.0. The HRC-adopting companies will take the manufacturing industry to the next level in a few years.

AI remains one of the primary enablers for HRC to advance to the next level in Industry 4.0. Modern AI systems include capabilities for robots to get information from a given context, similar to what their human counterparts are capable of, and thus perform their duties efficiently while working through challenges effectively. Machine learning entails utilizing predictors and feedback to teach robots, and hence they can anticipate mistakes and also proactively monitor efficiency, thereby minimizing unnecessary time and assuring effectiveness in functional processes. Moreover, AI improves real-time decisions on large data sets for productivity and efficiency in smart factories. New technical approaches like computer vision allow the robot to understand human gestures, facial expressions, and voice control thus improving the natural flesh and blood interface as well as safety when interacting with robots.

IoT and Sensors are critical components of HRC in Industry 4.0. Some of the robots are fitted with sensors that can detect human presence, read environmental conditions, and perform more efficiently than humans. IoT links different robots, machines, and systems by integrating them in a way that information can be passed from one device to another. This connection provides mechanisms through which it is possible to coordinate the robots and the humans and make them work together in harmony as never before. For instance, a robot fitted with IoT sensors can report its state to a central system based on a priority list of tasks and scheduling of maintenance checks. A high level of reliability is achieved through IoT-based predictive maintenance which enables the absence of a machine failure disrupting the flow of production.

As with all changes that tend to be associated with the integration of HRC, there are important legal and ethical implications. Challenges that concern data privacy, safety of workers as well as the effects of automation on employment should be solved to gain good results. In any sector where accidents or malfunctions may occur in the operation of collaborative robots, legal frameworks that determine the amount of liability for each party must be drawn. Additionally, highly recognized moral dilemmas of delegation of decisions to robots especially in sensitive operations must be addressed. This kind of AI models and systems should be gradually developed by following ethical rules and regulations which then creates trust among the stakeholders as well as employees. To fix these problems, companies must develop internal regulations that respond to novel legislation and standards of integrity.

The shift to manufacturing by HRC entwists the need for a workforce that has developed skill sets appropriate for interacting with enhanced robotic systems. This calls for a change of the education and training paradigms. Employees need to be educated in not only how to deal with manufacturing robots as tools used in their profession but also in how to solve problems and how to adapt to change. To narrow the skill gap, educational institutions should provide courses on robotics, artificial intelligence, and generally—digital technologies. Moreover, the practice of developing a positive organizational learning culture through training and skill development, notes that there is a necessity of frequent changes and updates in the environment of the manufacturing business process for the workers and organizations.

HRC will maintain its evolution and continue to hold the power to transform the manufacturing sector all around the globe. However, such combined technologies like edge computing, blockchain, and quantum computing might bring a breakthrough when integrated with HRC systems when they reach the next stages of development. For instance, the idea of edge computing could be applied to increase decentralized decision-making at the level of factories, as for blockchain this could improve supply chain information tracking and analysis. Furthermore, the combination of AR/VR may create natural interfaces for the operators; thus, enhancing Face to Face Human Robot Interaction. Through applying these advancements all those involved in the different stages of production can establish versatile, sustainable, and precisely tailored environments for manufacturing, which will fit the requirements of the future more closely.

2 Literature Review

HRC is a fundamental element of a contemporary manufacturing system combining human adaptability and artificial intelligence with speed and accuracy. Current advancements in ML have made it possible for robots to develop cognitive models for dynamic interaction Semeraro et al. [3] pointed out that time-dependent algorithms are crucial in dynamic task adaptation while Mukherjee et al. [4] have successfully surveyed reinforcement and supervised learning as key to improving robot's adaptability in industrial activities. Safety continues to be an essential consideration in Arents et al. [5] pointed out that there is a lack of safety procedures and measures across the studies; Li et al. [6] disclosed the fundamental importance of pre-collision safety and post-collision safety. Digital Twins (DTs) have turned out to be a revolutionary tool as reviewed by Baratta et al. [7] since they provided actual time data required for establishing high levels of HRC strategies. Building on this, Wang et al.

[8] presented a DT framework that enhances the safety and reliability of this framework by integrating deep learning with convolutional neural networks to perform action detection and classification through semi-supervised learning. Taken together, these works support the idea of using high-performance computational methods such as ML, alongside safety metrics and DTs, to overcome HRC's difficulties and create novel, safer, more efficient, and adaptive manufacturing environments.

HRC and the application of sustainable manufacturing and Industry 5.0 concepts have been widely explored in the current literature as a means of improving performance, flexibility, and sustainability in manufacturing systems. Thus, using Siemens' Tecnomatix simulation tool, Ojstersek et al. [9] analyzed HRC in context to cost, time, scrap rate, and utilization in terms of social, environmental, and economic objectives. They highlighted the very high applicability of simulation modeling in assessing collaborative workplaces most especially in the case of labor deficits and the need to keep global competitiveness. Based on the above uncertainties of HRC, Zheng et al. [10] introduced a novel CI framework that builds a synergistic relationship between human and artificial intelligence to handle human, robot, and task uncertainties. To overcome the human uncertainties in this study, fine-grained human digital twin modeling was proposed and for the robotic task uncertainties, learning from the demonstration method was introduced and was shown to be viable in an example assembly task.

In studying Industry 5.0, Lou et al. [11] analyzed how human-cyber-physical systems could support the flexibility of disassembly planning in remanufacturing, which is a concern of sustainable development. There are two approaches one is a multi-objective sequential disassembly planning model that factors ergonomics and task complexity, and there is an improved multi-objective hybrid grey wolf optimization algorithm used to address Pareto-optimal plans. These findings do point to the fact that the proposed approach can cope better with an array of constraints and offer various and optimal disassembly strategies on the problem setting. Collectively, these investigations underscore the development of HRC when integrated with enhanced simulation, intelligence, and people's revisions to promote effective and effective production change that can deliver sustainable innovation quickly.

3 Principles of HRC

3.1 Harmony Between Humans and Robots

HRC is all about creating an effective working environment in which humans and robots complement each other's abilities. Robots do well in repetitive, dangerous, or extremely precise tasks that may be too stressful or impossible to bear for humans. Humans can think critically, they can solve problems while remaining flexible and decisive. This commutative collaboration ultimately results in higher efficiency, output, and flexibility, more so in environments characterized by complexity and dynamism.

3.2 Human-Centric Design

Users of HRC systems have to be able to operate them and hence regard design from a human's perspective. HRC systems focus on the ergonomics of the processes with the operator who is considered and incorporated in the design [12]. This includes developing and designing robots that are easy to manoeuvre and hence low training is required when implementing these robots into an organization. Workers are not ignoring their smart pieces of devices but instead, the devices incorporate their physical and psychological blast. A worker tends to become more efficient if robots help him/her rather than distract him/her.

3.3 Safety and Risk Management

Safety is one of the biggest issues in any robotic and human interaction territory. To ensure there are no incidents and human employees are safe, the use of intelligent robots comes with safety devices such as adaptive algorithms that can detect humans, additional safety measures that can halt the movement of the robot if there is a sensed danger, and non-linear incident management where a robot's actions may be changed in the course of operation precision of forecasted and executed operations with that of practical efficiency of each environment where interaction with humans occurs. All the operations involving the cooperation of workers and machines conform to the prescribed requirements of safety [13]. Assessment of potential risks is carried out in a proper manner whereby patients are observed and their interactions with machines are evaluated, risks are anticipated, and adjusted to maximize the safety of the working area.

4 Human–Robot Interaction (HRI)

The HRI follows multiple techniques to process the commands and perform the task. Figure 1 gives a basic workflow of the HRI. Strategies used for human and robot interaction are discussed below,

User Interface Design: HRC's success depends on an interactive system designed with easy-to-understand and easy-to-use interfaces so that human workers can comfortably interface with robots. This can be in the form of a physical interface such

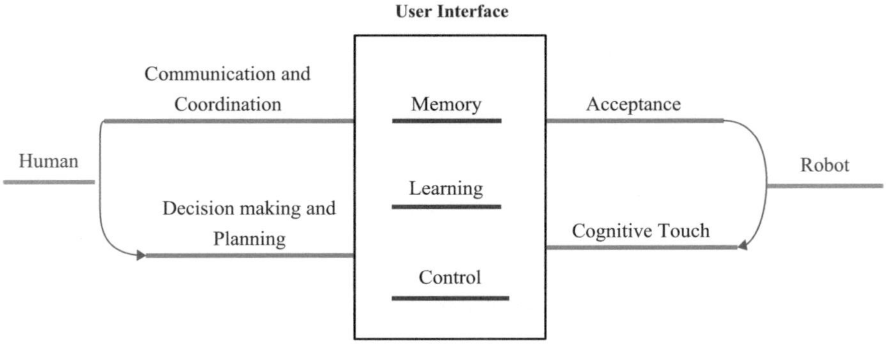

Fig. 1 Human–robot interaction methodology

as a touchscreen or software interface which could include programming languages or applications.

Trust and Acceptance in HRC Systems: This is important since HRC requires the trust of human workers with the robots. This entails making robots predictable in their actions and consistent with how they will interact with people in a particular setting as well as making sure that people obtain the information they need to know about a robot and its abilities and restrictions.

Communication and Coordination Strategies: That is why, direct communication with robots, generally addressed as human–robot cooperation, is critical to efficient HRC. This can be done through the use of words, sign language as well as other facial or body gestures. Thus, the coordination strategies make sure that each of the plans of the two parties is well understood with the least possibility of making mistakes or causing an accident.

Social and Cognitive Aspects of HRI: This is to assert that HRI is not only an aspect of physical touch, it is also a social and cognitive touch in collaboration. This counts for understanding human attitudes towards robots and how robots can effectively and appropriately incorporate themselves into human society [14].

4.1 Workflow Process of HRI

In Industry 4.0, the first step of the HRI process includes robots deploying various sensors to gather information from their environment. Such data is then transformed to help the robot in comprehension of human expectations so that proper decisions can be made. Then, the robot takes into consideration these decisions and comes up with an action plan, including communication methods, and correctly organizes the performance of activities. The robot interacts with users through user interfaces, and observes the environment for hazards, so that any unexpected events may be corrected instantaneously and without any delay. With each existent-bound communication, the performance of the robot keeps improving with the feedback considered. This way of performing HRI ensures that it is safe, efficient, and fortuitously applies to all areas of manufacturing and advanced technology in Industry 4.0.

5 Technological Foundations of HRC

The IoT integrates people, smart robots, machines, monitoring equipment, and other devices in a manufacturing setting and allows real-time data transfer as well as interaction. The IoT is a network that facilitates the reception and processing of data such as environmental and production sensor data, and data from human workers by the robots. With IoT, it enables the robots to alter their course of action depending on the information available at a certain point in time thereby optimizing interaction with the human stage workers even further which is shown in Fig. 2. For instance, it is possible to have an IoT system that will recognize the moment a certain machine reaches its failure point and order the robot to proceed and bn do preparations for maintenance such as cutting down other tasks to reduce the negative impact on the workflow. Table 1 illustrates some of the sample technologies that are used for various purposes in collaborating Humans with Robots.

Similarly, companies in this industry equip robots with **Artificial Intelligence (AI)** so that these devices can learn, adapt, and act reasonably based on given data [15]. Smartphone apps that require HRC as the area of interest only accept AI to be employed in contributing to the understanding of work processes by robots. Since AI equips the robots with these capabilities, they can explore their surroundings, detect patterns, and perform prognostication which enhances behavioural advancement [16,

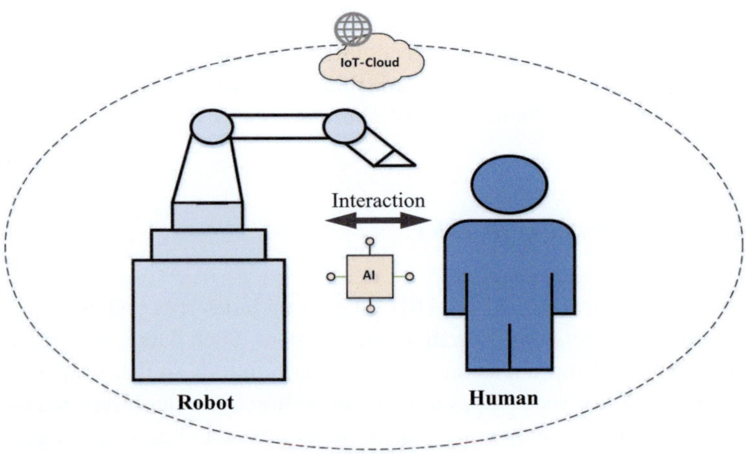

Fig. 2 Basic human–robot collaboration over IoT

Table 1 Technological uses in HRC

Technology	Purpose	Example
Natural language processing (NLP)	Enabling communication through speech or text	Voice commands for household robots
Computer vision	Understanding visual data from the environment	Object detection for robotic arms
Sensor integration	Detecting proximity, pressure, or motion	Safety mechanisms in co-bots
Artificial intelligence (AI) and machine learning (ML)	Enhancing robot adaptability and decision-making	Adaptive behaviour in dynamic environments
Human intention recognition	Predicting human actions and intentions	Gesture recognition in assistive robots

17]. For example, instead of standard actions and commands, AI-based machines can work with people, performing a variety of tasks: including recognizing gestures and voice commands, responding to changing situations under what they are subjected to, and even 'sensing' the needs of the human side.

Machine learning is a part of AI that helps robots get better at their jobs over time by learning from information and past events. Using machine learning algorithms, robots can look at huge amounts of data from sensors and other places, spot patterns, and fine-tune what they do. In HRC, machine learning lets robots work better with humans by always getting better at handling tricky jobs and working with human employees. To give an example, a robot might learn to match its moves better with a human co-worker, which cuts down on the chance of accidents and boosts output. This ability to adapt is key in places where jobs and conditions keep changing.

Cyber-physical systems (CPS) combine physical processes with digital tech letting robots interact with the real world as it happens. CPS links sensors, actuators, and computing parts to watch and control physical processes. In HRC, CPS helps robots react to real-world events and makes teamwork with human workers [18]. For example, a CPS might spot a sudden change in the production setting, like a shift in temperature, and tell the robot to change how it works. This instant back-and-forth is key to keeping human–robot teamwork precise and quick to respond.

Cloud computing gives robots and devices the computing muscle and storage space they need to crunch huge amounts of data. This lets them do smart things like analyse stuff in real time, predict when they will need repairs, and keep an eye on things. When it comes to humans and robots working together, cloud computing helps make things run smoother by looking at all the data and figuring out what is working well. Let's take this example where data from robots in different places can be looked at in the cloud to spot patterns and make the whole system work better. Along with that, cloud platforms make it easy to update all the connected gadgets at once so the robots are always using the newest and best features.

6 Types of HRC in Smart Manufacturing

Various types of HRC in Industry 4.0 depending on the application are explained below and given in Fig. 3.

6.1 Coexistence: Parallel Workflows

In the coexistence mode of HRC, robots and humans work in the very same environment but in different ways that they do not intermingle. This arrangement is usually applied where abilities or competence levels that correspond to responsibilities given to robots and humans are dissimilar. For example, in a manufacturing plant, robots may be used for boring welding of the parts, while human operators perform the quality control check or any other complex delicate operations which may not be possible for the robots but are repetitive as well.

A coexistence characteristic is defined as one in which there is limited interference between humans and robots and there is no likelihood of an accident. Pandemic prevention is still relevant, but since all the work is divided into tasks that are apart from each other, confrontation or mistakes because of interaction with robots is hardly possible. This mode of working increases the possibility of incorporating automation with little interference with physical human work.

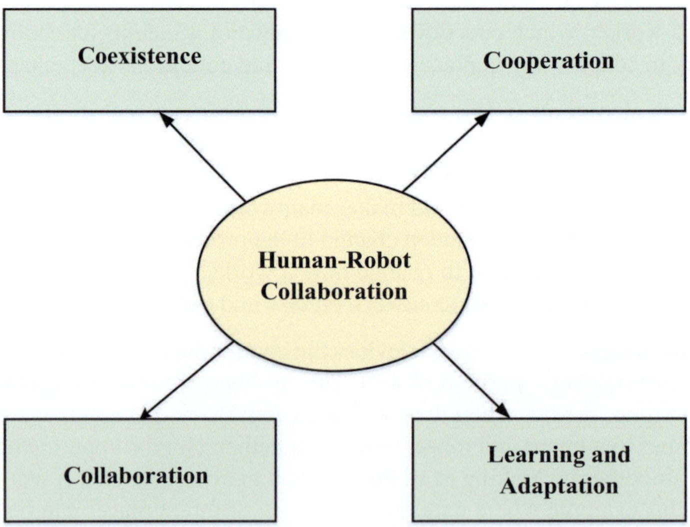

Fig. 3 Classification of HRC

6.2 Cooperation: Sequential Tasking

In cooperation, the human and the robot perform different subtasks of a given task at different moments, and not side by side. This approach is an example of the symbiotic combination of humans and robot's ability to accomplish a certain task. For instance, in the car manufacturing lines, a robot may lay down the framework of a car door perhaps a metal frame, while a human being can handle such details as installation of the wiring system or interior trims.

This type of HRC is useful in forms of processes that may need robotic precision but at the same time, human imagination and logic. By so doing, work is divided into certain stages which are usually handled by each respective party. Sequential tasking also ensures that the handover point between robots and human attendants is well defined to avoid loss of quality tasks such as food quality before they reach the customers.

6.3 Collaboration: Simultaneous Tasking

Amongst all the HRC solutions, collaboration is considered to be the most complex one and is also referred to as the ultimate form of HRC in which the human operator and the robot are performing a specific task in parallel [19]. This mode demands a great extent of contact and synchronization between the two. For instance, in an assembly line working jointly, a robot could position a part that a human being affixes by screwing. The effectiveness of this mode may be determined by the dependence on effective communication between the parties and their ability to respond to each other actions within the process.

Dominant coordination facilitates a greatly flexible way of working and can achieve difficult work that is awkward for both humans and robots to carry out. However, it entails the need for complex safety measures since people are in close contact causing high chances of an accident. Sophisticated sensor intelligence adjusting algorithms as well as constant surveillance equipment are a must to make the cooperation safe and effective. This duality makes this form of collaboration more useful with activities that need input from the human mind and output by a robot.

6.4 Learning and Adaptation in Collaborative Robots (Co-bots)

Co-bots are user-friendly robots that have been programmed to cooperate with people and have the feature of 'learning by doing' thereby acquiring new ways of performing tasks with time. In contrast with what traditional industrial robots have been like, co-bots are designed to allow them to interface with human colleagues more readily and naturally. They can watch the human activities around them, get some feedback on what they are doing, and then modify their movements or next tactics with the help of the human worker [20].

They learn from the task they are performing and through the use of artificial intelligence, refine the performance outcome accordingly. For instance, a co-bot may work alongside a human, and while it starts by observing the human in the course of the work, it slowly starts expecting the human to perform in a particular way to complement the human's methods. This not only increases efficiency but also makes the co-bot safer because it is so able to quickly react to changes in the surroundings or the actions of the human colleague.

Co-bots are helpful especially when work environments are complex and unpredictable since Co-bots may learn from one environment to another. They are flexible associates in Industry 4.0 thus they can adapt to new ideas and concepts in the market. Flexibility and adaptability are necessary for companies to meet the demands of the complex manufacturing processes in the modern world. Co-bots are also instrumental in creating a human-centered workplace as these are robotics that collaborate with man in the context of augmenting man's capacity without displacing him.

7 Applications of HRC in Industry 4.0

7.1 Smart Manufacturing and Production

The smart industry manufacturing concept is understood as the creation of highly automated, adaptive, and efficient production using modern technologies such as HRC [19]. Within smart factories, HRC enables effective collaboration between robots and human beings over shared production lines that are flexible enough to adapt to changes in product requirements like their volume or type.

Robots are also employed in assistive operations, for example, assembly, material handling, and quality assurance where efficiency is harnessed. For instance, robots could undertake the normal assembly process of attaching pieces, whereas human workers would be involved in the more complex processes of the sequence; troubleshooting, customization, or quality control. This cooperation ensures that productivity is at its peak by ensuring that every entity does what it does best. In addition, since robots are easy to change or to new tasks through programming, production lines will be modified with less time lost during conversion leading to improving the plant adaption towards market products.

7.2 Assembly, Inspection, and Quality Control

HRC is of great importance in assembly, inspection, and quality assurance operations where precision and uniformity aspects matter. In these scenarios, the tasks of high accuracy and repeatability such as fitting off complicated parts or carrying out quality checks on completed components are done using the robots. For example, it could be that the robots in the electronics industry would work on the circuit boards with so much precision placing very small components that humans could find it hard and take a lot of time. Even when the first extreme plastic parts have been fitted over the structural foam cores, robots could also perform a lot of inspections using several states-of-art sensors and imaging to check for any defects or discrepancies that could compromise the quality of the product. In the course of these procedures, humans are in control of the process, managing the entire workflow, and intervening whenever there is a need such as whenever a decision for rework is required, or the complexity of quality concerns exceeds the robotic capabilities.

It is these joint efforts of the logical robot processes together with appropriate human involvement that ensure the delivered quality of the products such as minimizing product defects and recalls. Besides, it rescues the human operators from mundane and more robotic work, enabling them to undertake core processes that deal with process enhancement and sustained improvement programs.

7.3 Logistics and Supply Chain Management

HRC is changing the logistics processes and supply chain management processes by relieving human resources from the repetitive and tedious processes of work to allow them to do more complex work [21]. In warehouses as well as distribution centres, the number of procedures carried out by robots increasing, such as picking, packing, sorting, and transportation.

Moving on, for instance, there are intelligent robots with sophisticated sensors and navigational capabilities that can transport themselves throughout the warehouse to fetch items from the shelves or bring the packed goods to the packing area. As a rule, this technology allows to minimize the time necessary to complete the orders, decreases the number of mistakes made during the work, and limits the physical work that needs to be done by people. However, the overall logistics process must be adhered to by people, which involves more interaction, thinking tasks like inventory, planning, and resolving investigation matters when incidents occur (through interaction, damaged items, hot items, etc.).

Delivering goods at faster speeds and with fewer errors is not the only benefit that this new technology will bring, rather it will also help to strengthen the supply chains' flexibility. Companies are now in a position to scale their activities and cope with peaks and troughs of demand for certain services as well as meet set customer deadlines due to the incorporation of automation in the dull details of the delivery.

Furthermore, HRC in logistics reduces the impact of labour shortages and cuts back on the costs of operations.

7.4 Healthcare and Service Industries

HRC also goes in tune with the healthcare segment, as well as other industries, which do not impede but further facilitate the use of robots to assist human workers in improving services provided for patients and consumers.

Medical contexts involve many different types of assistance from robots, but one of the most important ways that robots aid surgeons is in complicated procedures where many elements must be kept under control. A good example involves surgical robots, which have assisted in performing quite several minimally invasive surgeries and have thereby reduced the recovery period, continuing to improve patient outcomes. These are cases wherein human surgeons direct the procedures to make sure that expert judgment is made while the robots are used to perform the delicate parts of surgery. Besides being used in surgeries, robots are used in physical rehabilitation too wherein they help patients with post-operative therapy exercises, as well as in hospitals moving around consumables, drugs, etc.

In service sectors, customer care robots will be ready to work in the areas of customer service, hospitality, cleaning, and other related functions in highly demanding wear and others. For example, in restaurants, room service can be done by the robots or even the greetings and information to the guests can also be brought forth by the robots to save the employees the incidence of serving off the customers as they could instead interact more personally with other prospects. In retail shops, it is useful in restocking items, assisting in finding items, and even checkout of items in the cash registers.

8 Benefits and Challenges of HRC

8.1 Benefits

Enhanced Productivity and Efficiency: HRC enhances productivity by automating those tasks that would otherwise consume a lot of time to be accomplished by human labour.

Increased Flexibility and Adaptability: Intelligent robots are easy to re-program, thus making them easily capable of responding to various demands of production and manufacturing. This flexibility is mainly important in today's manufacturing sector because it is characterized by customization and quick response.

Improved Worker Safety and Ergonomics: Robots perform dangerous and strenuous jobs hence leading to the reduction of the risk of employee injuries and enhanced employee conditions.

Customization and Personalization of Products: HRC allows the manufacturing companies to produce small units of different products sequentially where the creation of products with personal characteristics has become a trend that was not easy to establish in traditional manufacturing systems since it would reduce production rates significantly.

8.2 Challenges in Implementing HRC

Organizational and Cultural Challenges: The interoperation of robots in environments that were initially designated for human tasks means that the culture of an organization needs to change. Several issues need to be discussed to protect people from getting fired from their jobs or being demoted and assure them that they will have to work with robots.

Technical Challenges include,

Interoperability and Integration: Technical challenges and closely related to one another and involve the ability of robots to interface with other robots and systems. Since HRC is mostly implemented in large and intricate manufacturing organizations, the issue of interoperability plays a significant role in making the concept successful.

Workforce Training and Skill Development: Again, the incoming of robots means that human workers will require new skills, especially in areas such as programming, repairing, and monitoring of robots. On the same note, retraining and education bear a close relation in ensuring that the human resource complements the aspects of technologies.

Ethical and Legal Considerations: HRC takes philosophical concerns into the realm of ethics and even the legal such as the responsibility of human replacement, especially in case of an accident, issue of accountability, and data protection are major concerns. These factors must be tackled to explain the success of HRC.

In Fig. 4, the advantages, difficulties, and opportunities of HRC across Industry 4.0 are outlined. This emphasis on how HRC helps in increasing production rates, elasticity, and safety, all along responding to the organizational, technical, and ethical issues that arise thereon when carrying out the application. The figure also presents various practical uses in large-scale industries and industries of the future such as smart manufacturing logistics and health care where HRC supports the enhancement of industrial productivity and development.

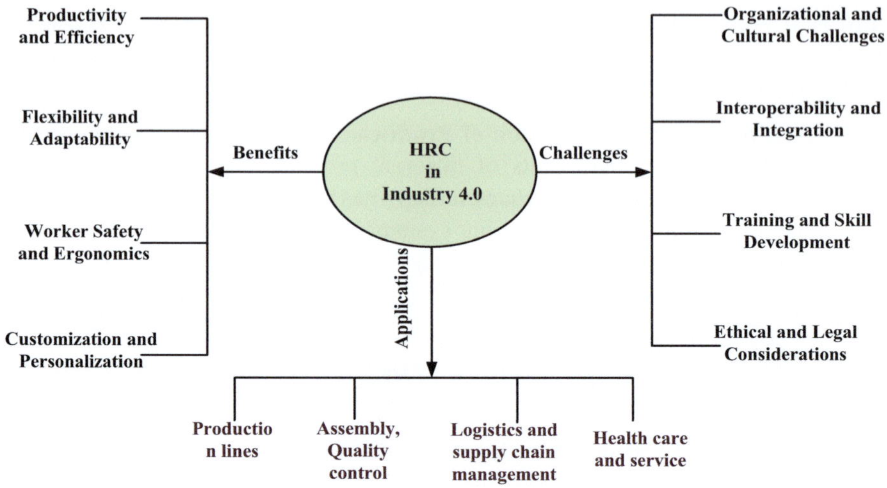

Fig. 4 Key benefits, challenges, and applications of HRC in modern industry

9 Advancements in HRC

9.1 HRC Collaboration with Digital Twins and Simulation in Industry 4.0

Digital twins are accurate replicas of the physical structures and processes that act as a precise manifestation of physical assets like mechanical, electrical, or even a manufacturing plant. Consequently, within the Framework of Industry 4.0, digital twins are employed to model different processes before these are incorporated into the real world. Digital twins used in HRC make it possible to improve the design, implementation, and evaluation of interaction in smart production halon [22]. In HRC, DTs provide a connection between the physical and cyber world rewarding efficiency in human/robot interaction. For instance, a DT could capture how a cobot cooperates with human operators within an assembly line which considers human motion characteristics, biomechanics, and dynamic changes in the robot's operations. This concerns the safety and effectiveness of physical robots working together with human ones.

Furthermore, the schematic has highlighted the fact that DTs are equally useful in enhancing the design phase of HRC systems. Various DTs can be used and adjusted by engineers or designers to test out different configurations, in regards to task distribution, and operational settings. Through these simulations, they are able to discover loops within the systems, understand optimal task schedulers, and improve the structure of the encompassing systems. This iterative process thus saves a lot of time and cost than can be spent in other trial and error methods. Digital Twins also have high-performance scores in the HRC implementation and assessment phase. They facilitate human–robot interactions' observation in real-time; the information being gathered originates from sensors installed in robots and smart gadgets. Such data can be used to compare the effectiveness of system performance, to discover deviations, and to apply corrective measures. Also, DTs support predictive maintenance since the patterns derived from the operational data help to predict the failure of A2212 equipment.

In smart production environments, DTs enhance the marketing decision by applying real-time data. For instance, when working on a joint project of accomplishing a task, a DT defines how the model changes with alterations in the physical environment such as shifts in the throughput demand of a manufacturing process or novel disturbances. Such flexibility makes it possible to keep both humans and robots effective in responding to changing circumstances and conditions to allow for continuous operations. In addition, DTs are combined with other technologies, such as AI and IoT, which improve their performance in the HRC context.

9.2 Data-Driven Decision Making in HRC

In Industry 4.0, data is the paramount secret to decision-making processes, management of organizational operations, and human–robot relationships. The use of data generated through HRC systems in decision making: Data-driven decision-making entails the use of data that is collected, analyzed as well as utilized in the process of enhancing efficiency, recognizing the probable time of need for maintenance together with making valuable decisions on the use of limited resources. Analytical decision-making in HRC is a key empowering influence in the manufacturing system of Industry 4.0 where inert real-time information is acquired from sensors, tools, and collaboration robots. All these data form a basis for improving productivity, managing resources adequately, and identifying when maintenance may be required. By adopting superior data manipulation and analysis methods of big data, organizations are able to use these insights in decision-making. The effectiveness of this data guarantees the transition from mere firefighting to planning, thereby preventing damage and improving productivity in totality.

Data-driven systems in particular improve communication by allowing robots to learn from the actions and preferences of their human counterparts in the course of HRC. For instance, real-time information can be applied to modify robotic behaviors as a way to match the human operators' schedules and rhythms. Further, the HRC-derived predictive analytics can be employed for discerning trends and patterns of future operational issues, such as equipment breakdowns or process slowdowns. On this account, this predictive capability is essential in building robust and reliable manufacturing systems. Additionally, HRC with a data-driven system will create efficient management of resources since it offers adequate control of energy consumption, materials, and time. This way, applied during the collaborative process, organizations can detect the patterns of waste and perform the corresponding interventions to minimize costs and environmental influence. This approach is in line with the sustainable strategy of Industry 4.0 which focuses on chip manufacturing factories hence reducing their impacts on the environment while increasing their output. The other evident benefit of employing data-driven decision-making in the area of HRC is the boost it will give towards boosting workplace safety and ergonomics. By integrating data emitted from wearable accessories in humans, gadgets mounted on robots, and senators implanted in the environment, it is possible to identify areas of risk within human–robot interaction and make real-time readjustments for the standard safety guidelines.

10 Real-Time Implementations

Examining the real-world applications, to gain valuable insights into best practices, technological advancements, and the impact of HRC on different sectors [23].

Warehouse Robots

Some of the specific application areas related to warehouse robots include the basic logistics industries and Supply chain management. These are used jointly with the human workforce in activities including order picking, inventory replenishment as well as materials transport. HRC in warehouses aims at increasing efficiency, decreasing the number of tasks solved with the help of human intervention, and increasing the accuracy of those tasks. The use of technologies such as robotics in warehouses has been pioneered by Amazon which has completely transformed its operations through automation. The use of robots by the company, particularly by Amazon Robotics previously known as Kiva Systems is one of the best examples of how HRC works to increase efficiency while lowering costs and improving the supply chain [24].

Agricultural Robots

HRC in the agricultural sector is revolutionizing techniques of farming by incorporating robots to undertake tedious chores like planting, harvesting, and analyzing crops [25, 26]. These are service robots meant to complement farmers in their work to make their systems efficient, reduce costs that are associated with labor, and give better yields.

Healthcare Robots

In the field of health care, HRC is a revolutionizing tool since it increases the accuracy, speed, and even the safety of operations and treatment or patient and hospital functions [27]. In recent years, the application of robots in healthcare has expanded across almost all medical fields with the function of assisting clinicians in enhancing patients' quality of life.

Military Robots

In the military, the focus is on the HRC to improve the state of affairs, the protection of the personnel, and the optimization of the system. Hiring robots is as follows; reconnaissance, bomb disposal, supply delivery, transportation, backup fighters among others.

Kitchen Robots

Leading to the improvement of HRC in the food preparation, and cooking sectors and meal services in both kitchens and households. Automated equipment is useful in performing monotonous and time-consuming activities, and is also useful in maintaining standard procedures, thereby making chefs and kitchen crew more inclined to come up with new ideas and more challenging tasks in the culinary arena.

Manufacturing Robots

In the manufacturing industry, HRC is already an innovative and effective way of working where both the efficiency of the human and the robot are integrated into the production line [28]. This makes it easier, adaptable, and safe for instance through the use of robots to undertake repetitive, hazardous, or delicate operations while on the other hand, manpower is dedicated to decision-making and oversight [29, 30].

11 Ethical and Legal Considerations in HRC

Thus, the ethical and legal concerns of HRC are important with the increased incorporation of robots in the workplace. Some issues relate to re-training, as employment has to be preserved while relying on automation, another issue is the proper use of AI, where the emphasis should be on making the process open and fair while keeping humans in control [31]. Robotic-associated legal issues have to do with apportioning blame in accidents involving robots essential requiring, regulation and risk management covering robots; with the need for specialized insurance. Data protection is also another consideration whereby the information of these workers should be protected to avoid cases of information leakage and cyber-crimes.

However, HRC raises several social concerns, which companies have to address: who will benefit from automation of tasks? Having rules and ethical codes of conduct is crucial in establishing and dealing with such problems and it is a great need for the organizations to develop Hoyt and human rights-compliant mechanisms in HRC to provide fair, safe, and ethical practices. Thus, by solving the above-mentioned ethical and legal dilemmas, HRC is capable of being installed as a tool that does not violate human dignity and safety, and at the same time, contributes to the improvement of productivity and innovation in Industry 4.0.

Further, the ethical and legal considerations of HRC have to be tamed with the help of transparency and accountability. It will be possible to come up with the guidelines on the robot's actions, and decisions needed so that there can be gradual development of trust between human and the robotic systems which are involved. Ethical implication also has to be applied on designing and deploying the decision in aspect of work force diversity. In addition, promotion of an ongoing conversation among the stakeholders involved, including employees, employers, regulators and AI developers can aid in dealing with new developments in HRC. In this way, organizations can reach positive outcomes for human interest as well as for technological influence, by introducing HRC as an ethical solution.

12 Future Enhancements in HRC

The future advancements of HRC will push the boundaries ahead by making robots more intelligent, responsive, and capable of working along with humans in a wide range of applications in our daily lives. Some future enhancements using various technologies are listed below,

Improved Sensing and Perception Technologies: In HRC, as the technology progresses, it is anticipated that there will be drastic improvements regarding sensors, thus facilitating robots to have better perception and cognition of space. More advanced sensors such as LiDAR, more advanced cameras as well as multi-modal sensors will be able to allow robots to get more precise and expansive information regarding their environment. This will also lead to enhanced situational awareness, better obstacle avoidance, and appropriate interactions with people, regardless of the complicated or ever-changing environments.

Improved HRI: Looking ahead, the human–robot interaction is likely to become less demanding and more intuitive and natural. Advanced gesture and speech recognition, together with haptic interfaces, will make it easier to interact with the robots. These interfaces will slowly become user-centered, changing based on how a specific user interacts with the system and will thus lower the barriers for HRC for other population groups such as disabled people.

Collaborative Autonomy: As a natural development, robots will be designed for more than simply facilitating decision-making by AI systems. It will mean that robots will function autonomously but also consider other people's course of action and intentions. Robots will provide a mature form of HRC wherein robots will assess human intention and respond accordingly, adapt to human action or interaction sequence, and modify behavior without extensive instruction.

Blockchain Technology: People should look for solutions to improve the safety and operability of HRC systems and one of these solutions is by introducing the blockchain concept in HRC systems. Through the use of the blockchain, data obtained from human–robot interaction can also be recorded and tracked in an immutable form. This would be especially effective for fields where it is as necessary to keep trust and integrity of the data such as finance, healthcare, legal, and so forth.

Augmented Reality (AR) and Virtual Reality (VR): Using these technologies in interactions between humans and robots will allow users to interact with the robotic process beforehand by creating a virtual model which can be changed as much as necessary before the process can take place outside the high-level simulation. This is going to be critical for mentorship, teamwork over a distance, and planning of complex tasks in cold environments with fewer risks to humans and moving robots and the accompanying attachments.

13 Conclusion

The incorporation of Human Robot Collaboration into the general framework of Industry 4.0 can be viewed as the evolutionary enhancement of supply chain management. With the help of innovative technologies like AI, Digital Twins, and CI after presenting exemplary experience in productivity improvement, safety management, and problem solution including the acute issues like lack of workforce and environmental issues, HRC has revealed the great opportunity. The combination of human agility, and robotic operation results in complementarity which brings about results that are optimized for the modern manufacturing environment.

As a result, introducing HRC has its advantages and disadvantages, mainly ethical and legal factors, workforce issues, and the requirement of numerous safety measures. Solving these problems is possible only through the convergence of such values as technology, openness of management decisions, involving various parties, and an orientation to the person. Moreover, IT-support for knowledge work and the use of Digital Twins for the creation of realistic models of HRC environments to allow for the further optimization of cooperative tasks, and maintaining the flexibility required to accommodate fluctuating production schedules.

As for the future work, the contingent evolution of HRC relies on the creation of massive frameworks which integrate the values that human beings treasure while considering the application of these technologies as supplements rather than replacements for human-centered careers. With mention of uncertainties, ethical behavioral protection, and innovation, HRC is in a position to set the path for sustainable and inclusive future in industrial technology domain, which can be aligned with Industry 4.0 and recently introduced Industry 5.0.

References

1. Baratta, A., Cimino, A., Gnoni, M.G., Longo, F.: Human robot collaboration in Industry 4.0: a literature review. Proc. Comput. Sci. **217**, 1887–1895 (2023)
2. Soori, M., Dastres, R., Arezoo, B., Jough, F.K.G.: Intelligent robotic systems in Industry 4.0: a review. J. Adv. Manufact. Sci. Technol. **4**, 20240070 (2024)

3. Semeraro, F., Griffiths, A., Cangelosi, A.J.R.: Human–robot collaboration and machine learning: a systematic review of recent research. Robot. Comput. Integr. Manufact. **79**, 102432 (2023)
4. Mukherjee, D., Gupta, K., Chang, L.H., Najjaran, H.J.R.: A survey of robot learning strategies for human–robot collaboration in industrial settings. Robot. Comput. Integr. Manufact. **73**, 102231 (2022)
5. Arents, J., et al.: Human–robot collaboration trends and safety aspects: a systematic review. J. Sens. Act. Netw. **10**(3), 48 (2021)
6. Li, W., Hu, Y., Zhou, Y., Pham, D.T.: Safe human–robot collaboration for industrial settings: a survey. J. Intell. Manufact. **35**(5), 2235–2261 (2024)
7. Baratta, A., Cimino, A., Longo, F., Nicoletti, L.: Digital twin for human–robot collaboration enhancement in manufacturing systems: literature review and direction for future developments. Comput. Ind. Eng. **187**, 109–764 (2024)
8. Wang, S., et al.: A deep learning-enhanced digital twin framework for improving safety and reliability in human–robot collaborative manufacturing. Robot. Comput. Integr. Manufact. **85**, 102608 (2024)
9. Ojstersek, R., Buchmeister, B., Javernik, A.: Human–robot collaboration, sustainable manufacturing perspective. In: International Conference on Flexible Automation and Intelligent Manufacturing, pp. 71–78. Springer (2023)
10. Zheng, P., Li, S., Fan, J., Li, C., Wang, L.: A collaborative intelligence-based approach for handling human–robot collaboration uncertainties. CIRP Ann. **72**(1), 1–4 (2023)
11. Lou, S., Zhang, Y., Tan, R., Lv, C.: A human-cyber-physical system enabled sequential disassembly planning approach for a human-robot collaboration cell in Industry 5.0. Robot. Comput. Integr. Manufact. **87**, 102706 (2024)
12. Lorenzini, M., Lagomarsino, M., Fortini, L., Gholami, S., Ajoudani, A.: Ergonomic human–robot collaboration in industry: a review. Front. Robot. AI **9**, 8139–8207 (2023)
13. Proia, S., Carli, R., Cavone, G., Dotoli, M.: Control techniques for safe, ergonomic, and efficient human–robot collaboration in the digital industry: a survey. IEEE Trans. Automat. Sci. Eng. **19**(3), 1798–1819 (2021)
14. Winkle, K. et al.: Feminist human–robot interaction: disentangling power, principles and practice for better, more ethical HRI. In: Proceedings of the 2023 ACM/IEEE International Conference on Human–Robot Interaction, pp. 72–82 (2023)
15. Asaad, H., Askar, S., Kakamin, A., Nayla, F.: Exploring the impact of artificial intelligence on human–robot cooperation in the context of industry 4.0. Appl. Comput. Sci. **20**(2), 138–156 (2024)
16. Darwish, D.: Human-AI collaboration in industry 5. In: Human–Machine Collaboration and Emotional Intelligence in Industry 5.0, pp. 44–70. IGI Global (2024)
17. Jayadharshini, P., Priya, C.S.R., Lalitha, K., Santhiya, S., Keerthika, S., Abinaya, N.: Enhancing retailer auctions and analyzing the impact of coupon offers on customer engagement and sales through machine learning. In: ICCEBS. Sri Sairam Engineering College, Chennai (2023). Malathy, S., Vanitha, C.N., Rajesh Kumar, D., Lalitha, K.: Augmented reality based medical education. In: ICCEBS. Sri Sairam Engineering College, Chennai (2023)
18. Piardi, L., Leitão, P., Queiroz, J., Pontes, J.: Role of digital technologies to enhance the human integration in industrial cyber–physical systems. Annu. Rev. Control. **57**, 10093–10094 (2024)
19. Vysocky, A., Novak, P.: Human–robot collaboration in industry. MM Sci. J. **9**(2), 903–906 (2016)
20. Taesi, C., Aggogeri, F., Pellegrini, N.: COBOT applications—recent advances and challenges. Robotics **12**(3), 79 (2023)
21. Ponnambalam, S., Chang, Q., Zhong, R.Y., Kucukkoc, I., Janardhanan, M.N.: Human–robot collaboration in the next generation manufacturing and logistics system. Flex. Serv. Manufact. J **35**(4), 975–978 (2023)
22. Müller, M., Ruppert, T., Jazdi, N., Weyrich, M.: Self-improving situation awareness for human–robot-collaboration using intelligent digital twin. J. Intell. Manufact. **35**(5), 2045–2063 (2024)

23. https://www.forbes.com/sites/bernardmarr/2022/08/10/the-best-examples-of-human-and-robot-collaboration/
24. https://www.aboutamazon.com/news/operations/how-amazon-deploys-robots-in-its-operations-facilities
25. Yerebakan, M.O., Hu, B.: Human–robot collaboration in modern agriculture: a review of the current research landscape. Adv Intell Syst **6**, 2300823 (2024)
26. Pal, A., Leite, A.C., From, P.J.: A novel end-to-end vision-based architecture for agricultural human–robot collaboration in fruit picking operations. Robot. Auton. Syst. **172**, 104–567 (2024)
27. Giallanza, A., La Scalia, G., Micale, R., La Fata, C.M.: Occupational health and safety issues in human–robot collaboration: state of the art and open challenges. Saf. Sci. **169**, 106–313 (2024)
28. Vahedi-Nouri, B., Tavakkoli-Moghaddam, R., Hanzálek, Z., Dolgui, A.: Production scheduling in a reconfigurable manufacturing system benefiting from human–robot collaboration. Int. J. Prod. Res. **62**(3), 767–783 (2024)
29. Fu, Y., Chen, J., Lu, W.: Human-robot collaboration for modular construction manufacturing: review of academic research. Autom. Constr. **158**, 105–196 (2024)
30. Duan, J., Zhuang, L., Zhang, Q., Zhou, Y., Qin, J.: Multimodal perception-fusion-control and human–robot collaboration in manufacturing: a review. Int. J. Adv. Manufact. Technol. **132**(3), 1071–1093 (2024)
31. Liang, C.-J., Le, T.-H., Ham, Y., Mantha, B.R., Cheng, M.H., Lin, J.J.: Ethics of artificial intelligence and robotics in the architecture, engineering, and construction industry. Automat. Constr. **162**, 105–369 (2024)

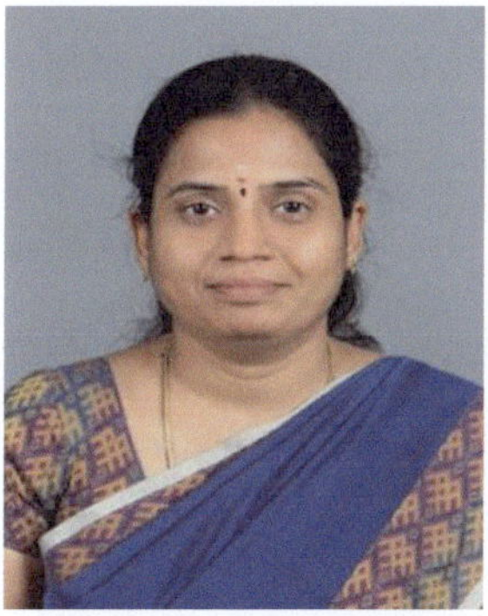

Dr. C. N. Vanitha is a Professor in the Department of Information Technology at Karpagam college of Engineering, Tamil Nadu, India. She received her Ph.D. degree in Computer Science and Engineering from Anna University, Chennai, India, in 2018, M.E. degree in Computer Science and Engineering from Anna University, India in 2008. She obtained her M.Phil., M.Sc. and B.Sc. degree in Computer Science from Bharathiar University in 2004, 2002 and 1999 respectively. She is a life member of Indian Society for Technical Education (ISTE) and International Association of Engineers (IAENG). Published more than 38 articles and papers in high impact factor SCI, SCIE, SCOPUS, WEB OF SCIENCE and ESCI Journals, 30 papers in international conferences indexed with ACM Digital Library, Springer, and IEEE Xplore. She has published 4 books, patent and contributed 11 book chapters to the Springer, Elsevier, Web of Science indexed books. Wireless Sensor Networks, Networking, Security and Machine Learning are her research interests. Received Best Faculty Award and Best Teacher Award. She is acting as a reviewer in Springer—Wireless Networks, Science Direct and editorial board member in many international conferences.

Ms. P. Anusuya is a Research Scholar (Ph.D.) in the Department of Information Technology at Karpagam college of Engineering, Tamil Nadu, India. She received her M.E. degree in Communication Systems at Kongu Engineering College, Erode, India in 2018 and B.E. degree in Electronics and Communication Engineering at Nandha College of Technology, Erode, Tamil Nadu, India in 2016. Areas of Research Interest were Wireless Sensor Networks, IoT and Wireless Communication. Published journal article in PeerJ Computer Science and MethodsX indexed as SCIE. Also, Publication in Asian Journal of Applied Science and Technology (AJAST) and International Journal of Engineering Research and Technology (IJERT) and Published conference article in IEEE Xplore.

Dr. Rajesh Kumar Dhanaraj is a distinguished Professor at Symbiosis International (Deemed University) in Pune, India. Prior to this, he served as a Professor at the School of Computing Science and Engineering at Galgotias University in Greater Noida, India. **His exceptional academic and research contributions have placed him among the top 2% of scientists globally, an honor recognized by Elsevier and Stanford University**. Dr. Dhanaraj completed his B.E. in Computer Science and Engineering from Anna University Chennai in 2007, followed by an M.Tech. from Anna University Coimbatore in 2010. He earned his Ph.D. in Computer Science from Anna University, Chennai, in 2017. His prolific career includes authoring and editing over 90 books on advanced technologies and holding 27 patents. He has published over 170 articles in esteemed journals and international conferences, including four papers in IEEE Transactions.

As a mentor, Dr. Dhanaraj has guided four Ph.D. candidates to completion, with eight more currently under his supervision. He is renowned for delivering insightful tech talks on disruptive technologies and has established valuable collaborations with professors from top QS-ranked universities globally.

Dr. Dhanaraj's research interests include Machine Learning, Cyber-Physical Systems, and Wireless Sensor Networks. His expertise in these areas has led to numerous research talks at esteemed institutions. He is a Senior Member of the IEEE, and a member of the CSTA and IAENG. Additionally, he serves as an Associate Editor and Guest Editor for several prestigious journals and is an Expert Advisory Panel Member of Texas Instruments Inc., USA.

Distributed Training of Neural Networks in Smart Manufacturing Systems

P. Jayadharshini, S. Santhiya, S. Keerthika, N. Abinaya, R. Ahalya, and V. N. Shree Nandhini

Abstract Distributed neural network training for intelligent manufacturing systems is perhaps the most significant player in the field of Industry 4.0 regarding its function and application. The biggest power users in the industry of the era, such as IoT, big data analytics, and AI, set up intelligent, automated, and capable production settings. On their part, roles in doing data analysis, predictive maintenance, quality control, and process optimization seem to play a pretty significant role in these systems. Actually, it starts with presenting a short outline of the smart manufacturing systems and the importance of neural networks. The paper heavily focuses on the fact that parallel training is the best approach toward these enormous amounts of data; however, it further points out that the real advantages of this technology are upgraded scalability and decreased training time. To begin with, some overview concepts of distributed deep learning are made, referring to differences of basic and distributed training, and also to the role of components within the distributed systems-that are computing nodes, communication protocols, and synchronization mechanisms. Taking it forward, this paper takes up the distributed neural network

P. Jayadharshini (✉) · S. Santhiya · S. Keerthika
Assistant Professor, Department of Artificial Intelligence, Kongu Engineering College, Perundurai, Tamil Nadu, India
e-mail: jayadharshini.ai@kongu.edu

S. Santhiya
e-mail: santhiya.cse@kongu.edu

S. Keerthika
e-mail: keerthika.ai@kongu.edu

N. Abinaya
Assistant Professor, Department of CSE, Hindustan Institute of Technology, Coimbatore, India
e-mail: abinaya.ai@kongu.edu

R. Ahalya · V. N. Shree Nandhini
Student, Department of Artificial Intelligence, Kongu Engineering College, Perundurai, Tamil Nadu, India
e-mail: ahalyar.21aim@kongu.edu

V. N. Shree Nandhini
e-mail: shreenandhinivn.21aim@kongu.edu

L. Krishnasamy et al. (eds.), *Distributed Deep Learning and Explainable AI (XAI) in Industry 4.0*, Information Systems Engineering and Management 55,
https://doi.org/10.1007/978-3-031-94637-0_9

training technology that uses knowledge sharing such as data parallelism, model parallelism, and hybrid parallelism along with case studies on different projects. The building blocks of distributed training like cloud computing, fog computing, edge computing, and networking are looked at. In this section, implementation strategies are discussed in association with a list of popular frameworks and tools describing the process involved in setting up a distributed training environment. Challenges of distributed training including data management, synchronization, communication overhead, scalability, and fault tolerance are explained. The final section of the course shows how performance might be optimized, including efficient data sharding, gradient compression, asynchronous training, and utilization of specialized hardware.

Keywords Industry 4.0 · Distributed deep learning (DDL) · Real-time monitoring · Predictive maintenance · Internet of things (IoT) · Neural networks

1 Introduction

Actually, smart manufacturing is innovative in its fusion of the most high-tech information and manufacturing technologies in very intelligent, automated, and adaptable production environments. The ideas of Internet of Things, big data analytics, and artificial intelligence are now concentrated on integration with the latest technologies that work in conjunction with each other to produce significant enhancements in productivity, quality, and flexibility of manufacturing processes [1, 2]. Among other subgroups of AI, the one on neural networks will feature prominently in the smart manufacturing domain. These algorithms are well-developed and very efficient for data patterns. They will enable one to find correct predictions or facilitate sophisticated decision-making processes for complex situations [3]. Applications of such algorithms encompass different domains of a production process, especially quality control, predictive maintenance, and process optimization.

The quality control technique can also use artificial neural networks when detecting defects in a product or faulty materials. Such networks would, through sensors, cameras, and other means, pick abnormalities that do not match the predefined standards very quickly. Hence, interference only the best quality products would reach the market. Besides this, neural networks are used adequately in predicting maintenance. Such networks, by continuous monitoring and data analysis of machines and equipment, can predict problems or failures that may be possible even before their development stage [4]. This advanced pre-repair machinery not only minimizes the downtime but, at the same time, assists in stretching equipment lifespan thereby avert any type of financial losses from these companies in the near future. Key functionality that artificial neural networks have been imported into the development of smart manufacturing process includes their ability to identify various efficiencies, bottlenecks, and profitable areas which one could capitalize on. These networks identified through data collected at every stage of the production cycle and implemented this information about recommendations brought these to life, and so these operations resulted in increased productivity, reduced waste, and improved efficiency while operating.

Training neural networks for applications related to smart manufacturing is also quite a challenge, particularly due to big data values. At this point, distributed training has been harnessed. Distributed training is a process where the workload of training is split up on multiple nodes for computing for it to be done in parallel and fast. In fact, distributed training offers more scalability because it facilitates feeding really large datasets and complex models into real-time decision-making, making smart manufacturing more feasible. In a nutshell, the sheer volume and complexity of data that these networks need to process is why distributed training in neural networks for smart manufacturing is urgently required. Distributed training can reduce training time and provide real-time insight in decision-making. It is because of this reason that this is also a big part of modern smart manufacturing systems.

2 Foundations of Distributed Deep Learning

Distributed Deep Learning is a methodology through which deep-learning models are trained on multiple computing nodes that may or may not be geographically dispersed. The methodology is quite different from the traditional approach called centralized training where all the computations are carried out on a single machine [5]. It's easy enough to implement centralized training that only depends on the computing power and memory available on a single machine. The problem arises when scalability becomes a bottleneck, and the time for training expands if one is dealing with extensive data or complex models [6]. Distributed training takes advantage of the combined computing power of multiple machines to parallelize calculations. This means the training process becomes significantly faster in an opportunity to safely handle more comprehensive models and more extensive data [7]. Additionally, the distributed training provides for a scalable mode of training since one can add as many additional computation nodes as there are needs arising and the approach thus allows effortless scaling to address the growing complexity of related problems [8].

The important elements of a distributed deep learning system are:

1. **Computational Nodes**: These nodes are the machines with processor, memory, and storage. It is engaged in distributed training. These nodes work together to calculate the deep learning models.
2. **Communication Protocols**: Communication protocols are fundamental for data and information exchange between computing nodes. The commonly used communication protocols in distributed deep learning are Message Passing Interface (MPI) and gRPC, Google Remote Procedure Call.
3. **Synchronization Mechanisms**: Synchronization mechanisms are techniques used to coordinate and align the training process between different computational

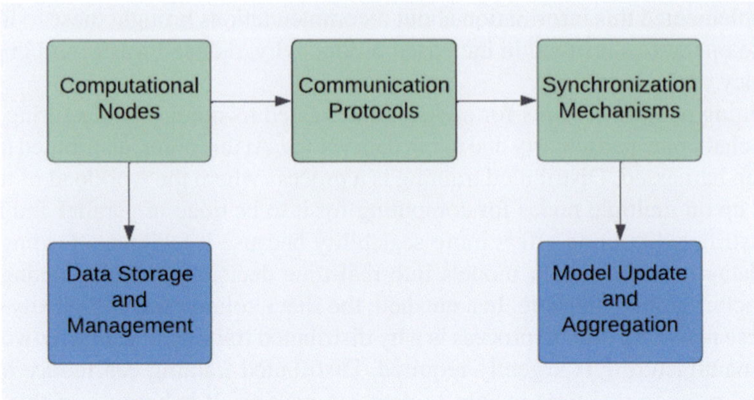

Fig. 1 Distributed deep learning systems

nodes. This can be done by including parameter servers and ring-allreduce algorithms for model updates and aggregations in order to provide highest efficiency and accuracy across distributed nodes.

The flow diagram below illustrates the components and flow of distributed deep learning systems:

Figure 1 shows the,

- Computational nodes are the machines used in distributed training.
- Communication Protocols enable data transfer among nodes.
- Synchronization Mechanisms Coordinate the training processes in the nodes.
- Data Storage and Management: stores data and retrieves data during training.
- Model Update and Aggregation manage the updating and aggregation of model parameters across nodes.

Organizations can utilize distributed deep learning systems with definite components and effective communications and synchronization to accomplish the mission of trained deep learning models and work with large-scale datasets and complex models in a very efficient way through distributed computing.

3 Architectures for Distributed Neural Network Training

3.1 Data Parallelism

In data parallelism, it breaks down the training dataset into smaller batches, and the nodes computers do them at the same time. In this case, forward and backward passes are conducted by nodes individually, hence producing the gradients of their batches, which are later weighed jointly to update the model's parameters [9]. This

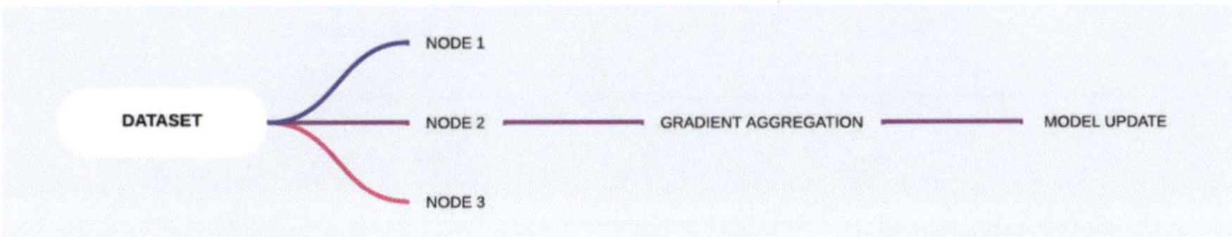

Fig. 2 Data parallelism

method is effective for simple tasks like the division of the dataset when doing image classification or NLP tasks (Fig. 2).

3.2 Model Parallelism

Model parallelism implies splitting the neural network model itself across different nodes. Each node must process its own part during both forwards and backward passes while still part of the larger model [10]. For example, if the model is so big that it can't fit into the memory of one node, this way gives the possibility to switch the parts of the model and do parallel processing of the model whole thereby making it possible to train even the most complicated architectures (Figs. 3 and 4).

Fig. 3 Model parallelism

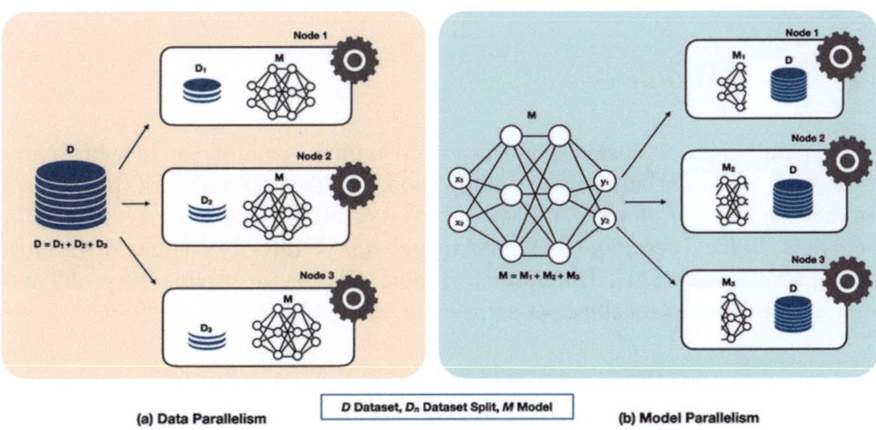

Fig. 4 Data parallelism versus model parallelism

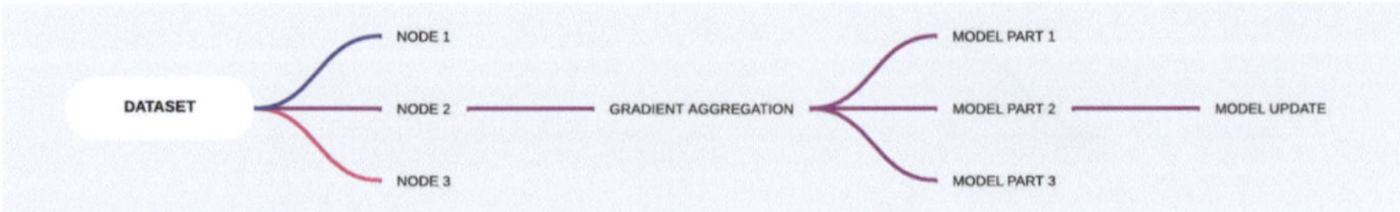

Fig. 5 Hybrid parallelism

Table 1 Comparison of types of parallelism

Aspect	Data parallelism	Model parallelism	Hybrid parallelism
Advantages	Scalability for large datasets	Memory efficiency for very large models	Balances workload and optimizes memory and computational efficiency
	Faster training convergence	Granular control over model segments	
	Simple implementation		
Challenges	Communication overhead due to frequent synchronization	Complexity in coordination across model segments	Implementation complexity due to combining data and model aspects
	Memory constraints on individual nodes	Scalability limits with very large models	Communication overhead during aggregation and synchronization
Suitability	Large datasets that can be partitioned easily	It cannot fit into single machine	Tasks requiring a balance between large datasets and complex models
	Distributed computing environments with sufficient resources	Specific parts of the model requiring intensive computation	Distributed environments with a need for optimal workload balance

3.3 Hybrid Parallelism

Hybrid parallelism includes single-data as well as model parallelism. In hybrid parallelism, training data is being divided among nodes for the purpose of data parallelism, and then each node follows a strategy of model parallelism with respect to the allocated data. In that scenario, this hybrid model will be directly compared with load sharing across nodes [11]. Therefore, it is more suitable for solving the problem of large sets or complex machines at a time (Fig. 5; Table 1).

3.4 Case Studies of Distributed Training Architectures

1. **Google's TensorFlow with Parameter Server Architecture**:

Google's TensorFlow framework follows data parallelism distributed training using a parameter server architecture. In such a setup, there are many computational nodes-worker nodes that compute on multiple batches of data in parallel, while there is a separate set of parameter servers that store and manage the model parameters. These parameter servers coordinate updates from computational nodes and distribute updated parameters back to the nodes. This architecture is very common for distributed training for large-scale deep learning projects.

2. **Facebook's PyTorch with Horovod**:

   Facebook's PyTorch framework uses Horovod, a distributed training framework, to aggregate gradients efficiently in data parallelism. Horovod uses the ring-allreduce algorithm for communication, which aggregates gradients optimally on different nodes during distributed training. This reduces communication overhead and further increases the scalability of deep learning tasks spread over distributed computers. Horovod is more useful in training deep models on large clusters of machines.

These case studies highlight the diverse approaches and tools available for implementing distributed training in deep learning. Each architecture and framework has its strengths and suitability for different types of tasks, datasets, and computational infrastructures. By understanding these concepts and leveraging appropriate technologies, organizations can effectively scale their deep learning training processes and achieve improved model performance.

4 Infrastructure for Distributed Training in Smart Manufacturing

4.1 Cloud Computing

1. Cloud Computing platforms such as AWS, Azure, and Google Cloud provide the facility of scalable resources for distributed training. Scalable and elastic resources on cloud computing platforms support pre-configured environments, virtual machines, and services like Kubernetes to handle distributed training workflows.
2. Centralized Data Storage: Cloud environments offer centralized data storage solutions in such as Amazon S3 or Azure Blob Storage [12], for example. Storage services make effective management and access possible for the distributed training processes (Fig. 6).

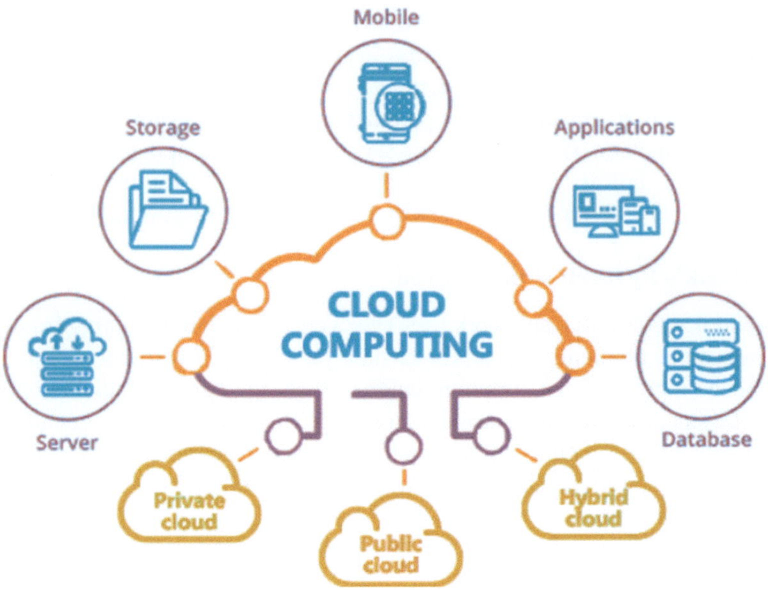

Fig. 6 Cloud computing

4.2 Edge Computing

1. Low Latency: A primary guiding principle of edge computing is to bring the computations closer to the source of data sources and from there, to minimize latency and thus the bandwidth usage. In smart manufacturing some edge tools such as sensors, actuators, and IoT devices work in real time performing analytics and computation on the edge of the network [13].
2. Real-time Processing: In the main, one of the edges of edge computing in smart manufacturing is a very time-consuming processing, in other words, immediate decision-making using sensor data for the optimization of production processes, anomaly detection, and efficient operativity of a factory (Fig. 7).

4.3 Fog Computing

1. In fog computing, it introduces an intermediary layer that can potentially carry out computations closer to its source through the expansion of the cloud computing network to the edge of the network but also uses the computation power of the cloud.
2. Edge-Cloud Integration: Fog computing integrates edge devices with cloud services, allowing seamless data flow, computation offloading, and resource management across the entire continuum of the edge-cloud model [14]. It allows

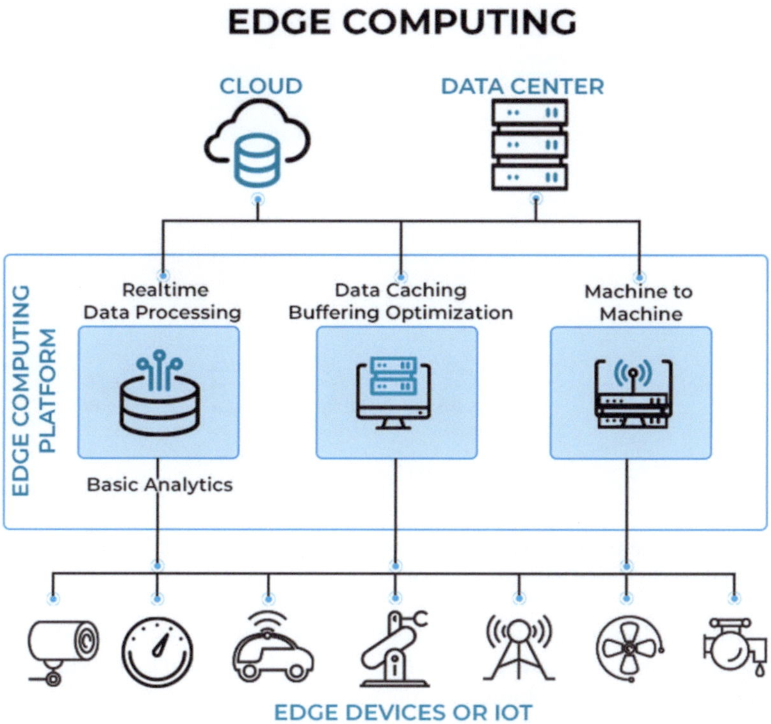

Fig. 7 Edge computing

edge devices to leverage cloud resources for computationally intensive analytics and machine learning operations within low latency real-time processing (Fig. 8).

4.4 Networking Requirements

1. High-Speed Data Transfer: An efficient high-speed data transfer is needed among nodes in distributed training to support strong networking infrastructure. High-bandwidth networks are Ethernet or wireless networks providing a high speed of data exchange and synchronization during the process of distributed training.
2. Low Latency Communication: Real-time processing of data and thus, decisions in smart manufacturing depend on low-latency communication. Major technologies responsible for fundamentally ensuring minimal latency of data transmission across edge devices, fog nodes, and cloud resources include 5G, edge computing, and optimized network protocols.
3. Reliable Connectivity Reliable connectivity is imperative to ensure consistent flow of data and also communication between distributed nodes. Redundant network connections, failover mechanisms and quality-of-service policies are

Fig. 8 Fog computing

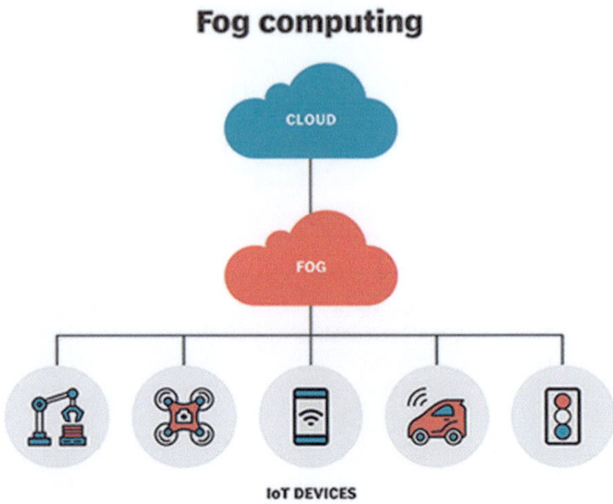

some of the means of ensuring reliable connectivity for the distributed training applications in smart manufacturing.

In Summary, the hybrid infrastructure of distributed training in smart manufacturing will combine scalable resources in cloud computing, low-latency real-time processing in edge computing, integration of edge and cloud computing in fog computing, and a robust networking infrastructure for the support of high-speed data transfer and communication. All these technologies will surely work together to enable efficient workflows in distributed training, real-time analytics, and decision-making in smart manufacturing environments.

5 Implementations of Distributed Training in Smart Manufacturing

5.1 Frameworks and Tools

5.1.1 TensorFlow

- TensorFlow provides several inbuilt strategies and architectures that support distributed training.
- There are several distributed training strategies offered by TensorFlow, including 'MirroredStrategy', useful for synchronous training on multiple GPUs; 'MultiWorkerMirroredStrategy', which works on synchronous training across a number of machines; and 'TPUStrategy', which provides a way to train on TPU pods.

- Parameter Server Architecture: TensorFlow tf.distribute.experimental. Parameter Server Strategy; scaling training across many nodes by leaving the management of the model parameters to parameter servers.

5.1.2 PyTorch

- PyTorch is a dynamic computing platform for machine learning, allowing distributed training; it is the most commonly used tool for research and development due to easy use.
- Horovod Integration: Horovod is the library, developed by Uber for doing distributed training with PyTorch, and it aggregates gradients efficiently across many nodes by employing ring-allreduce.
- DistributedDataParallel (DDP): PyTorch now has a module 'torch.nn.parallel.DistributedDataParallel', using which data parallelism can be implemented easily. It addresses the distribution of model replicas and gradient synchronization.

5.2 Setting up a Distributed Training Environment

5.2.1 Cluster Configuration

- Defining nodes: Identify and configure the nodes which are to be involved with the distributed training process [15]. Each node could be a GPU or TPU-equipped machine in a local cluster or a cloud-based instance.
- Roles and Responsibilities: Node roles may be defined to include worker nodes, which perform the computations, and parameter servers, that store the model parameters. Nodes should be configured to communicate properly with each other.

5.2.2 Data Distribution

- Sharding Data: Efficiently partition the dataset into smaller chunks (shards) and distribute them across the worker nodes.
- Balancing load: The data should be well-balanced, so that no node contains more data than the others as it would be a bottleneck.

5.2.3 Model Synchronization

- Gradient Aggregation Implements ways to accumulate gradients from all nodes during training [16]. For instance, TensorFlow can use parameter servers, while PyTorch leverages Horovod's ring-allreduce.

- Synchronous versus Asynchronous Training: In synchronous training, all the nodes get synchronized after each step, whereas, in asynchronous training, all the nodes operate independently and are synchronized at periods of time. The former one is easier to implement but slower, while the latter one is more complex but faster.

5.3 *Example Use Cases and Applications*

5.3.1 Predictive Maintenance

- Objective: Distributed training of sensor data from equipment in manufacturing to predict probable failures.
- Implementation: Collect data from various sensors, shard the data across multiple nodes, and train a neural network to recognize patterns indicative of potential failures.
- Distributed training ensures that large datasets from multiple machines are processed efficiently [17].
- Outcome: Since premature failure can be forecasted, downtime will occur much less, and maintenance can be scheduled ahead, improving overall efficiency and cutting costs.

5.3.2 Quality Control

- Quality Control: Objective: To implement distributed deep learning to detect defects in products in real time on the manufacturing line.
- Implementation: High Definition cameras and sensors to record the information of the products traveling on the assembly line.
- Shard that information and distribute it on nodes for the training of a CNN in order to identify defects.
- Outcome: Such defect detection in real time will immediately correct defects and limit defective products with high-quality products.

Flow Diagram: Setting up a Distributed Training Environment

Figure 9 explains the establishment of a distributed training environment follows through the following phases:

1. Cluster Configuration:
    - Identify and set up nodes.
    - Assign roles to nodes (e.g., workers, parameter servers).

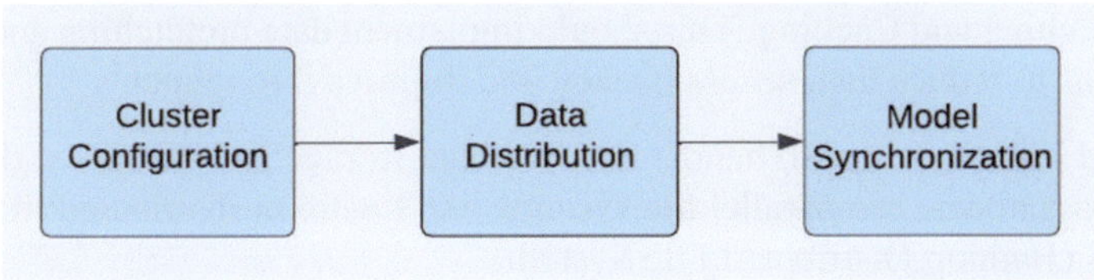

Fig. 9 Distributed training environment

2. Data Distribution:
   - Shard the dataset into smaller workable parts.
   - Distribute data shards across nodes.
3. Model Synchronization:
   - Implement gradient aggregation methods.
   - Decide on either synchronous or asynchronous training.

Each of these steps ensures a properly configured distributed training environment leading to scalable and faster training procedures application in smart manufacturing applications. Such configuration is primarily relevant for big datasets and intricate models, ensuring quick training of the neural networks for quality control, predictive maintenance, and other very important functions within the smart manufacturing systems.

6 Challenges and Solutions in Distributed Training

6.1 Data Management and Synchronization

Challenges:

Data Sharding and Distribution: Dividing and dispersing data across multiple nodes to ensure that there is a balanced workload.

Consistency and Accuracy-All nodes should have access to the last update of data and models to ensure consistency.

Data Storage: Managing the storage of large datasets across different nodes, especially when dealing with heterogeneous storage systems.

Solutions:

Automated Data Sharding: Use objects like TensorFlow's or PyTorch.DataLoader to automatically shard and spread the data evenly across nodes.

Data Prefetching and Caching: This should implement data prefetching and caching mechanisms to reduce data access latency and improve throughput.

Distributed File Systems: To handle massive data storage and guarantee data availability across nodes, use parallel file systems like Lustre or distributed file systems like HDFS (Hadoop Distributed File System).

6.2 Communication Overhead

Challenges:

Bandwidth Limitations: Network bandwidth is easily saturated, since nodes must very frequently exchange data with each other.

Latency: High latency in communication causes a significant degradation of the training speed, especially when using synchronous training and nodes are waiting for updates from others [18].

Message Passing Efficiency: Poor protocols for message passing will result in bottlenecks during data transfer.

Solutions:

Efficient Communication Protocols: Utilize efficient communication protocols such as gRPC, MPI (Message Passing Interface), or NCCL (NVIDIA Collective Communication Library) to reduce overhead.

Gradient Compression: Techniques such as quantization and sparsification can be utilized to reduce communication load.

Asynchronous Training: Allow nodes to work in isolation most of the time and synchronize less frequently-reduces the overhead of communication, but possibly improves the training.

6.3 Scalability Issues

Challenges:

Rising Data and Model Size: The computational and memory requirements increase with the volume of data and model complexity, which makes the scalable size of the training system harder.

Load Balancing: Ensure that all nodes make optimal use of resources and no node becomes a bottleneck because of uneven workloads.

Resource Allocation: Proper allocation of resources (CPU, GPU, memory) of the different nodes to tackle increased loads.

Solutions:

Horizontal scaling: Add more nodes to the cluster to send out computational load and handle larger data and models size [19].

Dynamic Resource Allocations: Orchestrations solutions, including Kubernetes, ensure that all of their nodes are used efficiently and make dynamic resource allocations based on the workload demand requirements.

Advanced Techniques for Parallelism: use hybrid parallelism to balance the computational load in efficiently exploiting the resources at hand.

6.4 Fault Tolerance and Recovery Mechanisms

Challenges:

Node Failures: In a distributed environment, node failures are inevitable, and they can disrupt the training process.

Checkpointing: Guaranteeing that the training process is checkpointed regularly so it can easily resume in case of failures from the last checkpoint made.

Error Detection and Correction: Quickly detecting errors and implementing mechanisms to correct them without significant downtime.

Solutions:

Regular Checkpointing: Regularly checkpoint the model parameters as well as the training state so that recovery could be done from the last saved state upon failure.

Redundancy and Replication: The data and model parameters should redundantly be replicated across multiple nodes to prevent complete node failure.

The fault detection system should design a robust fault detection system that rapidly identifies and isolates the faulty nodes, with a rerouting of tasks to healthy nodes to maintain uninterrupted training.

Self-healing systems: deploy self-healing mechanisms that will automatically restart failed nodes or containers and jump back to training from the last check-point [20]. The high-performance, scalable, and reliable design of distributed training systems in smart manufacturing can be achieved by means of developing effective solutions to these challenges so that neural networks are efficiently trained to meet the needs of modern manufacturing (Fig. 10).

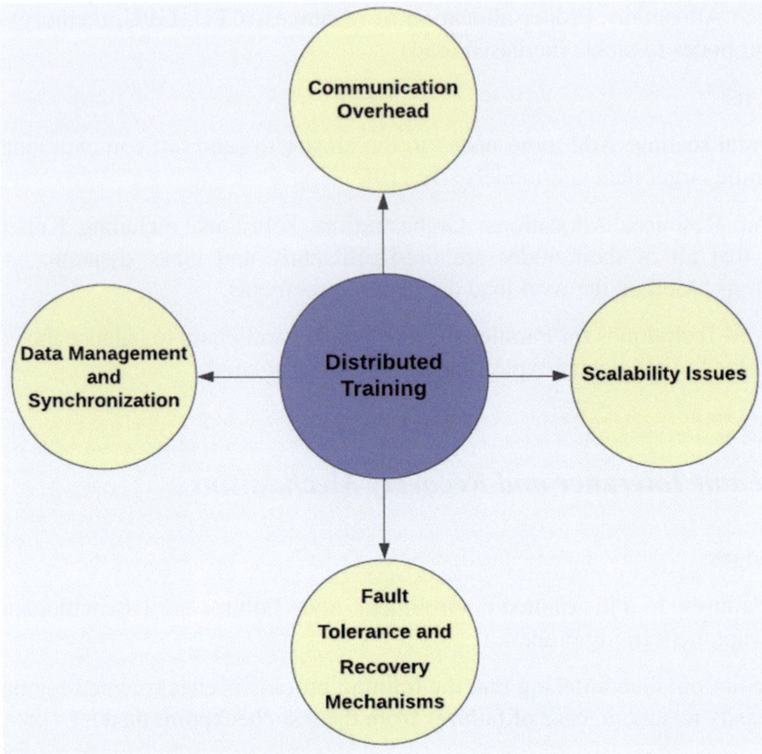

Fig. 10 Challenges in distributed training

7 Performance Optimization Techniques

7.1 Efficient Data Sharding and Distribution

Challenges:

- Uneven Workload Distribution: Inefficient data sharding can lead to some nodes having more data than others, causing an imbalance in the computational load.
- Data Locality: Ensuring that data is stored and accessed efficiently to minimize data transfer time between nodes.

Solutions:

- Automatic Data Sharding: It is done with frameworks like TensorFlow and PyTorch. The tools split the dataset into smaller pieces, all of equal size, to be distributed across the nodes.
- Data Locality Optimization Data stored in a way such that each node is processing data locally available on it; minimizing the data transfer time Data stored in

distributed file systems such as HDFS Ensures that data is stored in such a way that maximizes locality.
- Load Balancing: Implement load balancing algorithms that monitor and dynamically adapt the distribution of data to ensure workload balance is well taken between nodes.

7.2 Gradient Compression and Aggregation

Challenges:

- Communication Overhead: Frequent exchange of gradient updates between nodes can cause high communication overhead, especially with large models.
- Bandwidth Consumption: Processing large quantities of gradient data may consume too much bandwidth, hence delaying training [21].

Solutions:

- Gradient Quantization: Reduce the size of gradient data by quantizing gradients, which involves representing them with fewer bits. Techniques like 8-bit quantization can significantly reduce the data volume without drastically affecting model accuracy.
- Gradient Sparsification: Only send the largest gradients by zeroing out small gradients (sparsification). This decreases communication load because fewer gradients are exchanged between nodes.
- Aggregation Techniques Efficient aggregation techniques involve the use of ring-allreduce, which minimizes overhead communication by cutting down the number of exchanges needed for gradient aggregation across nodes. Lib Horovod and NCCL are examples of libraries that implement efficient allreduce operations.

7.3 Asynchronous Training Methods

Challenges:

- Synchronization Delays: Synchronous training makes nodes have to wait for one another to complete processing, thus causing delays whenever some nodes are slower.
- Staleness: Gradients in asynchronous training might become stale due to delays in updates, potentially affecting model convergence.

Solutions:

- Asynchronous SGD: Implement asynchronous stochastic gradient descent (SGD) where nodes independently compute gradients and update the model without waiting for synchronization. This would increase the speed of training since nodes do not idle while waiting for others.

- Elastic Averaging SGD (EASGD): A technique in which nodes perform local updates and periodically synchronize with a central parameter server [22]; this way, local computation and global synchronization balance each other.
- Delay Compensation Mechanisms may be devised to offset stale gradients. Then delayed updates will have meaningful contributions to model training. Techniques like gradient clipping and momentum can reduce the impact of staleness.

7.4 Use of Specialized Hardware

Challenges:

- Hardware Utilization: One should ensure the proper utilization of specialized hardware to get full acceleration using the GPUs and TPUs.
- Compatibility: Guaranteeing that the distributed training frameworks effectively utilize the capability of specialized hardware.

Solutions:

- Graphics Processing Units (GPUs): Leverage the high parallel processing capabilities of GPUs to significantly speed up common matrix computations. Frameworks provides support for GPUs enabling the efficient use of these processors in computations.
- TPUs (Tensor Processing Units): Make use of the TPUs that are specialized in deep learning to accelerate even further. TPUs provide high throughput for tensor operations and are very beneficial for large-scale model training.
- Hybrid Hardware Environments: Implement a hybrid environment, combining some elements of CPUs, GPUs, and TPUs so each can specialize in tasks at which it excels best [23]. Employ orchestration tools to manage the balancing of workload distribution across different hardware. Optimized Libraries Using optimized libraries for specific hardware, such as cuDNN and NCCL for GPUs for optimized deep learning primitives and efficient communication operations, respectively. TensorFlow compiler XLA will optimize computations for TPUs.

8 Case Study: Distributed Training for Predictive Maintenance

8.1 *Problem Definition and Requirements*

8.1.1 Predictive Maintenance in Smart Manufacturing

Predictive maintenance uses analytical tools and techniques in data analytics to predict when certain failures will be expected, hence capable of doing just-in-time maintenance to prevent unscheduled downtime [24]. Predictive maintenance promotes equipment reliability, decreases maintenance costs, and helps optimize production schedules within smart manufacturing.

8.1.2 Requirements for Distributed Training

Exponentially large volumes of sensor data: The wide spread of sensors and monitoring the performance, health, and environmental factors of the machinery in smart manufacturing environments generate a large volume of data.

Real-time Processing: Data needs to be processed and analyzed in real time to produce forecasts and alarms within the needed time.

Scalability: The solution needs to scale to accommodate larger volumes of data and more equipment as the manufacturing environment scales.

Accuracy: High predictive accuracy is crucial to ensure reliable maintenance scheduling and to avoid false positives/negatives.

8.2 *System Architecture and Implementation*

Data Collection: The monitoring equipment provided for manufacturing equipment has sensors installed on it to continuously collect data regarding a number of factors including temperature, vibration, pressure, and operational cycles.

Data Preprocessing: Filter out noise, normalization, feature extraction so that the raw sensor data is prepared for model training [25].

The preprocessed data is fed into a distributed training environment, wherein multiple neural networks, spread across the nodes, are trained to predict equipment failures.

A deployed predictive model monitors equipment in real time in order to predict faults and suggest maintenance schedules.

Implementation Steps:

Cluster Configuration: Configure a distributed computing cluster with nodes that are set up to perform certain tasks, such as data preprocessing, model training, and synchronization.

Data Sharding and Distribution: Efficiently shard the preprocessed sensor data across the nodes to ensure balanced workloads and minimize data transfer times.

Model Training Distributed learning with framework pytorch or TensorFlow should be used. Data parallelism, model parallelism, or their hybrid should be used as needed to get maximum training efficiency.

Model Synchronization: Using synchronizing mechanisms such as parameter servers or allreduce algorithms that help aggregate and update model parameters across nodes.

Deployment and Monitoring: They will deploy the trained models in the production environment to go on and monitor sensor data continuously; generate fault predictions, and consequently schedule maintenance activities accordingly.

8.3 Results and Performance Analysis

8.3.1 Evaluation Metrics

Accuracy of Prediction: Use accuracy metrics such as precision, recall, F1-score, and ROC-AUC to measure how accurate the predictions of fault are.

Training Time: Compare the training times of distributed versus centralized training setups to evaluate the efficiency gains.

Scalability: The ability of the system to scale when more equipment comes online as well as larger volumes of data [26].

Operational Impact: Determine the reduction in unplanned downtime, maintenance cost, and improvement on production efficiency.

8.3.2 Performance Improvements

Reduces Training Time: Distributed training vastly reduces the time necessary to train large-scale prediction models by parallelizing calculations across numerous nodes.

Better Prediction Capability: Advanced neural networks, with vast database training, make faults more accurately predicted, thus reducing false alarms and missed detection.

Scalability: The distributed approach enables the system to scale well with additional data sources and equipment while maintaining high performance.

8.4 Lessons Learned and Best Practices

8.4.1 Key Takeaways

Data Quality: Ensuring high-quality, consistent data from sensors is crucial for accurate model training and predictions.

Communication Efficiency: Reduce communication overhead among nodes. Gradient compression coupled with an effective synchronization mechanism should help improve performance in distributed training.

Hardware Utilization: This model's performance can be enhanced along with a significant decrease in training times by using specialized hardware like GPUs and TPUs.

Fault Tolerance: Implementing robust fault tolerance and recovery mechanisms is vital to maintain system reliability and resilience in case of node failures or network issues.

8.4.2 Best Practices

Incremental Model Updates: Instead of retraining the model from scratch, use incremental updates to predictive models to fetch the new data, which reduces the training time and computational load. Adaptive Learning Rates: Use adaptive learning rates to ensure stable and efficient convergence during distributed training. Regular Monitoring: Monitor the functioning of all the deployed models as well as the supportive infrastructure all the time and identify any problem emerging rapidly and resolve it. Combine the predictive maintenance models with collaborative filtering techniques and utilize knowledge from similar equipment to improve the predictive accuracy.

With these strategies and best practices in place, smart manufacturing systems can make full use of distributed training for predictive maintenance to enhance the reliability of equipment, better schedule maintenance routines, and otherwise optimize production operations.

9 Future Trends and Research Directions

9.1 Advances in Distributed Training Algorithms

9.1.1 Exploring New Algorithms

Distributed training is one field that is advancing fast with continuous ongoing development in algorithms meant to enhance efficiency and scalability [27]. Among the notable trends include:

- Federated Learning: This approach can train models on decentralized sources of data without transferring the data to any centralized server. This preserves data privacy, reduces bandwidth usage, and is highly relevant to smart manufacturing environments equipped with distributed sensor networks.
- Gradient Compression: New algorithms are in the works to compress gradients from model training so the communication overhead across nodes is greatly reduced. Techniques include sparsification, quantization, and encoding.
- Decentralized Training: Decentralized training algorithms ensure that the model parameters and training processes are more uniformly distributed on all nodes than traditional centralized parameter servers, having reduced bottlenecking and enhancing fault tolerance.
- Adaptive Synchronization: Making the synchronization frequency change dynamically during the training according to the training progress can help improve efficiency. This can be done by reducing synchronization during the initial phase of training when gradients are large and increasing it during later stages.

9.1.2 Impact on Smart Manufacturing

These advances in distributed training algorithms are likely to significantly improve smart manufacturing capabilities to process huge volumes of data, reduce training times, and improve model accuracy, thereby speeding up manufacturing by making processes more responsive and adaptive.

9.2 Other Industry 4.0 Technologies

9.2.1 Internet of Things (IoT)

The integration of distributed deep learning with IoT enables real-time collection and analysis of data from a vast array of connected devices. Synergy is for.

- Real-time Monitoring and Control: Distributed deep learning models monitor and analyze data from IoT devices that continuously observe industrial processes in real time to identify anomalies and optimize operations.
- Predictive Analytics: IoT sensors generate data that feed predictive maintenance models, enabling timely maintenance actions and reducing downtime.

9.2.2 Artificial Intelligence (AI)

Adding a distributed deep learning approach to other AI technologies enriches decision-making and automation capabilities:

- Enhanced Robotics: AI-powered robots equipped with distributed learning capabilities can adapt to changing conditions on the factory floor, improving efficiency and flexibility [28].
- Smart Supply Chain: Using distributed learning system data, the AI algorithm manages inventory, predicts demand, and reduces associated costs with logistics to optimize the supply chain.

9.2.3 Big Data Analytics

Distributed deep learning with big data analytics is an integration that provides powerful tools for:

- Data-Driven Decision Making: examining enormous volumes of production data in order to find patterns, gain insights, and make wise decisions.
- Process Optimization: Continuously optimizing manufacturing processes through advanced data analysis and model predictions.

9.3 Potential Impact on Manufacturing Efficiency and Productivity

9.3.1 Efficiency Improvements

- Reduced Downtime: Predictive maintenance using distributed learning is less likely to cause unforeseen equipment failures, ensuring smoother operations and higher equipment availability.
- Continuous monitoring, combined with real-time adjustments, ensures the optimization of manufacturing processes, a reduced level of waste and improvements in product quality.
- Faster Training and Deployment: Distributed training will speed the model's training and deployment. This acceleration can help to adapt to new manufacturing issues and changes in production demands quickly.

9.3.2 Productivity Gains

- Scalable Solutions. Distributed learning systems could be scaled with complexity in data or manufacturing [29] hence supporting smart manufacturing growth initiatives.
- Enhanced Decision Support: More accurate and faster predictive models will better help managers make proactive and informed decisions.
- Cost Savings: Optimized resource usage, reduced downtime, and optimal process efficiency result in considerable cost savings in manufacturing operations.

9.4 Open Research Challenges

9.4.1 Privacy and Security of Data

There are significant challenges in ensuring the privacy and security of data in training environments. Federated learning or secure multiparty computation are techniques being researched, but probably what is needed is something significantly more robust to be able to protect sensitive manufacturing data while still allowing efficient training.

9.4.2 Scalability and Efficiency

Manufacturing environments are getting more complex, and challenges in creating scalable and effective distributed training algorithms persist as the size of big datasets and complicated models continues to grow without sacrificing performance.

9.4.3 Fault Tolerance and Reliability

Designing fault-tolerant distributed training systems that can recover quickly from node failures and maintain high reliability is critical. Research is needed to develop advanced recovery mechanisms and redundancy strategies.

9.4.4 Energy Efficiency

Distributed training can be resource-intensive. Developing energy-efficient algorithms and hardware accelerators to minimize the environmental impact [30] and operational costs is an important area of research.

9.4.5 Interoperability

This requires the interoperability of different frameworks into distributed training, IoT devices, and manufacturing systems to ensure seamless integration and operation. Interoperability standards and protocols must be developed.

9.4.6 Real-Time Processing

A distributed training system should have real-time processing capabilities to facilitate instant decision making with responsiveness toward changing conditions in a manufacturing environment.

Through tackling these research issues and exploiting improvements in Industry 4.0 technologies, smart manufacturing systems can reach unprecedented heights in terms of productivity, efficiency, and creativity.

10 Conclusion

This chapter further continued to cover the distributed neural network training in the context of smart manufacturing. One of the major highlights was that intelligent and flexible production environments were to be created through the integration of cutting-edge technologies in the form of AI, big data analytics, and IoT. Because managing enormous amounts of data is always involved in training neural networks, the necessity of distributed training in terms of shortening training time and enhancing scalability is observed. Neural networks are useful for quality control, predictive maintenance, and process optimization. We described distributed deep learning, discussed its essential elements, and contrasted it with more conventional techniques. Real-world case studies and discussions of several architectures, including data parallelism, model parallelism, and hybrid parallelism, were included. An analysis of the required infrastructure—which included cloud, edge, and fog computing—showed how crucial strong networking is.

Challenges like data management, communication overhead, scalability, and fault tolerance were identified with solution provided. Performance optimization techniques included efficient data sharding, gradient compression, asynchronous training, and use of specialized hardware. A case study on predictive maintenance was provided as an illustration of practical applications and the benefits of this system, including significant performance improvements due to the distributed training process. Future Trends: Algorithm advancements, integration with other Industry 4.0 technologies, impacts on manufacturing efficiency and productivity. Conclusion This chapter emphasizes the importance of distributed training in developing smart manufacturing systems, including smart data processing, real-time decision-making capabilities, and the important aspect of scalability that handles increased data and model complexity. Second, integration with other technologies improves efficiency, quality, and productivity. Future developments into distributed training methods and the connection of these methods with new technologies will certainly introduce even more intelligent and autonomous manufacturing environments. Overcoming some of the key obstacles in this respect-due to data privacy, fault-tolerant operation, and energy efficiency-will most surely emerge during the innovation process.

References

1. Lee, J., Bagheri, B., Kao, H.-A.: A cyber-physical systems architecture for Industry 4.0-based manufacturing systems. Manufact. Lett. **3**, 18–23 (2015)
2. Zhong, R.Y., Xu, X., Klotz, E., Newman, S.T.: Intelligent manufacturing in the context of Industry 4.0: a review. Engineering **3**(5), 616–630 (2017)
3. Schmidhuber, J.: Deep learning in neural networks: an overview. Neural Netw. **61**, 85–117 (2015)
4. LeCun, Y., Bengio, Y., Hinton, G.: Deep learning. Nature **521**(7553), 436–444 (2015)
5. Abadi, M., Barham, P., Chen, J., Chen, Z., Davis, A., Dean, J., Devin, M., Ghemawat, S., Irving, G., Isard, M., Kudlur, M., Levenberg, J., Monga, R., Moore, S., Murray, D.G., Steiner, B., Tucker, P., Vasudevan, V., Warden, P., Wicke, M., Yu, Y., Zheng, X.: TensorFlow: a system for large-scale machine learning. OSDI **16**, 265–283 (2016)
6. Paszke, A., Gross, S., Chintala, S., Chanan, G., Yang, E., DeVito, Z., Lin, Z., Desmaison, A., Antiga, L., Lerer, A.: Automatic differentiation in PyTorch. In: NIPS Workshop (2017)
7. Dean, J., Corrado, G.S., Monga, R., Chen, K., Devin, M., Le, Q.V., Mao, M., Ranzato, M., Senior, A., Tucker, P., Yang, K., Ng, A.Y.: Large scale distributed deep networks. In: Advances in Neural Information Processing Systems, vol. 25, (2012)
8. Kraska, T.: MLbase: a distributed machine-learning system. In: Conference on Innovative Data Systems Research (CIDR) (2013)
9. Li, M., Andersen, D.G., Smola, A.J., Yu, K.: Communication efficient distributed machine learning with the parameter server. In: Advances in Neural Information Processing Systems, vol. 27 (2014)
10. Sergeev, A, Del Balso, M.: Horovod: fast and easy distributed deep learning in TensorFlow. arXiv preprint arXiv:1802.05799 (2018)
11. Ghahramani, Z.: Probabilistic machine learning and artificial intelligence. Nature **521**(7553), 452–459 (2015)
12. Goyal, P., Dollár, P., Girshick, R., Noordhuis, P., Wesolowski, L., Kyrola, A., Tulloch, A., Jia, Y., He, K.: Accurate, large minibatch SGD: training ImageNet in 1 hour. arXiv preprint arXiv: 1706.02677 (2017)
13. Goodfellow, I., Bengio, Y., Courville, A.: Deep Learning. MIT Press (2016)
14. Li, H., Kadav, A., Durdanovic, I., Samet, H., Graf, H.P.: Pruning filters for efficient ConvNets. arXiv preprint arXiv:1608.08710 (2014)
15. Varghese, A, Tandur, D.: Wireless requirements and challenges in Industry 4.0. In: International Conference on Contemporary Computing and Informatics (IC3I), pp. 634–638, (2014)
16. Preuveneers, D., Ilie-Zudor, E.: The intelligent industry of the future: a survey on emerging trends, research challenges and opportunities in Industry 4.0. J. Ambient Intell. Smart Environ. **9**(3), 287–298 (2017)
17. Deng, L., Li, X.: Machine learning paradigms for speech recognition: an overview. IEEE Trans. Audio Speech Lang. Process. **21**(5), 1060–1089 (2013)
18. Kumar, V., Grama, A., Gupta, A., Karypis, G.: Introduction to Parallel Computing: Design and Analysis of Algorithms. Benjamin/Cummings (1994)
19. Reagen, B., Adolf, R., Whatmough, P., Lee, S.K., Lee, H.J., Hsia, C., Johnson, M., Sun, D., Emer, J.S.: Minerva: enabling low-power, highly-accurate deep neural network accelerators. In: ISCA (2016)

20. Wang, Z., Qiu, X., Zhang, J., Chen, L., Zhu, H., Hu, S.: Towards a decentralized, privacy-preserving, and scalable framework for IoT applications. IEEE Internet Things J. **5**(4), 4241–4252 (2018)
21. Cao, Y., Chen, M., Shi, W.: Edge computing in smart healthcare systems: a case study. IEEE Internet Things J. **4**(5), 3274–3284 (2017)
22. Zhang, C., Li, J., Lu, J., Sun, X.: Computational intelligence techniques for predictive maintenance: a survey. IEEE Trans. Industr. Electron. **66**(3), 1818–1830 (2018)
23. Malawski, M., Figiela, K., Byrski, A.: Cost optimization of scientific workflow execution in cloud environment. Futur. Gener. Comput. Syst. **71**, 102–118 (2017)
24. Zhang, Q., Cheng, L., Boutaba, R.: Cloud computing: State-of-the-art and research challenges. J. Internet Serv. Appl. **1**(1), 7–18 (2010)
25. Yu, W., Liang, F., He, X., Hatcher, G.D., Lu, C., Lin, S., Yang, C.: A survey on the edge computing for the Internet of things. IEEE Access **6**, 6900–6919 (2018)
26. Ranaweera, P., Jayalath, M.D., Perera, I.: Real-time IoT solution for cost-effective smart water metering system. In: IEEE International Conference on Information and Automation for Sustainability (ICIAfS), pp. 1–6 (2016)
27. Stoica, I., Zaharia, M., Shenker, S.: A Berkeley view of systems challenges for AI. arXiv preprint arXiv:1712.05855 (2017)
28. Yao, S., Zhao, Y., Wang, Y.: Edge intelligence: paving the last mile of artificial intelligence with edge computing. Proc. IEEE **107**(8), 1738–1762 (2019)
29. Sze, V., Chen, Y.-H., Yang, J.E., Emer, J.S.: Efficient processing of deep neural networks: a tutorial and survey. Proc. IEEE **105**(12), 2295–2329 (2017)
30. Marx, V.: The big challenges of big data. Nature **498**(7453), 255–260 (2013)

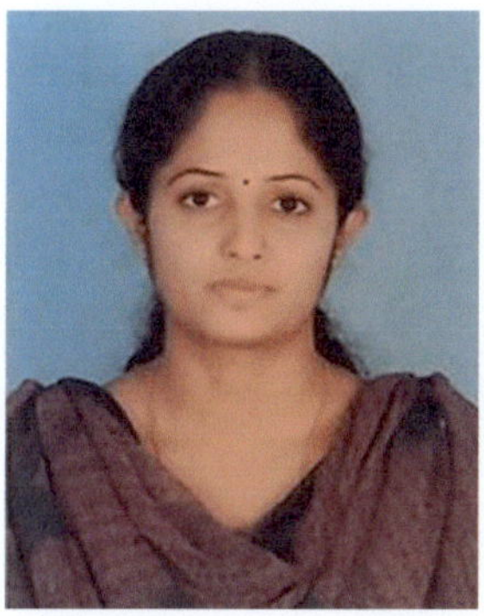

P. Jayadharshini

S. Santhiya

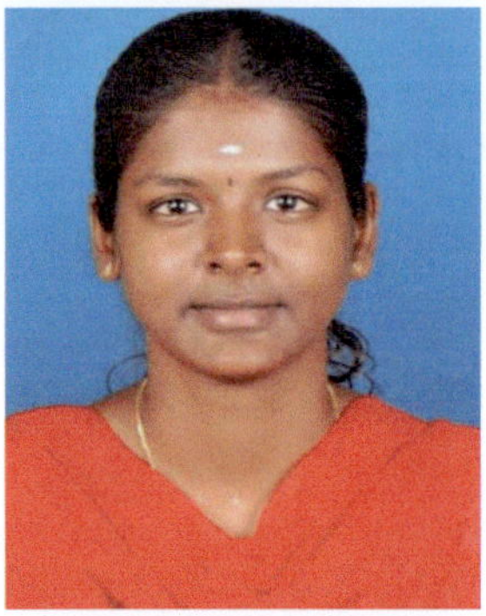

S. Keerthika

N. Abinaya

R. Ahalya

V. N. Shree Nandhini

Explainable Artificial Intelligence (XAI) for Enhancing Decision Making Processes in Building Industry 4.0

C. Kishor Kumar Reddy, Siramdas Sai Jaahnavi, R. Aarti, and Marlia Mohd Hanafiah

Abstract The Industry 4.0 rose and brought forth unrivaled technological combinations, where the industrial era is greatly shaped by artificial intelligence (AI). Although, there opacity issues inherited in AI systems pose a few challenges. To overcome this, Explainable (XAI) emerges as a solution for all, giving understandable justifications for AI systems through various techniques and interpretable models, that represent a transformative change toward creating transparent, understandable, and accountable AI systems. By using XAI, the space created between different complex problems to human understandable solutions is enhanced over transparency and trust towards AI decisions and processes, and this helps stakeholders to navigate and maintain their transformational era, innovations, and sustainability over various sectors. This chapter focuses on the basic principles of XAI and their relevant nature for enhancing decision-making processes in Industry 4.0. As we know, there is a huge increase in AI over industrial applications—ranging from predictive maintenance to complex automation and supply chain optimization, the need for explainability became most important to ensure trust and compliance with regulatory standards. XAI provides in depth conceptualization of the AI-driven decision making by elucidating the reason behind model outcomes, which in turn aids in debugging and improvising the models, fostering trust among the users. The principles of XAI's transparency, interpretability, fairness, and accountability are clear, focusing on their implementation in industrial applications. The chapter concludes with strategic recommendations for integrating XAI principles effectively

C. Kishor Kumar Reddy (✉)
Computer Science and Engineering, Stanley College of Engineering and Technology for Women (Autonomous), Affiliated to Osmania University, Hyderabad, Telangana, India
e-mail: ckishorkumar@stanley.edu.in

S. S. Jaahnavi · R. Aarti
Electronics and Communication Engineering, Stanley College of Engineering and Technology for Women (Autonomous), Affiliated to Osmania University, Hyderabad, Telangana, India
e-mail: aarti@stanley.edu.in

M. M. Hanafiah
Science and Technology, University Kebangsaan Malaysia, Bangi, Malaysia
e-mail: Mhmarlia@ukm.edu.my

L. Krishnasamy et al. (eds.), *Distributed Deep Learning and Explainable AI (XAI) in Industry 4.0*, Information Systems Engineering and Management 55,
https://doi.org/10.1007/978-3-031-94637-0_10

into Industry 4.0 initiatives to leverage AI's full potential by maintaining ethical and operational standards.

Keywords Explainable AI (XAI) · Artificial intelligence (AI) · Industry 4.0 · AI principles · XAI principles · Foundational design principles · Opacity in AI systems · XAI applications · AI applications

1 Introduction

Over the past centuries, industry has evolved, and major technological developments have transformed manufacturing and production processes. Agrarian economies gave way to industrialized society with the help of water and steam-powered machinery during the First Industrial Revolution, which began in the late 1700s. The Second Industrial Revolution allowed for the huge manufacturing of artifacts and the integration of consumer culture. The essence was intentionally revealed by the initiation of electricity and bulk manufacturing technologies during the latter half of 19th and early twentieth century. The widespread adoption of computers and digital technology in the latter twentieth century often called the third industrial revolution or the digital revolution, began by transforming the industry through increased automation and precision. Presently, the world is experiencing the Fourth Industrial Revolution, which can be understood as the coming together of digital gadgets adored as robots, the Internet of Things, and artificial intelligence. These technologies are revolutionizing several industries through improved data-driven decision-making, automation, and connection. The future industry trends are being shaped by rapid advancements in innovation that have expedited industrial mechanization, digitalization, automation, increased productivity, and global interconnection [1]. Significant impacts could be in the fields of Big Data and Analytics, IoT, Blockchain Technology, Augmented and virtual reality, and cyber security. Automation is the key that leads to future productivity and growth in the AI revolution.

The above Table 1 includes the specific technologies associated with each industrial revolution, providing a more comprehensive overview of their characteristics.

This Fig. 1 gives us the clear approach of what has been gone through over the years of development from past to present evolution.

This Table 2, highlights the revolutionary consequences of technological breakthrough in production and manufacturing processes, giving a succinct overview of each industrial revolution's influence and major industries.

Industry 4.0 and AI are like the dynamic duo of industrial innovation, bringing a wave of connectivity, super-smart solutions, and production to manufacturing and its processes. Imagine Industry 4.0 as the cool, contemporary thing on the block, also known as the Fourth Industrial Revolution. It's all based on the combination of digital and physical technology, shaping the lines between the virtual and real world in the coolest way possible. Industry 4.0 aims to characterize cyber-physical systems, self-governing equipment, and real-time data analysis. These technologies

Table 1 Evolution of industries over centuries

Industrial revolution		Period	Main technologies	Characteristics
First revolution	Industrial	Late 1700s to early 1800s	Water-powered and steam-powered machinery	Transition from agrarian economies to industrialized societies. mechanized production introduction
Second Revolution	Industrial	Latter 19th to the earlier half of twentieth century	Power generation and bulk manufacturing technologies	Massive of manufacturing commodities. Emergence of consumer culture
Third revolution	Industrial	Late twentieth century	Computers and digital technology	Increased automation and precision. rise of the digital revolution
Fourth revolution	Industrial	Present century	Robots, Internet of Things and Artificial Intelligence	Combining the digital technologies into various industries, Improved data-driven decision-making automation, global interconnection, and increased productivity

Fig. 1 Evolution of industries

Table 2 Overview of the impact and key industries associated with industrial era

Industrial revolution	Impact	Key industries
First industrial revolution	A change from agricultural to industrial economies, notable gains in economic growth, urbanization, and productivity	Textile manufacturing, coal mining and iron production
Second industrial revolution	The industrialization of commodities, the widespread use of energy, and the emergence of consumer culture	Automotive manufacturing, steel production, electrical engineering
Third industrial revolution	Digitization of processes, automation and precision in manufacturing, emergence of the knowledge economy	Information technology, tele-communications, electronics
Fourth industrial revolution	Convergence of digital technologies; enhanced connectivity, automation, and data-driven decision-making, reshaping of industries	Automotive, aerospace, healthcare and consumer goods

work together to develop intelligent, adaptable production systems. The ability of these systems to communicate autonomously, collaborate, and optimize processes enables unprecedented extents of productivity, customization, and efficiency.

The widespread application of artificial intelligence, which acts as the reasonable engine guiding intelligent automation and optimization processes, that are essential to realizing Industry 4.0's revolutionary vision. The unchangeable capacity of artificial intelligence to observe, adapt, and allow decision making on its own, makes it a promising tool for revolutionizing various aspects of the manufacturing valuing chain, including supply chain management, customer contact, predictive maintenance, and quality control [2]. Artificial intelligence powered quality control systems use pattern recognition, computer vision for identification of flaws in real-time, guaranteeing constant product quality and cutting down on waste. Additionally, AI-powered supply chain optimization algorithms improve demand forecasting accuracy, streamline logistics, and optimize inventory levels, allowing quick responses to market changes and cost savings. Artificial intelligence-driven personalization and predictive analytics also improve customer engagement by enabling businesses to provide customized goods and services based on customer preferences and behaviors.

This fusion of Industry 4.0 and AI holds immense promise for businesses navigating an increasingly digital and interconnected world. While AI-driven insights and automation help companies optimize operations, minimize downtime, and provide tailored goods and services at scale, this convergence also comes with significant hurdles. Yet they include technical challenges, data privacy, and upright considerations, as well as workforce disruptions. Solving these issues would be critical to enabling the full value of the industrial revolution and making its development and societal gains sustainable. The solution to providing the answers to those corporate

AI challenges came out with an AI solution called the Explainable AI which is a wide range of approaches, axioms meant to alleviate the opacity caused by many AI systems. As AI systems become more ubiquitous in all areas of life, such as health care, finances, and criminal justice, the need for oversight and accountability over AI decision-making processes has become evident. In general deep learning models-the black box, users are left less informed about the mechanics behind a model's judgments. Finally, the use of XAI systems can enable users to "see under the hood" of their AI system. This allows the user to establish by itself how the AI system is more understandable and explicable [3]. The intro of XAI played a major role to overcome many factors given by AI systems, a brief picture about this XAI is given in the further part of this section.

1.1 Explainable AI (XAI)

It is a wide range of approaches and axioms meant to alleviate the opacity caused by many AI systems. As AI systems become more ubiquitous in all areas of life, such as health care, finances, and criminal justice, the need for oversight and accountability over AI decision-making processes has become evident. Finally, the use of XAI systems can enable users to "see under the hood" of their own AI system. This allows the user to establish by itself how the AI system is more understandable and explicable [4].

Models that were created are naturally interpretable, including decision trees, rule-based systems, and linear models, is one of the ways to achieve explainability in AI. By directly mapping inputs to outputs in a legible manner, these models provide transparency and let users see how inputs affect the predictions or judgments made by the model. An alternative strategy involves utilizing post hoc explanation techniques on pre-existing AI models to produce explanations after the model's prediction or decision-making. These strategies include saliency maps, attention processes, model-agnostic methods like LIME (Local Interpretable Model-agnostic Explanations) and SHAP (SHapley Additive exPlanations) are examples of feature importance analysis. Post-hoc explanation techniques give consumers insights into the reasons influencing AI predictions or judgments on examining the models' behavior and emphasizing the most significant features or data points.

Furthermore, incorporating principles of human–computer interaction into AI systems can facilitate user understanding and interaction with the models. This includes providing intuitive visualizations, interactive interfaces, and natural language explanations that enable users to interrogate the model's decisions and provide feedback. Transparency and trust in AI systems are improved by human–computer interaction concepts, which involve users in the procedure of determining decisions. Furthermore, the concept of "Certified AI" involves developing AI systems with built-in guarantees on their behavior and performance. This includes bounds on prediction uncertainty and error rates, as well as assurances of reliability and safety. Certified AI approaches provide users with confidence in the credibility and accountability of AI systems, particularly in applications of safety–critical response [5].

XAI implementation presents several difficulties despite its potential advantages. Developing and implementing XAI systems has several challenges, chief among them being finding a balance between transparency, performance, and complexity; guaranteeing consistency and dependability of explanations across a range of contexts; and addressing societal biases and ethical concerns. XAI is an important initiative aimed at improvising transparency, accountability and interpretability of AI systems as they become more and more ingrained in our daily lives and decision-making processes. XAI improves confidence, accountability, and the ethical application of AI in a variety of fields and applications by offering intelligible justifications for actions made by AI [6]. Explainable Artificial Intelligence: AI system through humans, it has the possibility to retain intellectual oversight or methods for achieving it.

This Table 3 delves into the features and key principles of Explainable AI, highlighting its importance. It discusses the difference between white-box and black-box models, symbolic regression algorithms, and the role of human auditing in assessing AI system generalization to real-world data.

We will go through the synergies between Industry 4.0, AI and XAI in this chapter, we will explore the opportunities, challenges, and guiding principles for their combined deployment in industrial settings. Drawing on interdisciplinary research and practical insights from industry, we aim to elucidate the transformative potential of XAI within the context of Industry 4.0, while advocating for responsible and ethical AI development to ensure sustainable progress and societal benefit. Through this exploration, possibilities for contributing to the ongoing dialogue encapsulate the future manufacturing prospects and the roles of AI and Explainable Artificial Intelligence in shaping future industrial growth.

The following section's framework offers a methodical way to talk about how industry 4.0 has evolved and how important artificial intelligence has been at each stage of growth, and how XAI came into picture. It draws attention to the uses, and prospects of Industry 4.0 over explainable artificial intelligence.

Table 3 Features of XAI with initial aspects

Aspects	Explainable AI features
Key principles	Transparency, interpretability, explainability
White and black-box models	White-box provide understandable outputs, while black-box models are hard to explain
Application in various domains	Especially crucial in medicine, defense, finance, and law
Symbolic regression	Algorithm searches for the best mathematical model fitting a given dataset
Auditing rules	Human auditing helps assess the generalization of AI systems to real-world data

2 Background Work

A structured, systematic review of various challenges and future researches in XAI, were split into the following themes: broad issues and research in XAI, the machine learning life cycle, which uses a three-pronged approach of designing, developing, and implementing. This study aims to provide guidance in further exploring the XAI. Author delves into a clear concept of Explainable AI (XAI) taking into account the present day challenging obstacles and further chances highlighting Explainable AI in making artificial intelligence (AI) more transparent and accessible. This paper identifies provocations and future research superintendents in XAI, organized in two themes: general challenges and research directions, and challenges and research superintendents based on the instances of the machine learning life cycle. In a bid to increase the espousal of AI in crucial domains, it highlights the necessity for improved XAI tools and methodologies that would enhance the transparency and accountability of AI systems. The ability of XAI to solve ethical issues surrounding AI and to promote responsible AI. This study provides a snapshot of XAI research and analyzes the prospects and challenges facing the future development of this discipline [7].

The systematic literature review (SLR) on the most current advancements in XAI techniques and assessment metrics with regard to various application domains and tasks is presented in this work. This study examines 137 papers that were found using well-known bibliographic databases and published within the last few years. Many analytical conclusions were reached as a result of this methodical synthesis of research articles, Deep learning and ensemble models are being employed more than other forms of AI/ML models, textual elucidations are more palatable to final users, and reliable evaluation measures are being prospered to assess the quality of explanations. The majority of XAI tackles are developed for globally safety–critical areas [8].

The themes of argumentation and XAI together are covered in detail in this survey by going over all the significant approaches, research, and applications that employ argumentation to give AI explainability. More precisely, they demonstrate how Explainability can be made possible through Argumentation in order to solve a variety of issues related to discourse, opinion justification, and decision-making. We then dive more into the ways in which argumentation can be applied to create explainable systems across numerous application areas, such as security, the semantic web, robotics, medical informatics, law, and some general-purpose systems. Lastly, they offer methods for creating more interpretable predictive models by fusing argumentation theory with machine learning [9].

Through DARPA's XAI program, a wide range of innovative machine-learning approaches are being created and assessed. These include deep learning methods modified to learn explainable features, methods for "model induction," which use any black-box model and deduce into an explainable model, and approaches that grasp more organized, interpretable, and causal models. After a year of the XAI initiative, first technological demonstrations and findings suggest that these three main strategies warrant more research. They will also offer design choices to future developers that balance performance and explainability. To pinpoint the specific contributions of different tactics inside this trade space, the worth of the explanations offered by the development teams' XAI systems is being evaluated [10].

They surveyed current developments in XAI-powered surgical applications and medical diagnosis in this review. Following the inclusion of papers that satisfied the review's selection criteria, they took the pertinent data from the research and examined it. They also show how XAI can be used in medical XAI applications and offer an experimental presentation on breast cancer diagnosis. Finally, they give a summary of the XAI methods applied in medical applications, talk about the difficulties the researchers have encountered, and suggest future research avenues. According to the survey results, medical XAI is a potential area for future research. The purpose of this study is to provide medical specialists and AI scientists with a resource to use when developing medical XAI applications [11].

They provide a historical viewpoint on explainable artificial intelligence in this article. They address the current understanding of explainability, how it was primarily understood in the past, and how it might be understood in the future. They offer criteria for explanations in the article's conclusion, which they feel will be essential to the creation of systems that can be understood by humans. This article falls under the following categories: Artificial Intelligence > Explainable AI Technologies > Fundamental Concepts of Data and Knowledge [12].

This article offers a succinct analytical overview of the contemporary regarding artificial intelligence's explainability in light of recent developments in deep learning and machine learning. The study begins with a succinct historical introduction and taxonomy before outlining the primary explainability difficulties and basing them on the recently developed four explainability principles of the National Institute of Standards. Methods that have been published recently on the subject are then thoroughly examined and evaluated. Lastly, suggestions for further investigation are made. This article falls under the following categories: Technologies > Artificial Intelligence > Basic Data and Knowledge Concepts > Explainable AI [13].

By assembling all scientific studies using a hierarchical framework that categorizes concepts and ideas connected to the idea of explainability and the assessment methodologies for XAI methodology, this thorough study adds to the body of knowledge. This hierarchy's structure is based on a thorough examination of peer-reviewed scientific literature and current taxonomies. Additionally, they have offered a number of methods for determining the extent to which machine-generated explanations satisfy these requirements. This review ends by critically analyzing these shortcomings and defining future research objectives, emphasizing explainability as the foundational element of any artificially intelligent system [14].

With the use of a case study example, they gave readers a thorough understanding of the contemporary and new trends in this quickly developing field in this extensive study. The introduction of the paper provides background information on XAI, standard concepts, and an overview of recently proposed supervised machine learning algorithms in XAI. The review divides XAI approaches into four categories using a hierarchical classification approach: (i) data explainability, (ii) model explainability,

(iii) post hoc explainability, and (iv) assessment of explanations. They also include open-source tools, datasets with future research objectives, and assessment measures that are currently accessible. Researchers in various domains who are looking for effective XAI ways to confidently execute tasks and communicate the relevance of the findings, as well as XAI researchers who wish to increase the dependability of their AI models, are also intended audience members for this work [15].

They discuss the XAI frameworks in this article, emphasizing their features and IoT support. They demonstrate the popular XAI services for Internet of Things (IoT) applications, including Internet of Medical Things (IoMT), Industrial IoT (IIoT), and Internet of City Things (IoT), as well as security enhancement. Together with relevant examples, they recommend using XAI models rather than IoT systems for these applications' implementation and provide a summary of the main conclusions for further research. Along with important conclusions, they also showcase the state-of-the-art advancements in edge XAI structures and the support of sixth-generation (6G) connectivity services for Internet of Things applications. To put it briefly, this article is the first comprehensive compilation on creating XAI-based frameworks that are specifically designed to meet the needs of upcoming Internet of Things use cases [16].

The technique for transforming black box models into explainable (and interoperable) classifiers based on semantic rules is presented in this paper. Because our transformation method generates adversarial samples (often referred to as "corner cases") and incorporates progressive ignorance discovery, it assembles explainable rule-based classifiers with good performance and an efficient training process. Furthermore, a use-case scenario that protects user privacy and data using these explicable and interoperable classifiers has been established. It involves cooperative diagnostics, condition monitoring, and predictive maintenance of scattered smart industrial assets [17].

The current discussion's goal is to create such a framework, giving special consideration to the various explanatory needs of different stakeholders. This framework is based on a study of "opacity" from the philosophy of science and is based on cognitive science explanatory accounts. To allow various stakeholders to carry out their responsibilities within the machine learning ecosystem, the framework differentiates between the elucidating-seeking questions that are expected to be posed by various stakeholders and outlines the general manner in which these questions should be addressed. It is feasible to ascertain whether and to what extent approaches from Explainable Artificial Intelligence may be used to render whether opaque computer technologies can be used to address the black box hitch depends on how transparent they are [18].

On the one hand, the order, dependencies, and sequence of the research activities that are now and frequently carried out by multiple researchers have been highlighted here. Typically, this process begins with input data that is subsequently utilized to model using knowledge-based paradigms such as connectionist data-driven learning or symbolic reasoning. Following the creation of a model, knowledge discovery and analysis are conducted using XAI techniques, which enhance the model's interpretability. For the aim of explainability, this step offers one or more explanations to the models' end users. Ultimately, very few academics have suggested methods for assessing this level of explainability, either by putting forth formal, objective metrics or incorporating model creators and end users in a human-centered evaluation process. In this situation, the explanation should be the main focus since it is what end customers will engage with in the end. The various characteristics associated with the psychological idea of explainability should be considered while designing explanators. Afterward, researchers can concentrate on the modeling stage, ideally utilizing both artificial intelligence's symbolic and connectionist paradigms. This will make it possible to create models that are inherently interpretable at every level of development and resilient in terms of accuracy. Ultimately, the last stage should concentrate on creating an interactive interface to augment the interpretability and explainability of the model's inference as well as assessing these models' explainability using a human-in-the-loop technique including both final users and designers [19].

This paper furnishes a thorough summary of the state of XAI in deep learning along with mathematical representations of core work. To create reliable, intelligible, and apparent deep learning models, they first proposed a hierarchy of concepts and grouped the XAI techniques according to their degree of use, explanation scope, and algorithmic methodology. The key ideas utilized in XAI research were then discussed, and a historical chronology of notable XAI studies from 2007 to 2020 was presented. The authors first go to great lengths to explain each type of algorithm and methodology. They then assess the elucidation maps assembled by eight XAI algorithms using image data, go over the drawbacks of this method, and offer possible future routes to enhance XAI evaluation [20].

Through the use of some explainable processes, they were able to increase the epistemic confidence in the model in this survey. Naturally, XAI's objective is to acquire information and understand the model in order to prove its legitimacy. Still, the field of XAI is not just about aiding with reliability. Explainability has the potential to lead to novel training procedures and measurements that ensure the resilience and confidence of even the most complex and abstract models since explainable approaches can provide insights. Approaches that only focus on technological factors are also far from achieving an end-to-end XAI system. XAI techniques ignore and do not apply the user and developer interplays necessary for the AI system to be certitude. There aren't many unbiased tools available to demonstrate the dependability of AI systems. Because of this, interactive systems that provide justifications and feedback might be a cutting-edge way to prove to the user and decision-maker that the AI system is trustworthy in an impartial and accurate manner [21].

They demonstrated common XAI techniques utilizing a shared case study or assignment (credit default prediction); they examined competitive advantages from local and global perspectives, provided incisive analysis concerning explainability quantification, and proposed paths toward superintendent or human-centered AI by employing XAI as a source medium. This work can be used by practitioners as a library to understand, measure and analyze the positive traits of XAI's popularity. Furthermore, since the acceptance of AI in high-stakes applications depends on it, this survey emphasizes the necessity of responsible or human-centric AI systems in the future [22].

Taking into consideration the corpus of research on XAI, this study argues that although XAI in education has certain requirements in common with the broader use of AI, it also differs in several important ways. Consequently, they initially introduced a framework known as XAI-ED that takes into account six important factors regarding explainability for researching, creating, and creating AI-based teaching aids. These important features center on the following: the stakeholders, the advantages, the methods of explanation, the classes of AI models that are commonly employed, the human-centered designs of the AI interfaces, and the possible drawbacks of explaining things in the context of education. After that, they provided four in-depth case studies that demonstrated how XAI-ED was used in four distinct educational AI products. The possibilities, difficulties, and areas in need of further research for the successful integration of XAI in education are covered in the paper's conclusion [23].

These different works by researchers lead us to the adaptability and grasp about AI and XAI principles, and how they delved into the various applications. The following section gives us the briefness about AI and XAI principles and their applicability in Industry 4.0.

3 AI and XAI Principles

3.1 AI Principles

The industrial sector is currently undergoing a digital change known as Industry 4.0, or the Fourth Industrial Revolution (4IR), which is being propelled by disruptive developments such as robotics, analytics, data and connectivity, and human–machine interaction. AI systems should be able to provide succinct rationales for their actions, ensuring transparency and accountability. AI explanations should be intelligible and relevant to people, particularly non-experts, in order to foster confidence and aid in decision-making. AI systems must give precise explanations that emphasize the factors that have the greatest influence on the decision-making process. AI systems should operate under specified parameters, be mindful of their limitations and uncertainties, and express the degree of confidence in their forecasts [24].

3.1.1 Ethical Considerations

AI systems ought to be developed and applied with consideration for morality and human values. It includes fairness, transparency, accountability, privacy, and avoiding harm to individuals or society.

3.1.2 Explainability and Transparency

AI systems ought to be able to elucidate their choices and actions and be honest about them. It should be clear to users how AI algorithms operate and why particular results are produced.

3.1.3 Fairness and Bias Mitigation

AI systems should emerge to be fair and unbiased, treating all individuals and groups equitably. Developers ought to recognize and address biases in algorithms, data, and decision-making procedures.

3.1.4 Privacy and Data Security

AI systems should respect user privacy and protect sensitive data. Data collection, storage, and processing should adhere to legal and ethical standards, ensuring confidentiality and security.

This Fig. 2 gives us the picturization of AI principles that can be easily understandable with clear conscience.

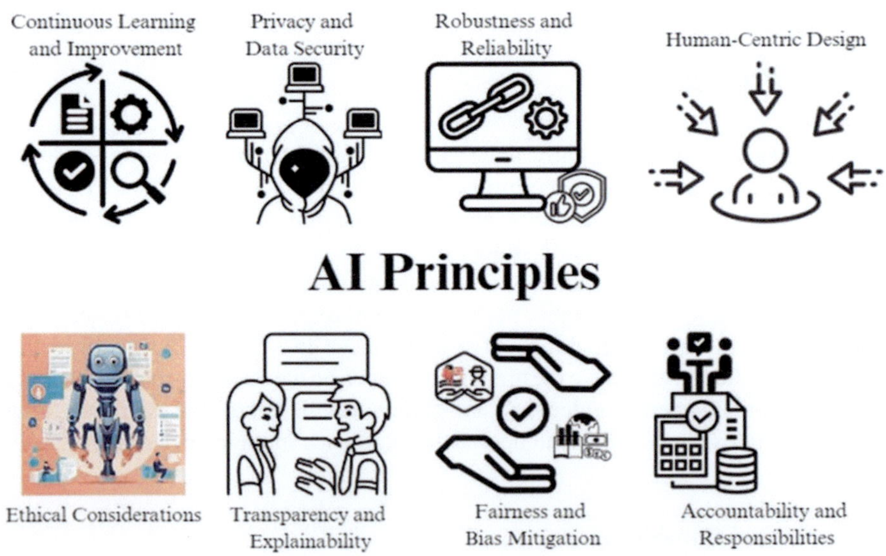

Fig. 2 Different AI principles that are essential for evaluating an application

3.1.5 Accountability and Responsibilities

Stakeholders and developers must assume responsibility for the results of AI systems. On-site protocols ought to be in place to address grievances, find errors, and ensure that AI technology is applied responsibly.

3.1.6 Robustness and Reliability

AI systems should be robust and reliable, performing consistently across different conditions and environments. They should be able to handle unexpected inputs, errors, and adversarial attacks without compromising performance.

3.1.7 Human-Centric Design

AI systems should emerge with the needs and preferences of human users in mind. User interfaces should be intuitive, accessible, and responsive, promoting effective engagement and collaboration between humans and machines.

3.1.8 Continuous Learning and Improvement

AI systems should be prone to learning from feedback and adapting to new information and circumstances. Initializing ongoing monitoring and assessment is necessary to assess performance and identify opportunities for enhancement.

Developers and other stakeholders may assure that artificial intelligence technologies advance and apply in a way that is ethical, responsible, and advantageous to society by abiding by these principles, which also help to reduce risks and unfavorable outcomes. The fundamental rules that underpin the moral-upright and responsible creation, application, and use of artificial intelligence technology are known as AI principles. These principles encompass various critical aspects that address the multifaceted nature of AI systems. They include ensuring that AI systems assist humans in decision-making while maintaining human oversight to override system decisions when necessary.

Moreover, AI systems should emerge to be predictable, safe, and reliable, promoting trust and confidence among users. Quality management is essential, with AI providers ensuring compliance with rigorous quality standards to maintain system performance and integrity. Privacy and data protection are paramount, with AI systems designed to prioritize the security and confidentiality of user data. Responsible data governance ensures that datasets used to train AI systems are correctly and accurately fostered, fairness, and accountability. Transparency is crucial and necessary to collaborate with AI providers to provide accurate information about system capabilities, limitations, and data sources to promote understanding and trust. Equity and non-discrimination principles mandate that AI systems emerge to avoid bias and discrimination, promoting diversity and equity in outcomes. Furthermore, AI systems must support general societal objectives and collide with social responsibility and sustainability concepts [25].

3.2 Opacity Issues of AI

The opacity issues of AI, as discussed in the provided sources, revolve around the opaqueness and inexplicability of AI systems, particularly in machine learning algorithms. Opacity in AI can stem from intentional corporate or institutional secrecy, technical illiteracy, and the inherent characteristics of machine learning algorithms that make them difficult to interpret. This lack of transparency can have significant implications for tactics of classification and ranking that have social consequences, including search engines, spam filters, and credit card fraud detection and more. Opacity in AI systems can lead to challenges in understanding how decisions are made, potentially resulting in issues related to inequality, discrimination, and unfairness.

As a way to maintain accountability, avoid harm, and encourage transparency in the operation of AI models across multiple domains and artificial intelligence systems' decision-making processes, it is imperative to address opacity in AI. These systems are often ramified as black boxes, which makes it stimulating for users to understand how they arrive at specific conclusions or assessments. This is especially true of systems based on sophisticated algorithms like deep learning. This opacity can be problematic in various contexts, particularly in high-stakes applications where the reasoning behind AI-driven decisions needs to be comprehensible and justifiable.

Opacity in AI systems can arise from several factors, including the complexity of the underlying algorithms, the use of large and opaque datasets for training, and the inherent unpredictability of machine learning models. As a result, users may struggle to trust AI systems or verify their outputs, leading to concerns about reliability, accountability, and ethical implications. To tackle the opacity issues in AI, researchers and practitioners are focusing more and more on the creation of Explainable AI (XAI) solutions. By offering explanations for their decisions and acts that are understandable to humans, XAI strives to assist in augmenting the transparency and understandability of AI systems. These justifications can aid users in comprehending the operation of AI models, seeing any biases or mistakes, and eventually fostering certitude in AI-related technologies. By improving transparency and interpretability, XAI can augment the accountability, reliability, and upright use of AI systems across several domains, from healthcare and finance to criminal justice and autonomous vehicles. But attaining transparency in AI is still a difficult and continuous task that calls for multidisciplinary research projects and creative thinking to guarantee that AI systems are robust, yet they're also transparent, accountable, and in line with human values and preferences [26, 27].

3.3 XAI Principles

The end goal of XAI, or explainable AI is to develop systems that can give concise, intelligible justifications for their decisions, hence promoting openness and confidence. XAI principles aim to fulfill regulatory requirements, facilitate accountability, mitigate bias, and promote fair outcomes in AI applications. XAI enables stakeholders to collaborate on improving AI systems, adhere to regulatory compliance, detect and correct biases, and hold systems accountable for decisions. XAI empowers users to understand and potentially contest AI decisions, promoting user advocacy and interests. XAI seeks to demystify AI operations by making algorithms interpretable, transparent, and justifiable, ensuring reliability and trustworthiness [24].

This Fig. 3 gives us the picturization of XAI principles that can be easily understandable with clear conscience.

3.3.1 Transparency

XAI places an enormous value on its role in allowing users to consider the decision-making process in AI systems. It entails giving precise justifications for AI activities, suggestions, and forecasts.

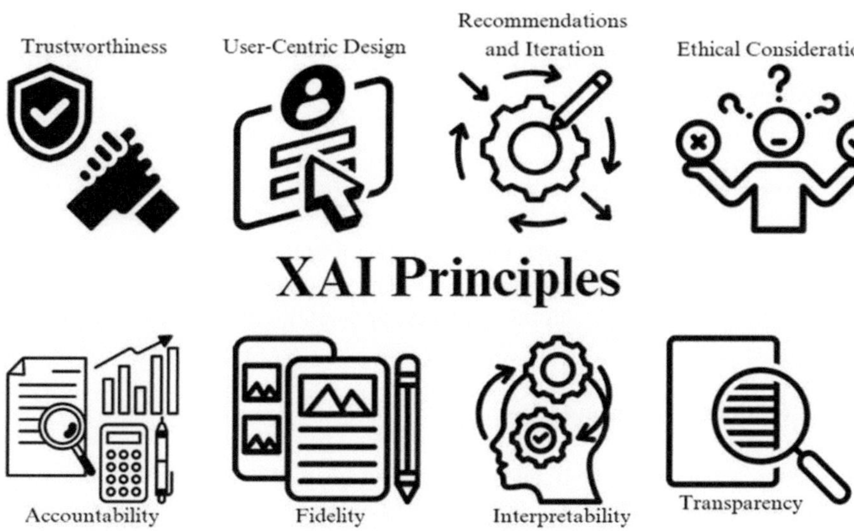

Fig. 3 Different XAI principles that are essential for evaluating an application

3.3.2 Interpretability

This helps consumers comprehend the elements influencing their judgments, and this AI model needs to be comprehensible. It entails the usage of models and algorithms that yield outcomes that are simple for people to comprehend and analyze.

3.3.3 Fidelity

XAI systems should provide accurate and faithful explanations for AI predictions and decisions. Explanations should accurately reflect the underlying reasoning and logic of the AI model, avoiding misinterpretations or distortions.

3.3.4 Accountability

XAI makes users and developers responsible for the results of AI systems. It includes clinching that AI decisions are fair, unbiased, and aligned with ethical principles and that developers can be held responsible for any errors or biases in the system.

3.3.5 Trustworthiness

XAI systems need to be trustworthy, inspiring confidence and trust in their predictions and recommendations. It ensures that AI systems are reliable, robust, and free from hidden biases or errors.

3.3.6 User-Centric Design

XAI systems should emerge with the needs and preferences of users in mind. It includes providing explanations in formats that are understandable and useful to users, such as visualizations, natural language explanations, or interactive interfaces.

3.3.7 Recommendations and Iteration

XAI systems ought to be able to gather user feedback and apply it to their decision-making procedures. It makes it possible for AI models to be continuously enhanced and improved upon in light of actual user experiences and usage.

3.3.8 Ethical Considerations

The value of upright considerations for the creation and use of AI systems is emphasized heavily in the XAI principles. It entails maintaining accountability, privacy, and fairness as well as refraining from using AI in ways that can injure people or violate their human rights.

Developers may promote trust and confidence in AI technology while reducing the risks of bias, inaccuracy, and misuse by following these guidelines to construct transparent, interpretable, and responsible XAI systems. Explainable AI (XAI) principles are foundational guidelines that aim to elevate the accountability, interpretability, and transparency of artificial intelligence systems. These principles emphasize the importance of furnishing clear and understandable explanations for the decisions and actions performed by AI algorithms. Transparency, or making AI systems' internal workings understandable and accessible to humans, is an indispensable component of the XAI concepts. The ability of AI models to provide outcomes that are simple for humans to comprehend and interpret is known as interpretability, and it is another crucial factor. A further emphasis on accountability is the need for AI system developers and users to accept accountability for their choices and results. To promise that AI systems adhere to social norms and values, XAI principles prioritize user-centric design, fairness, and ethical considerations. In a variety of fields and applications, stakeholders can advance the ethical, dependable, and trustworthy implementation of AI technologies by upholding these values [27, 28].

This Table 4 provides a clear and accurate comparison between AI principles and XAI principles, outlining their respective focuses and goals in the prospering and deployment of AI and XAI technologies.

3.4 Applications

The applications of Artificial Intelligence (AI) and Explainable AI (XAI) can significantly enhance productivity across various domains, albeit in different ways. AI productivity refers to the efficiency and effectiveness gained from the application of AI technologies in automating tasks, making predictions, and optimizing processes. In fields like manufacturing, healthcare, finance, and logistics, Large volumes of AI algorithms have the propensity to investigate data and they can spot trends, and make data-driven decisions at speeds and scales beyond human capabilities. This results in streamlined operations, improved resource allocation, and enhanced decision-making, leading to increased productivity and competitiveness. However, the productivity gains from AI can be hindered by concerns related to trust, transparency, and accountability. In dire circumstances where AI judgments affect people's lives or have broad societal repercussions, the "black box" cosmos of AI models can raise questions about their reliability and fairness, this is an instance of Explainable AI (XAI) [30]. XAI productivity complements AI productivity by focusing on making AI systems transparent, interpretable, and accountable. XAI techniques aim

Table 4 Comparison of aspects between AI and XAI principles [29]

AI principles	XAI principles
Focuses on the moral-upright and liable blooming and application of AI technologies	Focuses on rudimenting AI systems transparent, interpretable, and accountable
Emphasizes transparency, clear furnishing explanations for AI decisions and actions	Lays a strong emphasis on interpretability to make sure AI models and algorithms generate outcomes that are simple for people to comprehend and interpret
Includes principles related to fairness, transparency, accountability, privacy, and reliability	Places a strong emphasis on accountability, holding both users and developers responsible for the results of AI systems and guaranteeing the impartiality and fairness of artificial intelligence judgments
Focuses on prospering AI systems that are robust, reliable, and secure	Focuses on clinching that AI systems are trustworthy, inspiring confidence and trust in their predictions and recommendations
Emphasizes the value of taking user input into account and continuing to learn and grow	Advocates for user-centric design, ensuring that AI explanations are understandable and useful to users, such as visualizations, interactive interfaces, or natural language explanations
Addresses ethical considerations related to fairness, privacy, accountability, and avoiding harm	Considers moral-upright considerations in the design and deployment of XAI systems, including fairness, privacy, and accountability, and ensuring that XAI systems are used in ways that align with ethical principles

to provide clear explanations for AI decisions, enabling users to understand how and why specific outcomes are reached. By increasing transparency and interpretability, XAI fosters certitude in AI systems and empowers users to affirm the accuracy and fairness of AI-generated outputs. This, in turn, enhances the usability and acceptance of AI technologies across different applications [27].

AI and XAI are used together to enhance productivity, efficiency, and transparency in various processes.

3.4.1 Predictive Maintenance

AI Application: AI systems can analyze data from manufacturing equipment, such as electrical current, vibration, and sound, to predict signs of equipment failure well before they occur. This predictive capability helps improve maintenance efficiency, reduce overall maintenance costs, and minimize downtime by enabling timely interventions based on data-driven insights

XAI Application: XAI complements AI in this scenario by furnishing clear explanations for the predictions made by the AI system regarding equipment failure. XAI enables stakeholders, including maintenance technicians and decision-makers, to understand the rationale behind the AI's maintenance recommendations. This transparency enhances trust in the AI system's predictions and allows users to validate and act upon the perspectives provided by the AI model.

By integrating AI for predictive maintenance with XAI for explainability, manufacturing companies can optimize their maintenance processes, reduce costs, and enhance operational efficiency while ensuring that stakeholders have a clear understanding of the AI-driven maintenance recommendations. This collaborative approach between AI and XAI in manufacturing exemplifies how transparency and trustworthiness can be achieved in AI applications, leading to improved decision-making and performance in industrial settings [31].

3.4.2 Anomaly Detection

AI Application: AI systems can analyze real-time sensor data from manufacturing equipment, such as vibration, temperature, and electrical current, to detect anomalies that may indicate potential equipment malfunctions or deviations from standard operating conditions. By identifying these anomalies early, manufacturers can prevent unexpected breakdowns and optimize their maintenance schedules.

XAI Application: XAI complements the AI anomaly detection system by giving concise justifications why definitive data points are flagged as anomalies. This transparency allows operators to understand the reasoning behind the alerts, enabling them to take appropriate corrective actions. The explanations from XAI can include information about the specific sensor readings, historical patterns, and the algorithms used to identify the anomalies, empowering operators to validate the AI's findings and make informed decisions [31].

3.4.3 Supply Chain Optimization

AI Application: AI algorithms can scrutinize historical sales data, market trends, and demand forecasts to boost inventory levels, streamline logistics, and improve overall supply chain efficiency. By leveraging predictive analytics, the AI system can anticipate changes in demand and proactively adjust inventory, transportation, and distribution plans to minimize costs and maximize customer satisfaction.

XAI Application: XAI is used to provide transparent explanations for the supply chain optimization recommendations made by the AI system. This includes detailing the factors considered, such as lead times, supplier performance, and seasonal fluctuations, as well as the decision-making logic behind inventory adjustments or shipment prioritization. By understanding the rationale behind the AI's suggestions, stakeholders can validate the recommendations, make informed adjustments, and ensure alignment with the organization's supply chain goals [31].

3.4.4 Energy Management

AI Application: AI systems can optimize energy consumption in manufacturing facilities by analyzing real-time sensor data, meteorological predictions, and production schedules. The AI algorithms can identify opportunities to adjust equipment settings, shift production schedules, or implement other energy-saving measures to reduce overall energy usage and costs, while maintaining production targets.

XAI Application: XAI is used to give concise justifications for the AI system's energy management choices. This includes detailing the factors considered, such as energy prices, equipment efficiency, and production requirements, as well as the reasoning behind specific adjustments, such as shifting production to off-peak hours or temporarily powering down non-essential equipment. By understanding the explanations, operators can validate the AI's recommendations, ensure alignment with energy efficiency goals, and make informed adjustments as needed [31].

3.4.5 Medical Diagnosis

AI Application: AI systems might assist medical personnel diagnose diseases and ailments by analyzing medical imagery, patient data, and electronic health records. For example, AI algorithms can help radiologists discover possible health risks by spotting abnormalities in MRIs, CT scans, and X-rays.

XAI Application: XAI techniques complement AI by providing transparent explanations for diagnostic decisions. By offering clear perspectives into the features and patterns used by AI models to make diagnoses, XAI helps clinicians understand and trust the AI-driven diagnostic recommendations. This transparency enhances the dependability and precision of medical diagnosis while empowering healthcare professionals to make informed treatment alternatives [32].

3.4.6 Fraud Detection

AI Application: AI algorithms can scrutinize immense amounts of financial transaction data to identify trends, such as odd spending patterns or questionable transactions, that point to fraudulent activity. By automatically flagging potential fraud cases, AI systems help financial institutions mitigate risks and protect against financial losses.

XAI Application: XAI enhances transparency in fraud detection by furnishing interpretable explanations for the determinations that AI models undertook. For instance, XAI techniques can elucidate the factors and indicators considered by the AI system when flagging transactions as potentially fraudulent. This transparency enables fraud analysts and compliance officers to understand the rationale behind the AI's fraud detection alerts and take appropriate actions, such as further investigation or validation [33].

3.4.7 Personalized Recommendations

AI Application: Artificial intelligence-powered recommendation engines look at customer behavior, prior purchases, and preferences to make personalized product recommendations. These systems help retailers increase sales, enhance customer satisfaction, and improve user engagement by offering tailored suggestions based on individual preferences.

XAI Application: XAI methods furnish transparent explanations for the recommendations made by AI algorithms. By highlighting the factors and data points considered in generating each recommendation, XAI enables customers to understand why certain products are suggested to them. This transparency builds trust and confidence in the recommendation system, encouraging customers to make informed purchasing decisions [34].

3.4.8 Autonomous Vehicles

AI Application: AI algorithms power autonomous vehicles by processing sensor data, such as LiDAR, radar, and cameras, to sense and comprehend the surroundings. These systems enable self-driving cars to maneuver autonomously, recognize impediments, and traverse roadways while making judgments in real time.

XAI Application: XAI techniques furnish explanations for the decisions made by autonomous vehicle systems, such as lane changes or braking maneuvers. By offering clear perspectives into the factors considered and actions taken by the AI algorithms, XAI enhances trust and safety in autonomous driving technology. This transparency is essential for gaining regulatory approval and public acceptance of autonomous vehicles [35].

3.4.9 Flexible Education Resources

AI Application: AI-driven adaptive learning systems examine student performance information., learning preferences, and educational content to personalize the learning experience. These platforms provide activities, tests, and learning materials that are specially designed to meet the unique needs and learning styles of each student.

XAI Application: XAI techniques provide explanations for the recommendations and feedback provided by adaptive learning algorithms. By elucidating the reasoning behind the personalized learning suggestions, XAI helps students understand their strengths, weaknesses, and learning progress. This transparency fosters student engagement, motivation, and self-directed learning [34].

3.4.10 Predictive Maintenance in Power Plants

AI Application: AI systems analyze sensor data from power plant equipment, such as turbines and generators, to predict maintenance needs and prevent unplanned downtime. By detecting anomalies and patterns indicative of equipment failure, AI-driven predictive maintenance algorithms enable proactive maintenance scheduling and resource optimization.

XAI Application: XAI methods provide justifications for the maintenance recommendations generated by AI algorithms. By transparently highlighting the factors considered in predicting equipment failures and scheduling maintenance tasks, XAI boosts the trust and reliability of predictive maintenance systems. This transparency enables maintenance engineers and operators to understand and validate the AI-driven maintenance decisions [31].

The above mentioned Table 5, gives demonstration about case studies, the diverse applications of XAI in Industry 4.0 settings, ranging from predictive maintenance and quality control in manufacturing to energy efficiency optimization in smart grids and supply chain optimization in logistics. By deploying the power of XAI, organizations can unlock valuable perspectives from their data, optimize operations, and drive innovation in industrial processes. However, successful XAI implementations require careful consideration of domain-specific challenges, collaboration between data scientists and domain experts, and continuous monitoring and refinement of XAI models to ensure alignment with business objectives and operational requirements.

4 Comparisons Based on Explainable AI (XAI) in Industry 4.0: Principles, Models, Techniques, and Case Studies

This Table 6 illustrates how the productivity of AI systems can be impacted by XAI and AI concepts., with XAI generally emphasizing higher productivity by focusing on transparency, interpretability, accountability, trustworthiness, user-centric design, feedback, iteration, and ethical considerations.

This Table 7 provides an overview of the trade-offs between performance and explainability associated with different AI models commonly used in Industry 4.0 applications. Practitioners can select the best AI model that achieves the ideal balance between explainability and performance based on the particular requirements of the application and the significance of interpretability.

The aforementioned Table 8 gives a taxonomy/framework which categorizes various XAI techniques based on their characteristics and provides examples of their suitability for different Industry 4.0 applications. By choosing the best XAI methods depending on the objectives and particular requirements of each application, stakeholders can revamp the certitudeness, interpretability, and transparency of AI-driven systems in a variety of industrial contexts.

Table 5 A table summarizing case studies or real-world examples of successful XAI implementations in industry 4.0 settings

Case study	Description	Benefits	Lessons learned
Predictive maintenance in manufacturing	An automotive manufacturing company implemented an XAI-powered predictive maintenance system to anticipate equipment failures and optimize maintenance schedules	Reduced downtime by 30% through proactive maintenance interventions	Importance of integrating XAI with existing predictive maintenance systems to enhance transparency and interpretability of maintenance recommendations
Quality control in semiconductor manufacturing	A semiconductor manufacturing plant deployed an XAI solution for quality control, leveraging machine learning algorithms to identify defects in semiconductor wafers	Improved defect detection accuracy by 25%, resulting in fewer production defects and higher product yield	Need for domain-specific XAI techniques tailored to the unique characteristics and challenges of semiconductor manufacturing processes
Energy efficiency optimization in smart grids	A utility company utilized XAI techniques to optimize energy efficiency in smart grids, analyzing real-time IoT sensor data to predict energy consumption patterns	Achieved 15% reduction in energy consumption through optimized grid operations and demand-side management strategies	Importance of continuous monitoring and feedback loops to refine XAI models and adapt to changing environmental conditions and energy demands
Supply chain optimization in logistics	A logistics company employed XAI algorithms to augment supply chain operations, predicting demand fluctuations and optimizing inventory management	Increased supply chain efficiency by 20% through better demand forecasting and inventory optimization	Need for collaboration between data scientists and domain experts to ensure XAI solutions align with business objectives and operational constraints
Predictive analytics in healthcare	A healthcare provider leveraged XAI for predictive analytics, analyzing electronic health records (EHRs) to predict patient readmissions and identify at-risk populations	Reduced hospital readmissions by 18% through targeted interventions and proactive patient care management	Importance of data privacy and compliance with healthcare regulations when handling sensitive patient data in XAI applications

Table 6 Productivity differences between AI and XAI over principles [36]

Principle	Productivity of AI	Productivity of XAI
Transparency	May differ based on how intricate the data and AI model are	Generally high, as XAI emphasizes providing clear explanations for AI actions
Interpretability	May be moderate to low, especially for complex deep learning models	High, as XAI focuses on ensuring that AI models are interpretable by humans
Accountability	Moderate, since AI systems might not have accountability structures in place	High, since XAI places a strong emphasis on making users and developers responsible for AI results
Trustworthiness	Moderate, as trust in AI systems may be influenced by factors such as reliability and performance	Generally high, as making AI systems more dependable and trustworthy is the goal of XAI
User-centric design	Varies based on how much user demands are considered in the construction of AI systems	Generally high because XAI promotes the ideas of user-centric design
Feedback and iteration	May be moderate, as AI systems may not always incorporate feedback effectively	High, as XAI emphasizes the importance of incorporating user feedback and iteration
Ethical considerations	Varies, depending on the extent to which ethical considerations are prioritized in AI development	Generally high, as XAI places a strong emphasis on ethical considerations

This Table 9 highlights how XAI plays a crucial role in enabling collaboration between humans and AI systems in industrial settings, facilitating transparent decision-making, error detection and correction, adaptive decision support, intuitive human-AI interfaces, and the promotion of ethical and responsible AI practices. By leveraging XAI techniques effectively, organizations can harness the collective intelligence of humans and AI to achieve optimal outcomes across various industrial domains.

This Table 10 outlines the key challenges associated with scaling XAI systems in industrial settings and provides potential solutions to address these challenges. By overcoming scalability limitations, enhancing the interpretability of complex AI models, improving data quality and variability handling, and enabling real-time processing and decision-making, organizations can effectively deploy XAI solutions to extract actionable insights from large-scale industrial data and drive intelligent decision-making across various industrial domains.

This Table 11 highlights the potential synergies between XAI and emerging technologies like IoT, edge computing, and 5G/6G networks in industrial settings, as well as the associated challenges that organizations may encounter when integrating these technologies. By carefully overcoming these barriers and harnessing the combined capabilities of XAI and through the use of stemming technologies, businesses can

Table 7 Comparison of the performances and explainability trade-offs of different AI models commonly used in industry 4.0 applications

AI model	Performance	Explainability	Trade-offs
Decision trees	Ideal for managing data that is both classified and numerical. It can manage robust nonlinear relationships against outliers	Intuitive in decision-making process, easy to understand and interpret. Graphical representation facilitates the visualization	• Limited accuracy for complex datasets • Prone to overfitting with deep trees
Linear models	Simple and interpretable. Efficient training and prediction that is suitable with the large datasets	Coefficients provide direct perspective into feature importance. So, the decision-making process was transparent	• Limited ability to capture complex relationships in data • Assumption of linearity may not hold true for all datasets
Deep learning models	High predictive accuracy, ability to learn complex patterns from large datasets that are suitable for chaotic data like text, audio, and picture files	Black-box nature makes it challenging to interpret decisions. So, lack of transparency in internal representationsand decision-making processes	• Demands a lot of labeled data in order to be trained • Computationally intensive and resource-intensive • Prone to overfitting without proper regularization
Ensemble models	Improved predictive performance through aggregation of multiple models. Reduced risk of overfitting, versatile and flexible	Individual models within the ensemble may offer some interpretability, these methods can provide insights into feature importance and model consensus	• Increased complexity compared to individual models • Sacrifice some level of interpretability for improved performance

seize new chances for creativity, effectiveness, and competitiveness in Industry 4.0 environments.

The aforementioned Table 12 framework considers key factors such as fidelity, interpretability, human-centricity, actionability, robustness, and transparency to assess the efficacy and quality of XAI explanations. By assessing explanations based on these criteria, stakeholders can acquire a thorough comprehension of the strengths and limitations of XAI systems, enabling informed decision-making and fostering trust in AI-driven technologies.

Table 8 Comparison of different XAI techniques and their suitability for various industry 4.0 applications [37]

XAI technique	Description	Suitability for industry 4.0 applications
Interpretable models	Models designed to be inherently interpretable, such as decision trees, linear models, and rule-based systems	Well-suited for applications where transparency and understandability are paramount, such as quality assurance, process optimization, and predictive maintenance. Furnishing clear perspectives into decision-making processes, facilitating trust and verification of AI-driven recommendations
Post-hoc explanations	Techniques that provide explanations for AI model predictions after they have been made, including feature importance analysis, saliency maps, and SHAP (SHapley Additive exPlanations)	Useful for enhancing transparency in complex AI models, such as deep learning models, where interpretability may be lacking. Can aid in understanding the factors influencing AI predictions in applications like fault detection and anomaly detection
Visual explanations	Explanations presented in a visual format, such as heatmaps, graphs, or lively visualizations, to help users understand AI model outputs and decision-making processes	Particularly effective for conveying complex information in a user-friendly manner, making it suitable for applications where human–machine interaction is crucial, such as human–robot collaboration and intelligent monitoring systems. Enhance user engagement and comprehension by presenting AI insights in an intuitive and accessible format
Rule extraction	Process of extracting human-readable rules from complex AI models, enabling users to understand the logic behind model decisions	Ideal for applications where regulatory compliance and adherence to domain-specific rules are essential, such as safety–critical systems and regulatory compliance monitoring
		Provides actionable insights for decision-makers by translating AI model behavior into understandable rules and guidelines
Counterfactual explanations	Explanations that describe what changes to provided features would result in different model predictions, offering insights into how to achieve desired outcomes	Valuable for applications requiring actionable insights and scenario analysis, such as supply chain optimization and production planning
		Offering consumers the ability to investigate alternative courses of action and comprehend the influence of various variables on the forecasts of AI models

5 Discussion and Conclusion

Explainable AI (XAI) stands as a pivotal resolution to the opacity issues inherent in many AI systems, particularly within the backdrop of Industry 4.0. As highlighted by recent research, XAI has a requisite function in honing the transparency and accessibility of AI, addressing challenges across different phases of the machine learning

Table 9 A table discussing the functionality of XAI in accrediting human-AI partnership and decision-making in industrial settings

Function of XAI in decision-making and human-AI partnership	Description	Example applications
Transparency and interpretability	XAI techniques provide transparent explanations for AI-driven decisions, enabling humans to understand and trust AI recommendations	Predictive maintenance: XAI explanations clarify the factors influencing equipment failure predictions, allowing maintenance technicians to validate and act upon the perspectives provided by AI systems
Error detection and correction	XAI facilitates error detection by identifying instances where AI systems may make incorrect or biased decisions, enabling humans to intervene and correct these errors	Quality control: XAI techniques detect anomalies in manufacturing processes, alerting operators to potential defects or deviations from quality standards, which can be corrected to maintain product integrity
Adaptive decision support	XAI systems provide adaptive decision support by continuously learning from user feedback and adjusting their recommendations to better align with human preferences and objectives	Inventory management: XAI algorithms flexibly modify inventory levels in response to shifting market conditions and demand trends, incorporating feedback from supply chain managers to optimize stock levels and minimize costs
Human-AI interface design	XAI facilitates the design of user-friendly interfaces that enable effective partnership between humans and AI systems, allowing users to interrelate with AI models intuitively	Energy optimization: XAI-powered interfaces enable facility managers to visualize energy consumption data and adjust operational parameters in real-time, fostering collaboration between humans and AI in augmenting energy usage
Ethical and responsible AI	XAI promotes the blooming and deployment of ethical and responsible AI systems by making certain that judgments made by AI adhere to ethical standards, conventions, and human values	Supply chain management: XAI explanations justify AI-driven decisions in supply chain operations, ensuring compliance with ethical guidelines and regulatory requirements, such as fair treatment of suppliers and environmentally sustainable practices

Table 10 The challenges and potential solutions for escalade XAI systems to handle large-scale industrial data and complex AI models

Challenges	Potential solutions
Scalability of XAI techniques	Develop scalable XAI techniques that can efficiently process and scrutinize large volumes of industrial data without sacrificing performance
	To spread the computational load across several nodes or clusters, use distributed computing and parallel processing techniques
	Utilize cloud computing platforms and scalable infrastructure to provide on-demand resources for XAI instructions and inference of models
Interpretability of complex AI models	Developing XAI techniques specifically designed to interpret complex artificial intelligence models such as deep neural networks make sense and in an intuitive way
	Use model distillation or approximation techniques to simplify complex AI models into more interpretable forms while preserving performance
	Incorporate ensemble methods or model explanation frameworks that provide perspectives into the collective behavior of multiple AI models
Data quality and variability	Implement robust data preprocessing and cleaning pipelines to address data quality issues, like missing values, outliers, and noise
	Apply data augmentation and synthesis techniques to generate additional training data and reduce variability in industrial datasets
	Leverage domain-specific knowledge and contextual information, with the goal to elevate the XAI models' interpretability and generalizability
Real-time processing and decision-making	Optimize XAI algorithms for real-time processing by reducing latency and improving inference speed through algorithmic optimizations
	Deploy edge computing solutions to perform XAI inference locally at the edge devices, minimizing latency and bandwidth requirements
	Develop adaptive XAI systems that can continuously learn and update their explanations based on streaming data and evolving industrial contexts

Table 11 A table showcasing the potential benefits and challenges of amalgamate XAI with emerging technologies like IoT, edge computing, and 5G/6G networks in industrial settings [38]

Technology	Potential benefits	Challenges
IoT	Augment data collection: IoT gadgets make it possible to gather immense amounts of data in real time from industrial equipment and processes, providing valuable insights for predictive maintenance and process optimization	Interoperability and data synchronization: combining/synchronizing data from various IoT devices can be complex, requiring standardized protocols and robust data management systems
Edge computing	Decreased latency: edge computing reduces latency by moving closer to the data source and enables analysis and decision-making at the edge of the network	Resource boundaries: edge devices might only have a small amount of processing power posing challenges for running complex XAI algorithms and processing large datasets locally
5G/6G networks	High-speed connectivity: 5G/6G networks offer ultra-fast data transmission rates and low latency, facilitating seamless communication between IoT devices, edge nodes, and centralized AI systems	Infrastructure deployment: building and deploying 5G/6G infrastructure requires significant investment and may face regulatory and logistical challenges in industrial environments
XAI integration	Enhanced decision-making: integrating XAI with IoT, edge computing, and 5G/6G networks enables AI-driven insights and recommendations to be generated and acted upon in real-time, improving operational efficiency and productivity	Complexity of integration: integrating XAI with emerging technologies requires expertise in both AI and the specific technology domains, as well as careful consideration of interoperability and compatibility issues
	Improved transparency and interpretability: XAI provides clear justifications for AI-driven decisions and actions, enhancing trust and understanding among stakeholders and enabling effective partnership between humans and AI systems	Ethical and regulatory considerations: integrating XAI with emerging technologies raises moral dilemmas related to data privacy, fairness, accountability, and bias mitigation, requiring robust governance frameworks and regulatory compliance

life cycle. XAI facilitates responsible AI deployment in safety–critical domains and builds confidence and accountability by offering transparent and intelligible explanations for AI decisions. Building on this base, subsequent developments in XAI methods and evaluation measures highlight important patterns such as the inclination toward visual justifications assessing the worth of models for deep learning. Moreover, the integration of argumentation with XAI offers promising avenues for enhancing explainability in various domains, highlighting the interdisciplinary nature of XAI research. Initiatives such as DARPA's XAI program drive the development of novel machine-learning techniques aimed at creating more interpretable models. By exploring different strategies within the performance-explainability trade-off,

Table 12 A table that shows the methodology for evaluating the efficacy and caliber of XAI explanations

Evaluation factor	Description	Considerations
Fidelity	The accuracy to which the behavior and decision-making process of the underlying AI model are accurately reflected by the XAI explanation	The explanation should faithfully capture the relationships between input features and model predictions
		It should be consistent with the model's decision boundaries and reasoning
Interpretability	The ease with which users can understand and interpret the XAI explanation	The explanation should be presented in a clear and intuitive manner
		It should use familiar concepts and terminology accessible to non-experts
Human-centricity	The degree to which the XAI explanation is tailored To cater to end users' demands and preferences	The explanation should address specific user requirements and use cases
		It should be customizable, allowing users to adjust the level of detail or focus based on their preferences
Actionability	The extent to which the XAI explanation provides actionable insights that can inform decision-making or lead to improvements	The explanation should offer actionable recommendations or suggestions for improving outcomes
		Users should be able to leverage the explanation to refine their strategies, optimize processes, or mitigate risks
Robustness	The stability and reliability of the XAI explanation across different contexts and scenarios	The explanation should be consistent and reliable under varying conditions and inputs
		It should have been validated or tested to ensure its robustness and generalizability
Transparency	The degree to which the XAI explanation reveals the internal mechanisms and assumptions of the AI model	The explanation should give illustrations on how the AI model works and ensues at its decisions
		It should be transparent about any biases, limitations, or uncertainties inherent in the model

researchers aim to provide design alternatives that balance transparency with performance in AI systems. However, challenges remain, including the need for robust XAI techniques and the integration of ethical considerations into medical AI applications.

With the potential to increase production, efficiency, and connection, the incorporation of Artificial Intelligence (AI) into the Industry 4.0 framework serves as a substantial advancement in industrial innovation. Novel gadgets adored as robotics, cloud computing, IoT (Internet of Things) and big data analytics are driving this revolutionary evolution, which will alter manufacturing processes and bring in intelligent, adaptable production systems. Machine learning and predictive analytics, in particular, are unquestionably going to be major contributors to the optimization of supply chain management and quality control, among other aspects of the manufacturing value chain.

By leveraging AI's capabilities, companies stand to streamline operations, enhance decision-making, and offer personalized products and services on a scalable basis. However, the opacity inherent in many AI systems poses significant challenges regarding trust, transparency, and accountability. The opacity issues stem from the inherent lack of transparency and explainability within AI systems, peculiarly in complex machine learning algorithms. This opacity can arise from intentional corporate secrecy, technical intricacies, and the enigmatic "black box" traits of numerous AI models, thereby obscuring the decision-making process. Consequently, this inadequacy of transparency raises concerns over the fairness, equality, and potential biases embedded within AI systems (Fig. 4).

This gives us the growth analysis of AI and XAI over the years passed, this graph shows how much the XAI applicability has increased over the years.

This Table 13 outlines the significance of incorporating HCI principles and UX design in developing XAI systems for industrial applications, emphasizing the importance of intuitive presentation, usability, accessibility, and collaboration facilitation.

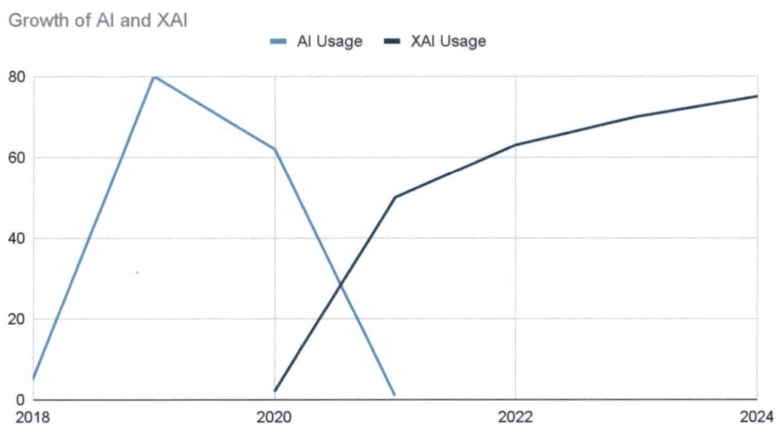

Fig. 4 Growth of AI and XAI as per growing years

Table 13 Role of HCI principles and UX design in developing intuitive XAI systems for industrial collaboration

Aspect	Description
Role of human–computer interaction (HCI) principles	HCI principles focus on designing interfaces and interplays between humans and computers to optimize usability and user experience. In the context of XAI systems, HCI principles guide the development of intuitive interfaces for presenting explanations
User experience (UX) design	UX design confine the process of enhancing user satisfaction by improving the usability, accessibility, and overall interaction with a product. In the context of XAI systems, UX design ensures that explanations are user-friendly and easy to understand
Importance of intuitive presentation	XAI explanations should be presented in a manner that is intuitive and easy to comprehend for users with varying levels of technical expertise. Intuitive presentation fosters trust in the AI system and encourages effective partnership between humans and machines
Enhancing usability and accessibility	Incorporating HCI and UX design principles in XAI systems enhances usability and accessibility, making explanations accessible to a wider audience. This approach promotes inclusivity and ensures that all users can benefit from the insights provided by the AI system
Facilitating effective collaboration	User-friendly XAI explanations facilitate effective collaboration between humanology and AI terminology by enabling users to interpret and act upon the insights provided. HCI and UX design principles serve as vital in promoting trust and cooperation in industrial settings

To address these challenges, Explainable AI (XAI) emerges as a vital resolution. XAI techniques are designed to furnish clear and understandable explanations for AI decisions, thereby enhancing transparency and certitude in AI systems. By rendering AI algorithms interpretable and transparent, XAI facilitates partnership between humans and AI systems, empowering users to comprehend how and why specific outcomes are reached. Moreover, the incorporation of principles from human–computer interaction and user experience design into XAI systems further augments usability and accessibility, ensuring that XAI explanations resonate with end-users.

The significance of incorporating HCI principles and UX design in developing XAI systems for industrial applications cannot be overstated. HCI principles, centered around optimizing usability and user experience, guide the development of intuitive interfaces for presenting explanations in XAI systems. Similarly, UX design, aimed at enhancing user satisfaction and interaction, ensures that explanations are user-friendly and easily comprehensible. Intuitive presentation of XAI explanations fosters trust in the AI system and encourages effective partnership between humans and machines, thus enhancing usability and accessibility for a wider audience.

However, despite the potential benefits of XAI, its implementation poses several challenges. These include striking a balance between transparency and performance, ensuring the consistency and reliability of explanations, and addressing societal biases and ethical concerns. Resolving these issues is essential to utilizing XAI to its most potential in Industry 4.0 environments, enabling the morally upright and liable application of AI. It will take multidisciplinary cooperation and ongoing research to advance XAI methods and promote accountable and transparent AI systems in Industry 4.0 and beyond.

A new era of industrial innovation is being heralded by the integration of Artificial Intelligence (AI) into Industry 4.0, which offers unmatched potential for increased productivity, efficiency, and connection. However, harnessing the full potential of this convergence necessitates tackling the challenges surrounding trust, transparency, and accountability within AI systems. Explainable Artificial Intelligence (XAI) is an important solution to these problems, providing clear and understandable AI decision-making and facilitating partnership between humans and AI systems. By embedding XAI principles into Industry 4.0 applications, companies can bolster transparency, trust, and accountability in AI-driven processes, thereby facilitating responsible and ethical AI deployment. Moreover, the integration of human–computer interaction and user experience design principles into XAI systems further enhances usability and accessibility, ensuring that XAI explanations resonate with end-users. This approach not only cultivates trust but also promotes effective partnership between humans and machines, maximizing the avails of AI in Industry 4.0 settings. However, the implementation of XAI presents challenges such as balancing transparency with performance, ensuring the consistency and reliability of explanations, and addressing societal biases and ethical concerns. Overcoming these challenges is vital for accomplishing XAI's full potential in Industry 4.0 and ensuring responsible and ethical AI deployment. Given the future, continued research and interdisciplinary collaboration will be indispensable for advancing XAI techniques and fostering transparent and accountable AI systems in Industry 4.0 and beyond. By tackling these obstacles and leveraging the transformative potential of XAI, we can unlock new frontiers of innovation and propel industrial progress toward a more transparent, trustworthy, and ethical future.

References

1. Xu, M., David, J.M., Kim, S.H.: The fourth industrial revolution: opportunities and challenges. Int. J. Financ. Res. **9**(2), (2018)
2. Raja Santhi, A., Muthuswamy, P.: Industry 5.0 or industry 4.0S? Introduction to industry 4.0 and a peek into the prospective industry 5.0 technologies. Int. J. Interact. Des. Manuf. (IJIDeM), Springer **17**, 947–979 (2023)
3. Stepin, I., Alonso, J.M., Catala, A., Fariña, M.P.: A Survey of Contrastive and Counterfactual Explanation Generation Methods for Explainable Artificial Intelligence. IEEE (2021)
4. Došilović, F.K., Brčić, M., Hlupić, N.: Explainable Artificial Intelligence: A Survey. IEEE (2018)

5. Ghassemi, M., Oakden-Rayner, L., Beam, A.L.: The false hope of current approaches to explainable artificial intelligence in health care. Lancet Digit Health **3**, e745–50 (2021)
6. Adadi, A., Berrada, M.: Peeking Inside the Black-Box: A Survey on Explainable Artificial Intelligence (XAI). IEEE (2018)
7. Saeed, W., Omlin, C.: Explainable AI (XAI): a systematic meta-survey of current challenges and future opportunities. Sci. Direct. **263** (2023)
8. Islam, M.R., Ahmed, M.U., Barua, S., Begum, S.: A Systematic Review of Explainable Artificial Intelligence in Terms of Different Application Domains and Tasks, MDPI (2022)
9. Vassiliades, A., Vassiliades, N., Patkos, T.: Argumentation and Explainable Artificial Intelligence: A Survey. Cambridge University Press (2021)
10. David Gunning, D.W.A.: DARPA's Explainable Artificial Intelligence Program. AI Magazine (2019)
11. Zhang, Y., Weng, Y., Lund, J.: Applications of Explainable Artificial Intelligence in Diagnosis and Surgery. MDPI (2022)
12. Confalonieri, R., Coba, L., Wagner, B., Besold, T.R.: A Historical Perspective of Explainable Artificial Intelligence. Wiley (2020)
13. Angelov, P.P., Soares, E.A., Jiang, R., Arnold, N.I., Atkinson, P.M.: Explainable Artificial Intelligence: Analytical Review. Wiley (2021)
14. Vilone, G., Longo, L.: Notions of Explainability and Evaluation Approaches for Explainable Artificial Intelligence. Elsevier (2020)
15. Alia, S., Abuhmed, T., El-Sappagh, S., Muhammad, K., Alonso-Moral, J.M., Confalonieri, R., Guidotti, R., Del Seri, J., Díaz-Rodríguez, N., Herrera, F.: Explainable Artificial Intelligence (XAI): What We Know and What's Left to Attain Trustworthy Artificial Intelligence. Elsevier (2023)
16. Jagatheesaperumal, S.K., Pham, Q.-V., Ruby, R., Yang, Z., Xu, C., Zhang, Z.: Explainable AI Over the Internet of Things (IoT): Overview, State-of-the-Art and Future Directions. IEEE (2022)
17. Terziyan, V., Vitko, O.: Explainable AI for Industry 4.0: Semantic Representation of Deep Learning Models. Elsevier (2022)
18. Zednik, C.: Solving the Black Box Problem: A Normative Framework for Explainable Artificial Intelligence. Springer (2019)
19. Vilone, G., Longo, L.: Explainable Artificial Intelligence: A Systematic Review. Arxiv, Cornell University (2020)
20. Das, A., Rad, P.: Opportunities and Challenges in Explainable Artificial Intelligence (XAI): A Survey. IEEE (2020)
21. Rojat, T., Puget, R., Filliat, D., Del Ser, J., Gelin, R., Díaz-Rodríguez, N.: Explainable Artificial Intelligence (XAI) on Time Series Data: A Survey, Arxiv. Cornell University (2021)
22. Islam, S.R., Eberle, W., Ghafoor, S.K., Ahmed, M.: Explainable Artificial Intelligence Approaches: A Survey, Arxiv. Cornell University (2021)
23. Khosravi, H., Tsai, Y.-S., Shum, S.B., Kay, J., Knight, S., Chen, G., Martinez-Maldonado, R., Gašević, D., Sadiq, S.: Explainable Artificial Intelligence in Education. Elsevier (2022)
24. Arrieta, A.B., Díaz-Rodríguez, N., Del Ser, J., Bennetot, A., Tabik, S., Barbado, A., Garcia, S., Gil-Lopez, S., Molina, D., Benjamins, R., Chatila, R., Herrera, F.: Explainable Artificial Intelligence (XAI): Concepts, Taxonomies, Opportunities and Challenges Toward Responsible AI. Elsevier (2019)
25. Bellini, V., Cascella, M., Cutugno, F., Russo, M., Lanza, R., Compagnone, C., Bignami, E.: Understanding basic principles of artificial intelligence: a practical guide for intensivists, national library of medicine. Acta Biomed (2022)
26. Burrell, J.: How the Machine Thinks: Understanding Opacity in Machine Learning Algorithms. Sage (2016)
27. Vaassen, B.: AI, Opacity, and Personal Autonomy. Springer (2022)
28. Phillips, P.J., Hahn, C.A., Fontana, P.C., Yates, A.N., Greene, K., Broniatowski, D.A., Przybocki, M.A.: Four Principles of Explainable Artificial Intelligence, NISTR 8312, U.S. Department of Commerce (2021)

29. Alicioglu, G., Sun, B.: A Survey of Visual Analytics for Explainable Artificial Intelligence Methods. Elsevier (2021)
30. do Silveira, G.N., Viana, R.F., Lima, M.J., Kuhn, H.C., Crovato, C.D.P., Ferreira, S.B., da Rosa Righi, R.: I4.0 Pilot Project on a Semiconductor Industry: Implementation and Lessons Learned. MDPI (2020)
31. Chen, T.-C.T.: Explainable Artificial Intelligence (XAI) in Manufacturing Methodology, Tools, and Applications. Springer (2023)
32. Chaddad, A., Peng, J., Xu, J., Bouridane, A.: Survey of Explainable AI Techniques in Healthcare. MDPI (2023)
33. Mill, E., Garn, W., Ryman-Tubb, N., Turner, C.: Opportunities in real time fraud detection: an explainable artificial intelligence (XAI) research agenda. Int. J. Adv. Comput. Sci. Appl. **14**(5), (2023)
34. Conati, C., Barral, O., Putnam, V., Rieger, L.: Toward Personalized XAI: A Case Study in Intelligent Tutoring Systems. Elsevier (2021)
35. Atakishiyev, S., Salameh, M., Yao, H., Goebel, R.: Explainable Artificial Intelligence for Autonomous Driving: A Comprehensive Overview and Field Guide for Future Research Directions, Arxiv (2024)
36. Božić, V.: Explainable Artificial Intelligence (XAI): Enhancing Transparency and Trust in AI Systems. ResearchGate (2023)
37. Kishor Kumar Reddy, C., Anisha, P.R., Khan, S., Hanafiah, M.M., Lavanya, P., Madana Mohana, R.: Sustainability in Industry 5.0: Theory and Applications. CRC Press, Taylor & Francis (2024)
38. Kishor Kumar Reddy, C., Anisha, P.R., Hanafiah, M.M., Doss, S., Lipert, K.J.: Intelligent Systems and Industrial Internet of Things for Sustainable Development, Sustainability in Industry 5.0: Theory and Applications. CRC Press, Taylor & Francis (2024)

C. Kishor Kumar Reddy currently working as Associate Professor, Dept. of Computer Science and Engineering, Stanley College of Engineering and Technology for Women, Hyderabad, India. He has research and teaching experience of more than 10 years. He has published more than 60 research papers in National and International Conferences, Book Chapters, and Journals indexed by Scopus and others. He is an author for 2 text books and 2 co-edited books. He acted as the special session chair for Springer FICTA 2020, 2022, SCI 2021, INDIA 2022 and IEEE ICCSEA 2020 conferences. He is the corresponding editor of AMSE 2021 conferences, published by IoP Science JPCS. He is the member of ISTE, CSI, IAENG, UACEE, IACSIT.

Siramdas Sai Jaahnavi is currently pursuing a Master's in Embedded Systems at Stanley College of Engineering and Technology for Women, Abids, Hyderabad, India. She has an impressive academic and research background, with three publications to her credit—one national and two international. Additionally, she has presented her project at a prestigious national seminar, demonstrating her commitment to innovation and academic excellence.

R. Aarti received her B.Tech degree from Aurora's Technological and Research Institute, Hyderabad, Telangana and M.E. degree in Digital Systems (ECE) Hyderabad, Telangana, she is pursuing her Ph.D. from JNTU Hyderabad. She is currently working as an Assistant Professor, Dept. of Electronics and Communication Engineering, Stanley College of Engineering and Technology for Women and has a teaching experience of 3 years, has been an active participant in various FDP, Workshops and several MOOCS courses. Her research interests includes Digital System Design, Machine Learning and Artificial Intelligence in biomedical and emerging technologies. She is a budding researcher and has contributed to a few publications and Book Chapters. She is a member of IAENG.

Marlia Mohd Hanafiah is a Professor at the Faculty of Science and Technology, Universiti Kebangsaan Malaysia (UKM). She is also a Head of Centre for Climate Change System, Institute of Climate Change, UKM. She received her Ph.D. Degree in Life Cycle Assessment (LCA) from Radboud University Nijmegen, the Netherlands in 2013. She has supervised/co-supervised 60 masters and 22 bachelor theses and currently she supervises 30 Ph.D.'s in the areas of LCA, wastewater treatment, green technology and sustainability. She is the coordinator of several postgraduate courses at her department. Her research projects focus on modelling potential environmental impacts of multiple stressors in an LCA context and exploring potential bio-based and nanomaterials for recycling and treating wastewater towards sustainability and circular economy. She has published over 90 peer-reviewed papers, books, books chapters, technical reports and serves as peer reviewer for several high impact journals. As a project leader, she has received several international and national grants from various funding agencies with a total amount of more that RM 2 million. Since 2013, she has been involved in consulting and reviewing Criteria Document and Standard for companies, organizations and governments. She is currently conducting transdisciplinary research with a strong interest that integrates various scientific disciplines such as environmental engineering, industrial ecology, toxicity, mathematical modelling and environmental science. Her ultimate goal is devoted to introduce Life Cycle Thinking as a holistic view to assess environmental performance of products and technologies to solve fundamental and unprecedented society challenges, especially those related to sustainability.

An Effective Explainable AI-Based Discrete Swarm Herd Optimization Model for Intrusion Detection in Industry 4.0 Networks

T. Saravanan, S. Maheswaran, Saigurudatta Pamulaparthyvenkata, P. Preethi, and N. Indhumathi

Abstract The increasing integration of artificial intelligence in Industry 4.0 networks makes explainable models crucial for ensuring the transparency and trustworthiness of security protocols. This paper proposes a particular Explainable AI-Based Discrete Swarm Herd Optimization model for intrusion detection as a solution to the distinct security challenges with Industry 4.0 environments. The proposed system reduces the number of false positives around network anomalies and threats, using swarm intelligence for optimal feature selection and classification. The proposed model utilizes explainable AI techniques so that the operators can understand how the system works, thereby allowing better responses towards security threats. The DSHO algorithm is characterized as discrete, thus ensuring an efficient handling of high-dimensional data in industrial networks while being computationally efficient. Further, a fuzzy logic-based approach for prioritizing intrusion risks is integrated to further refine response actions. This method improves continuously by interacting with real-world network data and thus gives a dynamic and adaptive intrusion detection solution. The usage of explainable AI will provide transparency, and this would enable the administrators to understand the results appropriately for improved overall network security when compared to other existing works. The model will be up to date with periodic updations and allows effective prevention of

T. Saravanan
Department of CSE, GITAM School of Technology, GITAM (Deemed to be University), Bengaluru, Karnataka, India
e-mail: tsaravcse@gmail.com

S. Maheswaran · N. Indhumathi
Department of Electronics and Communication Engineering, Kongu Engineering College Perundurai, Erode, TamilNadu, India

S. Pamulaparthyvenkata
Bryan, TX, USA

P. Preethi (✉)
Department of IT, Kongunadu College of Engineering and Technology, Trichy, India
e-mail: preethi1.infotech@gmail.com

L. Krishnasamy et al. (eds.), *Distributed Deep Learning and Explainable AI (XAI) in Industry 4.0*, Information Systems Engineering and Management 55,
https://doi.org/10.1007/978-3-031-94637-0_11

constantly evolving cyber threats. It has been seen to contribute immensely towards Industry 4.0 network security from sophisticated attacks.

Keywords Industry 4.0 · Artificial intelligence · Discrete swarm · Fault prioritization · Production improvements · XAI

1 Introduction

Future manufacturers will employ microarray technologies [1, 2], and quality will always be the most important consideration in all manufacturing methods, irrespective of the kind of procedure or the goods being produced. That's where the phrase "Industry 4.0" originated, referring to the industrial sector's "fourth The Industrial Period (4IR). It speaks of the "technological revolution" that businesses employing cutting edge technologies are able to achieve. This discipline attempts to enhance product quality by novel methods and state-of-the-art computational technologies [3, 4]. Although there are numerous social constructivisms and trends, understanding social studies at Grade 4.0 is the primary challenge. The multiplicity of logical structures and mental models makes this challenging. Reducing manufacturing waste and increasing efficiency are two of the most important things modern business can do to maintain competitive pricing. This is one of the more important objectives that should be followed. Efficient managing faults and prompt mistake correction in manufacturing lines are required to achieve this goal [5].

However, it is dependent upon identifying and categorizing challenges that have occurred before. Prioritizing defects could hasten repairing shortcomings but this is not a guarantee. Systems that are data-driven could provide problem control. The increased use of instruments in manufacturing processes to measure the basic health of equipment has culminated in a considerable rise in both the amount of information and its sophistication. The synthetic neural network algorithms that allow fault control employ this data to perform their functions. The goal of this research is to offer a summary of the requirements for issue prioritizing, identifying damage techniques, and specific criteria, along with the circumstances in which those strategies function. A review of the appropriate literature is also included in this research, with a focus on offering solutions for different stages of fault mitigating. The results of the study that was made public show that there is insufficient research on fault prioritization that addresses the various learning modalities, which emphasizes the need for expert judgment [6, 7]. Numerous diagnostic methods that automate fault diagnosis have been created recently. None, however, were sufficient for our production challenge in cement, as seen below. The majority of fault diagnosis systems described in printed materials analyzed the differences in the procedure (mean or variance) by evaluating one management chart, usually an X-bar or R (range) chart. However, in many operations, it is necessary to combine the two graphs since many recognized reasons may exist simultaneously. Due to the possibility of numerous plausible convertible causes, this is the case.

A more reliable diagnosis, however, can come from seeing odd patterns and having a deep understanding of the procedure. However, no prior Industrial Organizations recognition programs have been able to synthesize this combo, which could be beneficial for diagnostic. However, the effectiveness of the template was not examined when these methods were developed inside the framework of an actual case study. The incapacity to identify a variety of distinct and contemporaneous manufacturing processes, in addition to a substantial percentage of false identification, is often noted as a hurdle during these studies. Despite control charts showing trends, most deep learning and machine learning methods for recognizing control automating lack specific knowledge regarding trends and pivotal moments. This data is necessary to conduct a practical assignable cause analysis, which in turn speeds up the implementation of appropriate corrective actions [8, 9] A neural expert system that can perform intelligent ongoing surveillance and forecasting, corrective in nature, and remedial diagnostics of process oversight in plaster manufacturing was developed as a result [10–14]. In order to develop the recommended approach for issue detection by the team of experts along with offering input on the present projections, we will be tackling each of the crucial model elements: identifying and assessing any irregularities in the X-bar and R charts while concurrently viewing and assessing a variety of industrial processes, both single and simultaneous, both genuine and artificial.

1.1 The Main Contribution of the Research

Three major contributions of the proposed research are,

- This paper introduces a new variant of DSHO designed especially to address the needs of an Industry 4.0 network. It is an efficient and accurate approach towards identifying cyber intrusions with the aid of swarm intelligence for optimal feature selection and classification.
- The model integrates explainable AI methods for the enhancement of intrusion detection decision interpretability to be understandable and, hence, trustworthy and actionable to the security administrators who need to respond and act according to the system's outputs.
- The model proposes a fuzzy logic-based approach for rating identified security threats in terms of prioritization with a motivation toward ensuring crucial risks are considered within the earliest possible timeframe to result in a more responsive adaptive network security framework.

1.2 Novelty of the Research Work

In this research, a unique control chart-based surveillance and evaluation method for manufacturing processes is presented. The suggested technique may identify many different natural and artificial (single and concurrent) systems, while simultaneously monitoring and evaluating any anomalies in both the X-bar and R charts. This is in contrast to previous approaches that can only detect random fluctuations. A technique for estimating variables, different positions, and shifts in control diagrams that correspond to nonrandom structures is also presented in the study. Additionally, the method offers suggestions for preventive and/or corrective action in times of crisis, which makes it a crucial instrument for guaranteeing the reliability and caliber of industrial systems. Enhancing industrial systems' efficiency and productivity is anticipated, since this journal article makes a significant addition to commercial process management. On April 1, 2023, the Kaggle dataset was released and made accessible. This link leads to a page where you may enter model-explainability-in-industrial-image-detection. What makes this work unique is that it uses a conversation-driven design, incorporates natural language communication, uses double neural network convolutional neural scenarios, prioritizes faults using fuzzy logic, emphasizes regular updates, incorporates physical fix drivers with an automated administration system, and introduces the DSADRRFP algorithm for better performance. All of these elements work together to create an artificial intelligence (AI) system for business safety precautions that is more precise and effective. The remainder of this essay is organized as follows: The research methodology is summarized in Sect. 2, followed by the conceptual framework in Sect. 3, the system framework in Sect. 4, a comparative study, and real-world findings from the case study in Sect. 5. The paper concludes in Sect. 5.

2 Related Work

Using deep learning algorithms, researchers have made major progress in defect prediction for software systems on the industrial internet. Numerous investigations have examined the efficacy of various methodologies in augmenting fault prediction precision, recall, accuracy, and f-measure. A defect prediction model based on the fusion of the long short-term memory (LSTM) and locally linear embedding (LLE) algorithms was presented by [15]. Their model performed better than previous methods in use after it was trained on datasets taken from NASA's MDP dataset. The significance of proficient dimension reduction was underscored by the writers, who also stressed the advantages of using deep learning methodologies for software system failure prediction. Reference [16] conducted a research on the use of deep learning algorithms for bearing fault diagnostics. They used LSTM and other deep learning techniques to forecast software errors. According to their findings, LSTM and other deep learning techniques performed more accurately and efficiently than current models. Deep learning techniques were used by [17] to solve the problem of failure identification and exclusion in processes in industries. They reviewed widely used databases in software defect prediction, assessed performance metrics, and examined earlier research. Their study demonstrated the possibility for enhancing software malfunction prediction via the use of information mining, machine learning, and deep learning approaches. A further study suggested using deep learning to forecast the remainder of usable life (RUL) of industrial equipment when just a portion of the system health data is available [18]. The authors ignored non-discriminative components in favor of informative data with notable degrading traits by using a supervised concentration method. Their technique took into account the difficulties caused by imperfect data and real-world constraints in order to deliver efficient and trustworthy machinery evaluation and prognosis methods in contemporary industries. Furthermore, in a different work, the authors suggested a deep learning-based method for defect diagnosis and prediction in superconducting systems [19]. Improving the accuracy and efficiency of defect detection and prediction in magnetic technologies was the goal. Their work concentrated on using clever data-driven strategies to provide encouraging prognostic outcomes. These investigations demonstrate how deep learning algorithms, like LSTM, are becoming more and more popular for failure identification and prediction in software-intensive systems. The suggested models and techniques outperform current approaches in terms of efficiency, highlighting deep learning's potential to handle fault prediction difficulties across a range of domains [20].

2.1 Hybrid Models Based Fault Detection

After information about several common faults was gathered, a CNN classifier was able to categorize every mistake with a level of precision of 92.66% and recognize fault situations with a level of accuracy of 99.02% [21]. The results suggest that deep learning could be able to detect cold forging faults. There is no denying that systems are very important, and their impact can be seen in almost every aspect of modern life. On the other hand, it keeps growing since more and more services are switching to digital formats. For this reason, producing technology that can be trusted requires enhancing the processes used to create it and guaranteeing its quality [22]. A hybrid hidden Markov process and an AI framework are utilized to identify errors in the measurement database. In comparison to the traditional gas sensor array, our method performed much better when analyzing real-time information as well as various gas mixes. Our approach outperformed other current technology in terms of tracking possibly hazardous chemicals and spotting mistakes in sensor information. The combination of HMM and ANN defect identification techniques had many false positives of 0.01% and performed extremely well on the available data sets [23].

2.2 Alarming Data Analysis Based Fault Detection

A lightweight CNN model created for the diagnosis of audio faults in automobiles shows a considerable improvement in precision in classification when using the LFOA approach. This method limits the quantity of input characteristics taken from recordings of sound and lowers the cells in the DCNN's hidden layer. It is possible to develop a lightweight DCNN models that is appropriate for use on edge machines, such as mobile devices, through the use of the LFOA algorithms. The recommended model is an efficient research model for assessing the health condition of automobiles, as shown by studies that show it improves the accuracy of identifying the six problems to be found [24]. This study is interesting since it suggests a new AOC-ResNet50 network and shows that it can be effectively used to identify defects in wind turbines. A research comparing the identification of problems in wind turbine converters for power to competing convolutional neural network (CNN) models for deep learning supported this [25]. The results show that, even with different levels of deterioration, our artificial brain can predict the voltage levels of many different cells. Moreover, it drastically lowers a parametric model's forecasting error by 53%. This improvement allowed our network to predict a failure 31 h before it really happened, which is a 64% faster response time than the parametric model [26]. Next, the company and connection among previous alarm data are determined using an identity BiLSTM-CNN classification. This step comes after the one before it. The model is used for continuous identification of defects when the training period is over. Finally, the proposed model is applied to the analysis of the well-known Tennessee Eastman manipulate, and its result is shown. This remains true despite the sample's high level of inequality. It should be noted that the approach can still achieve this level of accuracy with certain restrictions serves as an example of this. The suggested approach may increase diagnostic accuracy by around 10% when when compared with currently accepted state-of-the-art techniques. Furthermore, when used with a complex nonlinear rubbing defect signal, it yields good fault precision in classification. [Reference required] [Reference required] This result suggests that the recommended structure is particularly suitable for use in the large industries that are present in the real world.

3 Proposed Taxonomy with DSHO

The first stage in the overall fault management process is gathering information. This stage involves managing characteristics and preparing the information before algorithms are trained for determining faults and identifying defects. Next, the discovered defects have priority and either the person in charge or the system itself executes the fault repair operation manually or mechanically. Because the stages are customized and some demand a certain degree of maturity, not all topics of inquiry are covered by this survey. White squares indicate research areas that are yet unexplored, while gray boxes reflect study topics that are actively being examined. The suggested model's architecture is seen in Fig. 1.

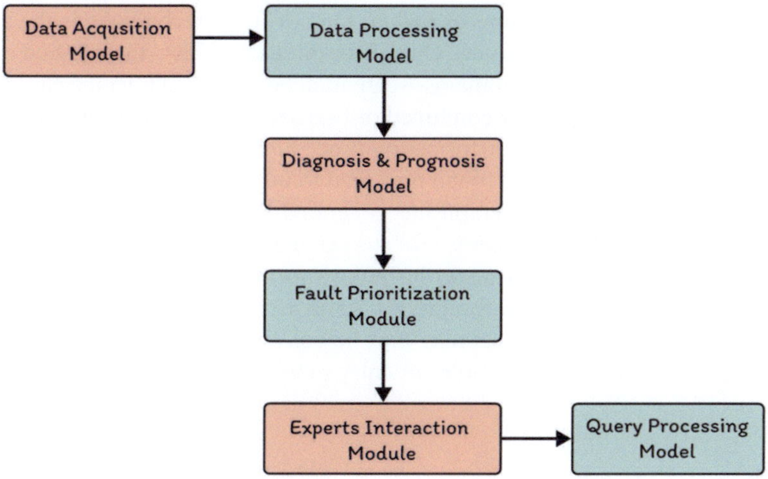

Fig. 1 Suggested model's framework

3.1 Data Acquisition Phase

After the data collection Phases is finished, any information gathered will be saved in the data storage facility, which will be utilized for diagnostic and prediction.

3.2 Data Pre-processing Phase

Let us consider the following scenario: throughout the procedure of knowledge discovery, there may be an overwhelming quantity of duplicate and unneeded data, such as noise or unreliable data. The course of action will be more challenging in such scenario. Therefore, it is essential that data preparation be done in some way before going on to the subsequent stage. Within the business community, this barrier can be referred to as BD. The most crucial steps in the data planning phase are often cleaning the information, integrating the information, data reduction, and information transformation. Among other tasks, this procedure involves settling conflicts, finding and removing outliers, normalizing data that is chaotic, and filling in the information that is missing. Combining data from many separate data repositories is the process of data integration. Carefully executed integration may help reduce and perhaps completely eradicate redundant and inconsistent data gathering that results from it. After data reduction, the data set is represented in a much smaller volume, yet it still has the ability to provide analytical findings that are almost identical to those of the original set of information. Different methods exist for reducing dimensionality. One of these approaches is as simple as using methods to extraction features on the information set. These techniques use preprocessed data and identify characteristics that indicate an imminent failure or defect. One of these techniques is the simple use of feature extraction techniques. One of the three domains—the a period of time frequency, or time–frequency domains—can often be used to extract the characteristics. The information are either combined or translated into formats appropriate for DM throughout the content transformation procedure. This makes the DM process more efficient, and the patterns found may lead to improved understanding.

Massive volumes of information are generated throughout the data collection process as a result of the advancement of storage media and computing power. Data pretreatment may effectively clean up the initial information, minimize the number of aspects in the data, and put the information back in the storage facility for information discovery. Large volumes of data may thus be converted into patterns or statistical parameters and then used as variables of entry in the DM algorithm.

3.3 Diagnostic Phase

We took use of Chen's approach to find relationships among the different data collected by sensors series. As an example, the information from the sensor and is associated with an ensemble of failures that are written as DTa = di1, di2,..., dim, while the sensory data b is associated with a clusters of failures that is written as DTb = dj1, dj2,..., djn. These two notes both allude to a comparable set of issues. The likeness of la and lb is then calculated using the parallels among DTa and DTb. LS is the sign for this connection (la, lb).

$$LS(I_a, I_b) = \frac{\sum_{i=1}^{m} \max_{1<j<n}\left(DS\left(d_{ai}, d_{bj}\right)\right) + \sum_{j=1}^{n} \max_{1<i<m}\left(DS\left(d_{bj}, d_{ai}\right)\right)}{m+n} \quad (1)$$

where the linguistic link between defect and both d_{ai} and d_{bj}, which separately pertain to DTa and DTb, is represented by the expression $DS(d_{ai}, d_{bj})$. The "mandn" numbers, which are enclosed in parenthesis, represent the greatest number of mistakes that the DTa and DTb have corrected. The quantity of L_{ij} is used to show the level of similarities among two different sensor information sets (li and lj), and matrices LR (n lnl) is used to show the parallelism among various information from sensors.

They will go over the diagnostic method for locating hidden sensor modeling and data-related linkages as well as foretelling data associated with malfunctions in the next part. The parallels, connections, and interactions among sensor information and issues were originally included in the set of characteristics that was created during the first step of the construction process. In order to utilize the set, this was done. This has led to the development of a capacity that utilizes segmentation and focus procedures in dual neural networks. While the proper side of the framework is in responsibility of learning the more important connecting ties among sensors in clothing and errors, the erroneous side of the network is in control of learning the general interpretation of an inertial imaging link. After that, a closely associated level and an extra inversion are used to combine these two models. The likelihood that a sensing set of information is connected to an error produced this representation's link grade. The detection sensor l1 and defect d2 will be used to show our dual CNN or diagnostic linkage approach.

3.4 Fault Prioritization Phase

In problem handling and leadership systems used in a variety of sectors, such as production and medical imaging, the error prioritizing component is essential. Prioritizing and swiftly identifying errors or malfunctions that need to be fixed right away is its main goal. The Phase may use data-driven techniques, such as machine-learning computations, to evaluate and understand massive amounts of information from sensors that are generated from manufacturing lines or imaging equipment. This evaluation of information helps to improve overall quality and efficiency, minimize production costs, and cut down on unavailability. By quickly locating and resolving essential issues, the fault prioritisation component effectively maximizes efficiency in operation while enhancing system dependability and efficiency.

3.5 Experts Interaction Phase

Fault prioritizing is essential in fault management methods in sectors like industrial and health care imaging. Prioritizing and swiftly identifying errors or faults that need to be fixed right away is its main goal. The course can employ data-driven techniques, such machine learning computations, to evaluate and understand massive amounts of data collected from sensors that are generated from manufacturing lines or imaging equipment. This examination of data helps to improve overall quality and efficiency, minimize production costs, and cut down on downtime. By effectively locating and resolving key issues, the error prioritisation Phases contributes significantly to the optimization of efficiency in operation and enhances systems functionality and reliability.

3.6 Query Processing Phase

The process of prioritizing errors or problems inside the Query Processor Unit entails ranking them according to importance and effect on the overall operation of the system. Ensuring effective utilization of resources and successful fault correction is the aim of the prioritization process. A variety of factors, including the extent of the issue, potential effects on system operation, frequency of occurrence, and labor required to fix it, can be considered when prioritizing defects. In order to decrease disruptions and enhance the network's reliability and security, the engineering team could be better off allocating resources in this way by prioritizing essential problems and analyzing and classifying defects according to these qualities.

3.7 Construction of Feature Matrix

The feature map that contrasts data from sensors set 1 with the fault data set 2 is created by combining three biological principles. First, if l1 and d2 show homology and connection linkages with more common information from sensors, there is a greater chance that l1 and d2 will be connected. When l1 and l2 conduct similar duties and d2 is linked to l2, it is probable that l1 is also tied to d2. Assuming that ×1 comprises all of the connections between l1 and the other RNAs in ln2, let us assume that ×1 represents the first row of L. We record the associations between each sensor data and d2 in the second column of D, designated ×2. After combining ×1 and × 2, a matrix of 2n dimensions is produced. Secondly, there is a greater chance that l1 and d2 are related if there are similarities between l1 and d2 diseases and links to more common problems. The letter ×3 in A's first row indicates the links between each fault and l1. The second row of D, ×4, lists the many similarities that may be found between d2 and these illnesses. Furthermore, ×3 and ×4 are combined, and the resultant matrix has dimensions of the second kind. Third, when l1 and d2 interact and establish linkages with the identical information from sensors, a linkage between them is feasible. The first row of Y, ×5, records the contacts that arise between l1 and the various data collected by the sensors, and the subsequent line of B, ×6, records the interactions between d2 and these information from the sensors. When ×5 and ×6 are integrated, a vector with dimensions of 2n*m is produced.

3.8 Convolutional Phase on the Left

To train generic depth symbols for l1 & d2, the multilayered element to the left of it receives the distinctive matrices P, which is made up of l1 and d2. It will utilize the first trained samples, the first dense level, to demonstrate how effectively both the single or the pooling operations function since it is simpler to grasp using examples. In order to learn as much as possible about P, we first create a new matrix, which we will refer to as P′, by filling it with 0.

3.9 Convolutional Layer

The height of a filter within the first convolution is represented by the number nf, while its width is represented by nw. Using Wconv1, $Rn_w \times n_f$ filtering to the matrices P 0 yields the cnn models Zconv1, Rn conv1 $(1, 4 - n_w + 1)$, $(1, n_t + 2 - n_f + 1)$, provided that nconv1 is the number of pixel. $P_{k,i,j}$ is a region that is contained within the filtering system and is reached once the kth filters slid to the place P 0 k, i, j. P 0 i, j is the side that is situated in the ith row and the jth item of P 0. The official definitions of $Z_{conv1,k}(i, j)$ and $P_{k,i,j}$ are as follows:

$$P'_{k,i,j} = P'\left(i : i + n_w, j : j + n_f\right), NP'_{k,i,j} \in R^{n_w x n_f} \tag{2}$$

$$Z_{conv1,k}(i, j) = f\left(W_{conv1}(k, :, :) * P_{k,i,j} + b_{conv1}(k)\right) \tag{3}$$

$$i \in [1, 4 - n_w + 1], j \in \left[1, n_t + 2 - n_f + 1\right], k \in [1, n_{conv1}] \tag{4}$$

where nt = nl + nd + nm, f is a relu operation, and b_{conv1} is a biased vector. The component located in the i th row and j th columns of the kth feature map called $Z_{conv1,k}$ is denoted as $Z_{conv1,k}(i, j)$. The dual-CNN, a that uses two CNN algorithms to do specific duties or improve the outcome of an assignment, is explored in Pooling Layer Fig. 2. Dual-CNN theory has been investigated in a number of fields, including denoising of pictures and caption. Dual-CNN algorithms were developed and taught for these investigations in order to solve particular issues and enhance the efficiency of tasks.

The Fig. 2 shows the Merged CNN for Fault detection. The process of flattening complex information into a singular vector, like maps of features in a CNN, is known as flatness. The input may now be supplied into a dense artificial intelligence layer, which is a fully connected layer, thanks to this change. Usually, flattening is done before of the thick coverings so that the network can perform job classification or make forecasts. In a neural network, a layer referred to as dense layers—also called completely linked layers—is one in which every neuron remains linked to every other neuron in the layer above it. These layers are essential for deciphering intricate patterns and generating predictions using the information that were retrieved. They enable the neural network to carry out tasks like regression as well as classification and to get high-level representations. A procedure or method that continues itself is called recursive; usually, the result of one iteration serves as the input for

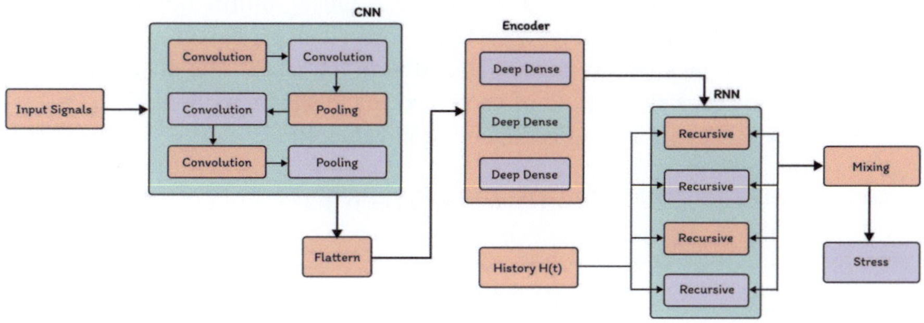

Fig. 2 Merged CNNs for fault identification

the subsequent one. Recursive techniques may be used in neural networks to analyze hierarchical structures and generate sequences. Recursive deep neural networks have been used for sentiment evaluation and parsing, among other machine learning applications. When neural networks are mixed, many methods or processes are used. When referring to CNNs, the term "mixing" may be used to describe actions like as concatenated or elementwise combination that bring together features or interpretations from various networking branches or layers. Mixing activities may boost the network's capacity to gather various types of data and boost efficiency. In this situation, stress has no special connotation relating to neural networks or profound learning. Stress management methods or treatments may be investigated in relation to human well-being in order to reduce stress while promoting general well-being and productivity.

In maximum pooling, 2D filter is slid across each channel in the characteristic map, summing the features that fall within its coverage. $Z_{\text{convpool1}}$, the dimension of the outcome of the pooling layer may be computed.

$$z_{convpool1,k}(i,j) = \max\left(Z_{conv1,k}\left(i : i + n_g, j : j + n_p\right)\right) \tag{5}$$

$$1\varepsilon\left[1, 5 - n_w - n_g + 1\right], j\varepsilon\left[1, n_t + 3 - n_f - n_p + 1\right] \tag{6}$$

$$k\varepsilon[1, n_{conv1}] \tag{7}$$

3.10 *Attention Phase on the Right*

According to our notion, the focus Phase are in charge of figuring out if features or link linkages are important for the representation of defect D2 and information from sensors L1. Thus, both the person-level method and the categorization stage suggested method are parts of the Phase. Most of the time, varied qualities within P contribute differently to different kinds and amounts of information from sensors as well as the defects that go along with them. For instance, information from sensors that have been demonstrated to be associated with a certain defect are often more important than the ones which weren't linked with the issue. The attention value_ ijF is assigned to every element xij of vectors xi in the square matrices P = × 1, × 2,..., × i,..., × 6. The following is the attention value_ijF.

$$\mathrm{s_i^F} = \mathrm{H^F}\tanh\left(\mathrm{W_x^F xi} + \mathrm{b^F}\right) \tag{8}$$

$$\alpha_{ij}^F = \frac{\exp\left(s_{i,j}^F\right)}{\sum_k \exp\left(s_{i,j}^F\right)} \tag{9}$$

where the regularity carriers in HF and W are indicated by xF, and the error message for the full function is represented by bF. The vector si^F = [si1^F, si2^F,…, sik^F,…, s (ini)^F] retains the attention scores the fact that indicate the worth of each attribute contained in xi, wherein ni is the length of xi and the value that x(in i) has been assigned. The adjusted concentration value for parameter xij is reflected by the number_ijF.

$$y_i = \alpha_i^F \otimes x_i \tag{10}$$

3.11 Attention at the Relationship Level

The interaction between sensor information and defects may take many distinct forms. Correlation relationships include similarities among circRNAs, similarities among detector information and faults, similarities among faults, interaction between various data types, and connections between defects and sensor information. The representation of sensor data-fault links varies based on the evaluated correlations. Therefore, you must apply a learning method to each variable yi independently in order to construct a complete focus description at the real level. The assessments of attention at the individual level are influenced by the following variables:

$$s_i^R = h^R \tanh\left(W_y^R y_i + b^R\right) \tag{11}$$

$$\beta_i^R = \frac{\exp\left(S_i^R\right)}{\sum_{j\in 6} \exp\left(S_i^R\right)} \tag{12}$$

where bR stands for the biased vectors and WyR for the set of weights. The standardized attention value for connection yi is called iR. The emotions at the part and relation layers accomplish the obtained and revealed concealed portrayal of relationship.

$$g = \sum_i \beta_i^R y_i \tag{13}$$

where the degree of the current relationship is indicated by the letter R. Let G be the grid that comes after it has been padded with zeros. It is therefore possible to construct the attention representations Z "att" by placing G into a practical and maximal pooling layer.

3.12 Final Phase

We will refer to the data obtained from the middle convolutional Phase as Zatt and the model that was learned from the left recurring component as Z glo. Everything that was discovered about focus will be Zatt. The sign Zcon, which is created by putting a first on top of a later and a latter below it, represents a mix of Z glo and Zatt (Fig. 2). An additional neural processing step is applied to the Zcon level with the goal to generate the final form Zfin. The line vector z0, which is produced by straightening the Z fin, is fed into the softmax algorithm and a convnet named W out to produce the value p.

$$p = softmax(W_{out}Z_0 + b_0) \tag{14}$$

Since pi is a linked random factor of C subclasses (where C = 2), it captures both the probability that an occurrence and sensor information are determined to have a partner and the probability that they do not.

Diagnosed with Loss of Connection.

According to our notion, L is the bridge loss connecting the test datasets dispersion and the likelihood of diagnosing a lncRNA-fault relationship, and L is defined as follows:

$$L = -\sum_{i} TX * \sum (j = 1)^{cZjlogpj} \tag{15}$$

where T represents the collection of training cases and zR2 represents classification label vectors. The vector z has an aspect of 0 for its first level and a level of 1 for its subsequent dimensions if l1 is linked to d2. In contrast, the first component of z is equivalent to 1 and the following aspect is equivalent to 0 if l1 is not linked to d2.

$$\min_{T} \theta L(\theta) = L + \lambda ||\theta||^2 \tag{16}$$

where is a variable that shows how the regularization term and the size of the initial data are traded off. For the goal variable to have the highest level of effectiveness, we use the Adam optimizing approach.

3.13 Data Acquisition and Signal Processing and Analysis

In information gathering structures, basic parts like LabVIEW program and analog-to-digital conversion cards (DAQ boards) are available to engineers. Engineers may develop and build their own bespoke data collecting systems with the help of these technologies. Engineers may easily and efficiently manage, analyze, and show information in immediate form on their computer's screen thanks to LabVIEW's features. The NI-DAQmx driver and LabVIEW work together to make it easier to create triggered gathering information apps. Four basic construction pieces make up an NI-DAQmx use, and trigger circumstances for those apps are defined in the settings area using the relevant LabVIEW VIs or callback functions. CNNs often use the pooling approach to minimize the geographic dimensionality of feature maps while maintaining essential data. It facilitates the removal of superfluous characteristics and lowers computing overhead. In a CNN design, layer pools like max or typical pooling are implemented after layer convolution.

3.14 Inner Fault Prioritization Phase

The prioritization process needs the input of informed experts. They must consider the several possible issues that may occur with the detected data. They also need to take into account the fact that machines rely on one another rather than working independently. Industry experts must thus take into account not only the significance of individual equipment but also the chain of failure and the ways in which machines affect other processes along the manufacturing line. Furthermore important is the expert degree of topic knowledge. The manufacturing processes and CPSs that are used are very innovative and provide a vast range of conceivable combinations. The lack of methods that do not depend on data makes fault prioritization challenging in an industrial setting. The strategy has been put to the test and shown to work in a range of industrial environments. However, this is an expensive option since it requires a lot of time and work and depends on statistics. An automatic problem prioritization mechanism has been attempted in a limited capacity. They do not, however, relate to Industry 4.0 in any way. We see a great possibility to apply knowledge from different disciplines and courses of study to the industrial sector. This is crucial since digitalization usually makes manufacturing plants more complicated, and the sole thing that might reduce supply disruptions is setting priorities correctly.

3.15 Inner Query Processing and Interaction with Experts/ Consumers

Classifying flaws is the next step that must be completed. To do this, the data must first be classified and issues must be found in accordance with their various types. The prioritization process needs the input of informed experts. They must consider the several possible issues that might come from the data that is detected. They also need to take into account the fact that machines rely on one another rather than working independently. Industry experts must thus take into account not only the significance of individual machines but also the chain of failures and the ways in which machines affect other processes along the production line. The CPSs and assembly lines in use both have a high degree of A rigorous computer technology called NLP, which stands for natural language processing, makes it possible to gather data on how humans use and comprehend language. Thanks in large part to AI, it has evolved to a new level of innovation, technical development, and economic success known as Industry 4.0. The German government originally proposed the idea of Industry 4.0 in 2011, and it describes an advanced industrialized economy that has garnered international prominence. IoT and the incorporation of artificial intelligence and NLP have made Industry 4.0 a reality for people worldwide, allowing for advantages and involvement. Through their knowledge of client needs, NLP and AI have affected customer decisions and enabled better communication as well as individualized products and services.

The research analyzes the context of the data, including attitude, intention, and fault representation, using pragmatic analysis. To accomplish considerable classification, industries use global optimization based on globalization. Industry 4.0 relies heavily on NLP applications to handle numerous forms of communication and improve operations and customer service for businesses. Businesses can learn more about their consumers, cater to their demands, and ascertain the needs of their clientele according to corporate objectives by combining AI and NLP of uniqueness and provide a wide range of potential arrangements. In an industrial context, fault prioritizing is challenging since there are few methods that do not depend on data. Fuzzy logic is a useful tactic that may be used in a variety of situations to mitigate the negative consequences of perceived shortcomings. The strategy has been put to the test and shown to work in a range of industrial environments. This tactic is labor- and time-intensive, depending on statistics, thus it's an expensive option. The process cannot be computerized since it requires a high level of skill and familiarity with the subject area. This is crucial because manufacturing facilities become more complicated as a result of digitalization, and the only way to reduce supply disruptions is to properly prioritized.

3.16 Integrating XAI with DSHO

Proposed architecture for precise industrial surveillance of networks is described in depth in this part. Pre-processing, feature selection, and classification are the three stages of the malware detection system.

Pre-processing involves cleaning and normalizing the data, which improves information quality. Subsequently, the KHOs technique is used in the process of feature choice to extract the noteworthy information characteristics. Lastly, the data is classified using the Proposedmethod, which correctly identifies the existence of incursions and classifications the information into groups. These subsections provide a detailed description of these processes. The simplified representation of the suggested systems for intrusion detection is shown in Fig. 3.

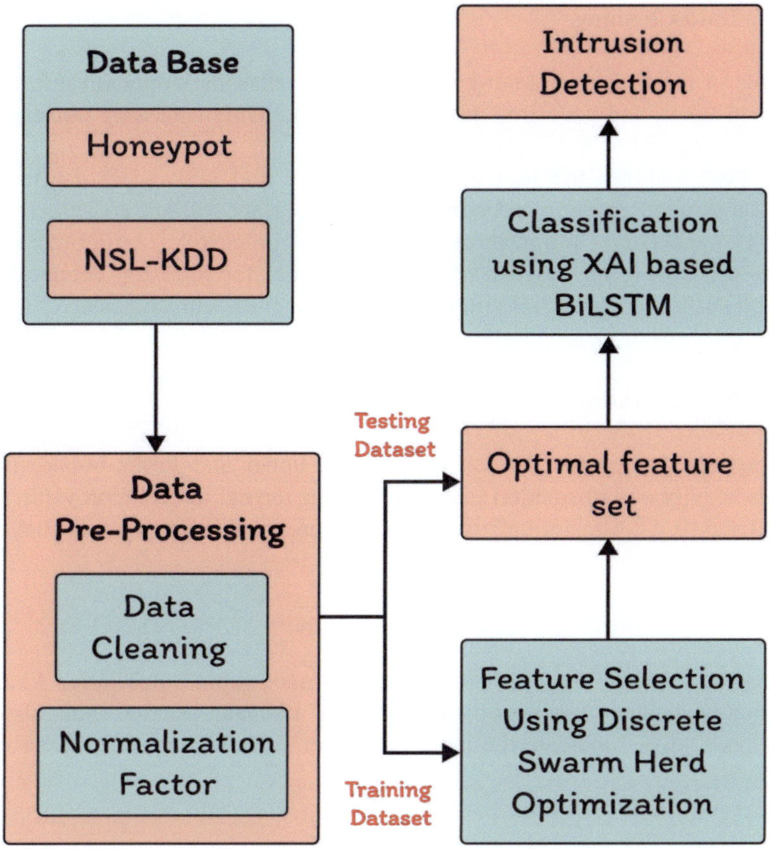

Fig. 3 The suggested security detection system's schematic frame

3.17 Data Pre-processing

To improve the capture and processing of knowledge, gathering data involves modifying the data ranges in an inventory. There is a significant difference compared to among the dataset's highest and lowest ranges. The standardized nature of information throughout this procedure make an algorithm less challenging. When using neural network methods for algorithmic categorization, normalizing the information becomes more powerful. The learning neural network will become completely efficient and increase training pace due to input standardization if it learns to apply the back-propagation approach.

3.17.1 Data Cleaning

Utilizing an information cleaning technique, the dataset's redundant information, noises, mistakes, and undesired data are eliminated. Only necessary information is allowed to proceed with this procedure.

The Fig. 3 shows the system uses Honeypot and NSL-KDD datasets, pre-processed through cleaning and normalization. Relevant features are selected using Discrete Swarm Herd Optimization, reducing data complexity. The optimal features are classified using an XAI-based BiLSTM model for intrusion detection. This approach ensures accurate, interpretable threat detection with enhanced efficiency.

3.17.2 Normalization Function

The normalized operation, which contains an upper and lower bound method, depends heavily on information scaling to change the net information value among [− 1, 1] and [0, 1]. The normalizing formula is provided in the formula above.

$$I = \frac{d - d_{MIN}}{d_{MAX} - d_{MIN}} \tag{17}$$

The term I in Eq. (1) represents the transformed input value, often known as the normalized value. Additionally, the word "d" indicates the real value; the terms "d_{MAX}" and "d_{MIN}" indicate the input parameter "d"'s highest and lowest values, respectively.

3.18 Feature Selection Process for IDS

The procedure of selecting features is becoming more and more popular as a means of choosing important feature subsets. Large multidimensional and numerous highlighted information influence the model's accuracy in detecting intrusions. Distinct features and inconsequential characteristics that might lead to a misclassification by the machine learning algorithm are included in the data characteristics that belong to distinct classes. Through the removal of irrelevant characteristics from the information set, the attribute choice method reduces the complexity of the information. In addition to removing redundant data, it lowers the percentage of false positive incursion detections. Additionally, it lowers calculation load while increasing detection performance. As a result, the choice of features has a greater influence on the model's potential for generalization and accuracy in detection. As a result, we presented the krill herd optimisation (KHO) method in this research for choosing features, which efficiently chooses the feature subsets by decreasing the information dimensionality and boosting the intrusion identification method's detection rate and precision. This KHO method's concise explanation is represented by the following model.

3.18.1 Discrete Swarm Herd Optimization Algorithm

While each KH member made an impact via movement, the KHO algorithm replicated the behavior of krill. When the krill people locate the feeding center, they choose the optimal option. The KHO method may perform mining and investigating in optimizing issues based on the lagrangian and evolutionary behaviour of each krill. Within this KHO method, the value of chance plays a crucial function. Three activities are used to determine the time-varying location of the particular krill: randomized dispersion, feeding action, and the krill individual's motion initiation. The d-dimensional search area of lagrangian architecture is used by the KHO method, which is represented by its formula below;

$$\frac{\partial Z_i}{\partial t} = Ok_j + Gk_j + Ek_j \tag{18}$$

The search agent's movement inductive methods, forage behavior, and randomized dispersion are denoted by the words Ojj, Gjj, and Ejj in this instance. Initializing variables such as the highest possible diffusing velocity (EMAX), krill position (Zj), maximum forage velocity (MG), maximal generated velocity (oMAX), and a maximum amount of repetitions JMAX and krill O numbers. Calculate each krill's movement individually. Single krills are motivated to relocate and make an effort to preserve the greater density due to the collective consequences. The direction of motion induction ones, denoted by rj, is assessed in relation to the local objective and the repelling swarm's density. Next, it is stated as;

$$Ok_j^{NEW} = o^{MAX}\sigma_j + x_0 Ok_j^{old} \tag{19}$$

The symbols for the inertia size, the last action, and the maximum generated velocity are xo, OjOLD j, and oMAX, respectively.

$$\sigma_j = \sigma_j^{LO} + \sigma_j^{t} \tag{20}$$

$$\sigma_j^{LO} = \sum_{k=1}^{o} k_{j,k} y_{j,k} \tag{21}$$

$$k_{j,k} = \frac{Y_k - Y_j}{||Y_k - Y_j|| + \alpha} \tag{22}$$

$$k_{j,k} = \frac{k_j - k_k}{k^{WO} - k^{BEST}} \tag{23}$$

jBEST and jWO represent for the most awful and greatest krill people, respectively. A stands for the bigger affirmative number. Oo indicates the number of neighbors. Additionally, the words jj and jk represent the suitability function of the jth krill and the kth others, respectively; Yk and Yj indicate the corresponding placements of the jth krill and the kth neighbor, respectively; and rsj and rLOj signify the objective directional effect and localized effect supplied by fellow citizens, respectively.

$$\sigma_j^t = D^{BEST} k_{j,BEST} Y_{j,BEST} \tag{24}$$

The following equation expresses DBEST, which stands for the krill person efficiency person via best efficiency;

$$D^{BEST} = 2\left(R + \frac{J}{J_{MAX}}\right) \tag{25}$$

$$E_{t,j} = \frac{1}{50} \sum_{k=1}^{0} Y_j - Y_k \tag{26}$$

where J is the present repetition, Et;j denotes detecting distance, and η represents a random number that falls inside the range [0,1] The definition of the foraging activity is based on previous observations and the precise location of the foodstuffs. Next, it is stated as;

$$Gk_j = v_g \eta_j + x_g Gk_j^{OLD} (27) \tag{27}$$

xg represents inertia mass with forage activity, gj the fitness worth for the jth krill, and gBEST the krill optimum goal.

$$\eta_j = \eta_j^{FOOD} + \eta_j^{BEST} \tag{28}$$

$$\eta_J^{FOOD} = D^{FOOD} K_{j,FOOD} Y_{j,food} \tag{29}$$

$$D^{FOOD} = 2\left(1 - \frac{J}{J_{MAX}}\right) \tag{30}$$

gBEST j determines the jth krill personal best aims, which are represented in the equation below.

$$\eta_J^{BEST} = k_{j,jbest} y_{j,jbest} \tag{31}$$

$$Y^{FOOD} = \frac{\sum_{j=1}^{o} \frac{1}{kj} Y_j}{\sum_{j=1}^{o} \frac{1}{kj}} \tag{32}$$

The diffusion velocity for maximal is used to characterize the actual diffusing motion, and the following equation expresses the random directional matrices;

$$Ek_j = E^{\max}\left(1 - \frac{J}{J_{MAX}}\right)\phi \tag{33}$$

which falls among − 1 and 1, represents the random orientation vector. Through KHO, the biological reproductive methods of mutation and crossing over are combined to improve KHO's efficiency. The crossover value for Yj0's nth element is represented by the formula below,

$$Y_{j,n} = \begin{cases} Y_{s,n}\ R_{j,n} < c0 \\ Y_{j,n}\ else \end{cases} \tag{34}$$

$$Y_{j,n} = \begin{cases} Y_{hBEST,n} + v\left(Y_{q,n} - Y_{r,n}\right)\ R_{j,n} < Nu \\ Y_{j,n} \qquad\qquad else \end{cases} \tag{35}$$

Nu represents the mutation the likelihood, which is set to Nu ¼ 0:05 The following equation is used to get the krill's location vector in the time frame.

$$Y_j(t + \Delta t) = Y_j(t) + \Delta t\frac{\partial Y_J}{\partial t} \tag{36}$$

3.19 Classification Using BiLSTM-Explainable Artificial Intelligence (BiLSTM-XAI)

The categorization job in intrusion detection programs (IDS) is a crucial step that divides the data being input datasets into two groups based on the presence or absence of computer breaches. BiLSTM-XAIs that precisely detects the existence of any illegal and irregular behavior in networks in order to get effective categorization results. The next subsections include more specific examples of these strategies.

3.19.1 Bidirectional Long Short-Term Memory (BiLSTM)

BiLSTM stores information forwards as well as backwards of the neural network's path. An encoded sequence of the attributes of the Inception classifier is sent to the LSTM classifier. The linear structural theory (LSTM) models are used for obtaining time data and features from spoken language movies. To recognize shared time-dependent trends in sequences of input derived from feature patterns that have been trained.

$$j_p = \mu\left(Z_j \cdot \left[d_{p-1}, g_{p-1}, y_p\right] + a_j\right) \tag{37}$$

$$e_p = \mu\left(Z_e \cdot \left[d_{p-1}, g_{p-1}, y_p\right] + a_e\right) \tag{38}$$

$$d_p = e_p \cdot d_{p-1} + j_p \cdot d_p \tag{39}$$

$$q_p = \mu\left(Z_0 \cdot \left[d_{p-1}, g_{p-1}, y_p\right] + a_0\right) \tag{40}$$

$$g_p = q_p \cdot \tanh\left(d_p\right) \tag{41}$$

The sequences of input, output, and the state of the memory at each given point in time, p, are represented by the symbols yp, gp, and thep. The cell stimulation are shown with d ~. The vector of inputs and these values have the same size. For nonlinear sigmoid functions, the sign l is used.A layer of layered LSTM cells may interact with one another and use weights that are comparable to those of another layer. These layers are used in the identification of long-term bidirectional relationships among time steps. Using BiLSTM results in output with characteristics from both past and future time steps.

3.19.2 Explainable Artificial Intelligence (XAI)

The topic of XAI study involves investigating various techniques that will enable autonomous intelligent systems to function in a way that is comprehensible and interpreted by humans. Human–machine interface is closing the gap among data science and social science, which advances AI technology and moves the field toward rational, transparent, and responsible AI. Generally speaking, a thorough explanation makes it easier to evaluate the advantages and disadvantages of learning frameworks and makes the model more reliable and intelligible. One kind of explain model called "post hoc explain" uses a black box model to extract facts in order to help decision-makers make better choices. By offering a prompt explanations in place of internal functioning, it primarily provides professionals and end users with vital data. Enhancing transparency of instructional techniques in decision-making about anticipated output is the primary goal of XAI.

Local Interpretable Model-Agnostic Explanations (LIME)

Every instance that the black box model generates is explained using the LIME method. The efficiency of the classifiers in determining the proximity of the information examples to be explained using the local proxy approach, which can be quantitatively shown in equation form as follows, determines how the LIME framework will be explained.

$$Explanation(y) = ArgMin_{G \in g} \int (F, G, \pi_y) + \varphi(g) \quad (42)$$

In this case, y stands for the data examples that need to be explained, R F; G; py {~ shows the fidelity term, G for the degree of complexity term, F for the black box approach, and G for the information instance that has been explained. In order to align the data instance proximity to the anticipated outcome, the local proxy modeling makes an effort. Initially, the LIME system generated data characteristics from the raw data by using the disturbance principle. In contrast, the LIME approach is employed differently by taking into account univariate distributed characteristics in order to ascertain the probability distribution of all characteristics. For categorical characteristics, selection is done according to the amount associated with the classes, whereas three options are used for numerical characteristics. Firstly, the unprocessed information is sorted into groups based on quantiles. A single bin is picked at at random, and samples are taken regularly among the highest and lowest bins allocated. Next, LIME uses a standard distribution to approximate the real variation in numeric characteristics. Next, the original pattern of numerical parameters is approximated using the kernel densities functional.

Shapley Additive Explanations (SHAP)

By evaluating a second person, the Shapley additive reasons (SHAP) are used to generate the rationales. For this investigation, the kernel SHAP method has been used. This method, which is model-agnostic, is used to assess SHAP values. It offers the greatest outcomes and precise SHAP values as well. SHAP values reflect how characteristics contribute to model output predictions. By displaying the relative contributions of each characteristic, the SHAP enhances the explainability of the learning method. The coalition's notion was used in this technique to calculate the SHAP values.

$$\varnothing_{\mathrm{k}}(\mathrm{y}) = \frac{1}{\mathrm{N}\sum_{\mathrm{n}=1}^{\mathrm{N}} \varnothing_{\mathrm{k}}^{\mathrm{n}}} \quad (43)$$

$$\varnothing_{\mathrm{k}}^{\mathrm{n}} = g\left(y_{+k}^{n}\right) - g\left(y_{-k}^{n}\right) \quad (44)$$

The Shapley features values of the projected y occurrences are described by the term $\varnothing_{\mathrm{k}}^{\mathrm{n}}$, mean margin contributions is denoted by /n k, black box predictions is indicated by $g\left(y_{+k}^{n}\right)$ without replacing the kth characteristic, and black box predictions is implied by $g\left(y_{-k}^{n}\right)$.

3.19.3 Combined Proposed Approach for Efficient Intrusion Detection

The suggested Proposed method for effectively categorizing incursions in a factory network is shown in Fig. 2. The following describes the Proposed approach's step-by-step process.

Phase 1: The input files, NSLKDD and Honeypot, are introduced into the BiLSTM structure, which categorizes the information's characteristics and looks for any anomalous characteristics in the network's behavior.

Phase 2: The BiLSTM architecture may have an incorrect loss functions in its features extracted.

Phase 3: The precision of detection is reduced as a consequence of the loss operations, leading to misinterpretation outcomes.

Phase 4: As a consequence, in order to shield the computer system from assaults in the future, comprehensible justifications with arguments for the incorrectly categorized outcome are required.

Phase 5: By implementing this method, the suggested system of intrusion detection becomes more transparent when deciding how to interpret forecasts.

Phase 6: The LIME and SHAP types of explicable artificial intelligence were presented in this work in order to make this occur.

Phase 7: The XAI techniques improve the efficiency of translation by recognizing the influence of harmful data.

Phase 8: As a result, the Proposedtechnique ascertains if any illegal or unusual network behavior—that is, intrusions—are present.

4 Empirical Results and Interpretations

The performance evaluation determined by the system to detect intrusions is provided in the next section. The models are run on a PC with an i3 CPU and 4 GB of RAM using MATLAB 2016 (a). The CICIDS2017 dataset is exploited in order to test the performances of the developed Proposedmodel on the identification of different categories of cyberattacks: DoS, Brute Force, and Infiltration. This section considers the achieved performance of each of the proposed models with regard to the main metrics: accuracy, precision, recall, specificity, and F1-score. Results For the detection of various types of intrusions, a substantial increase of the accuracy and other efficacy metrics is found. These datasets were split into two sub-classes: one for training and the other for evaluation, with a 70:30 split ratio. Table 1 shows the method's variable configurations.

Table 1 Parameter settings

Factors	Level
Population size	51
Min no. of repetition	101
Velocity of diffusion	0.07
Velocity of foraging	0.3
Rate of learning	0.02
Size of batch	65
Rate of dropout	0.6

4.1 Performance Metrics

The effectiveness of the suggested approach is assessed using a number of efficiency measures, including accuracy (A), precision (P), specificity (SP), F1-score, recall (R), and Matthews correlation value (MCC). Performance and acceleration duration are also calculated. Each metric's calculation is given below.

$$A = \frac{t_{pos} + t_{neg}}{t_{pos} + f_{pos} + t_{neg} + f_{neg}} \tag{45}$$

$$P = \frac{t_{pos}}{t_{pos} + f_{pos}} \tag{46}$$

$$SP = \frac{t_{neg}}{t_{neg} + f_{pos}} \tag{47}$$

$$R = \frac{t_{pos}}{t_{pos} + f_{neg}} \tag{48}$$

$$F1{-}score = \frac{t_{pos}}{\frac{f_{pos} + f_{neg}}{2 + t_{pos}}} \tag{49}$$

$$MCC = \frac{t_{pos} * t_{neg} - f_{pos} * f_{neg}}{\sqrt{(t_{pos} + f_{neg})(t_{pos} + f_{pos})(t_{neg} + f_{pos})(t_{neg} + f_{neg})}} \tag{50}$$

4.1.1 Receiver Operating Characteristic (ROC)

The real beneficial value at various thresholds is compared to the true positive value to create the ROC curve. Table 2 provides the comparison of proposed model against various attacks.

Table 2 Performance comparison of different attacks on CICIDS2017 dataset

Attack type	Accuracy (%)	Precision (%)	Recall (%)	Specificity (%)	F1-score (%)
DoS	95.8	96.1	94.7	95.2	95.4
Brute force	96.5	95.9	96.0	96.7	95.9
Infiltration	98.1	97.5	97.0	97.8	97.2
Web attack	94.6	93.8	94.2	94.9	94.0
Botnet	96.7	96.3	96.1	96.4	96.2

The Fig. 4 signifies the comparison result of various attack in graphical representation and Fig. 5 shows the different models true positive and false positive representation.

Applying the model to all types of intrusion detection concerning DoS, Brute Force, Infiltration, Web Attack, and Botnet attacks shows the highest accuracy in the Infiltration attack, which is 98.1% as in Fig. 4. There is outstanding precision with an F1-score of 97.2% for the mentioned more advanced and sophisticated type of attack. As in the case of Brute Force and Botnet attacks, both had reached high accuracy, above 96%, with good precision and recall values balanced and showed that the model is reliable in identifying and classifying them correctly. In parallel, the DoS attack had a good recall of 94.7%, yet it also scored accuracy of 95.8%. That means for such an attack, the model would not miss all true positives but would, at least minimize false positives. The lowest accuracy was attained by the category labeled Web Attack with accuracy 94.6%, yet at the same time, it scored a good F1-score, equal to 94.0%, which ensures continuity in the detection even from tougher attacks. The Proposed model maintains very high specificity, distinguishing the legitimate traffic from attacks with a minimum loss in terms of false negatives in all attack types as given in ROC of Fig. 5. Such a balance between precision, recall, and specificity

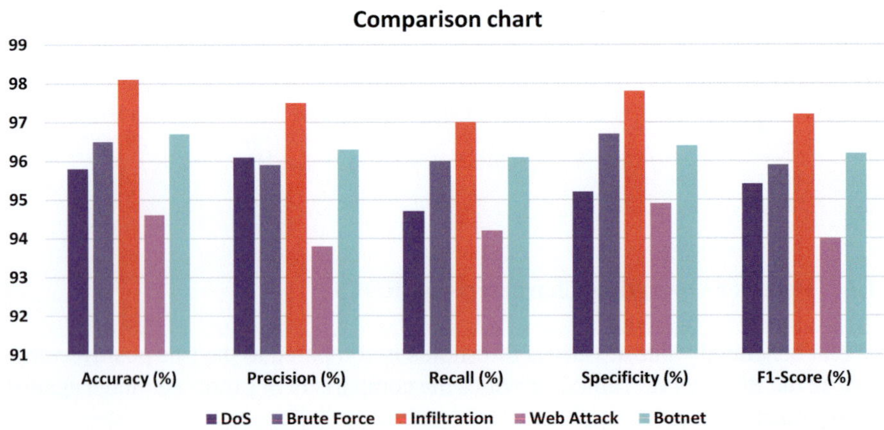

Fig. 4 Comparison chart for various attacks

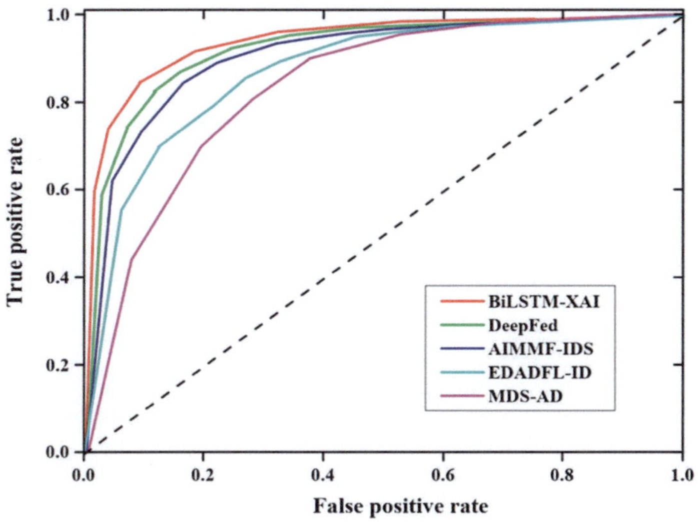

Fig. 5 Evaluation of ROC curve

underlines how the model performs well in varied attack types but gives out an overall high accuracy and efficiency about detection within Industry 4.0 networks.

5 Conclusion

In summary, the Explainable AI-Based Discrete Swarm Herd Optimization (DSHO) model proposed here works efficiently to overcome the existing challenges in cybersecurity with networks under Industry 4.0. The empirical results indicate a remarkable improvement in accuracy, precision, recall, specificity, and F1-score for all types of attacks like DoS, Brute Force, Infiltration, Web Attack, and Botnet attacks concerning the CICIDS2017 dataset. It is worth mentioning the fact that Infiltration attack achieved the highest accuracy of 98.1 percent, indicating that the model was well capable of identifying sophisticated threats. Also, the design of the model uses swarm intelligence for optimal feature selection such that performance enhances, and computational efficiency increases when applied with high dimensional data. Such integration of explainable AI techniques would enable the operators to have a better insight into the decision-making process that drives the model, trusting the process with greater implications on the response back in regard to security threats. Moreover, the prioritization based on fuzzy logic concerning intrusion risks by the system ensures a response based on the seriousness of the threats detected by the system. Hence, it makes the system adaptive to the dynamic cyber environment. In essence, the new model with great improvements in the security posture of Industry 4.0 networks effectively combats the looming risks from sophisticated cyberattacks while still maintaining an optimal balance between false positives and true positives. With continuous improvement through iterations based on real-world data interaction, this model places itself as a viable solution for proactive intrusion detection systems; hence, it opens doors to ongoing security measure development that may eventually be adapted for Industry 4.0.

References

1. Webert, H., Döß, T., Kaupp, L., Simons, S.: Fault handling in industry 4.0: definition, process and applications. Sensors **22**, 222 (2022)
2. Vogel-Heuser, B., Rösch, S., Fischer, J., Simon, T., Ulewicz, S., Folmer, J.: Fault handling in PLC-based industry 4.0 automated production systems as a basis for restart and self-configuration and its evaluation. J. Softw. Eng. Appl. **9**, 472–514 (2016)
3. Leitão, H.A., Rosso, R.S., Leal, A.B., Zoitl, A.: Fault handling in discrete event systems applied to IEC 61499. In: Proceedings of the 2020 25th IEEE International Conference on Emerging Technologies and Factory Automation (ETFA), pp. 1039–1042. Vienna, Austria (2020)
4. Preethi, P., Asokan, R., Thillaiarasu, N., Saravanan, T.: An effective digit recognition model using enhanced convolutional neural network based chaotic grey wolf optimization. J. Intell. Fuzzy Syst. **41**(2), 3727–3737 (2021)
5. Cinar, Z.M., Abdussalam Nuhu, A., Zeeshan, Q., Korhan, O., Asmael, M., Safaei, B.: Machine learning in predictive maintenance towards sustainable smart manufacturing in industry 4.0. Sustainability **12**, 8211 (2020)
6. Rekha, P., Saranya, T., Preethi, P., Saraswathi, L., Shobana, G.: Smart Agro Using Arduino and GSM. Int. J. Emerg. Technol. Eng. Res. (IJETER) **5** (2017)
7. Rousopoulou, V., Nizamis, A., Vafeiadis, T., Ioannidis, D., Tzovaras, D.: Predictive maintenance for injection molding machines enabled by cognitive analytics for industry 4.0. Front. Artif. Intell. **3**, 23 (2020)
8. El-Mahdy, M.H., Maged, S.A., Awad, M.I.: End-to-end fault tolerant control of discrete event system using recurrent neural networks. In: Proceedings of the 2022 2nd International Mobile, Intelligent, and Ubiquitous Computing Conference (MIUCC), pp. 266–271. Cairo, Egypt (2022)
9. Sandhya, N., Saraswathi, R.V., Preethi, P., Chowdary, K.A., Rishitha, M., Vaishnavi, V.S.: Smart attendance system using speech recognition. In: 2022 4th International Conference on Smart Systems and Inventive Technology (ICSSIT), pp. 144–149. IEEE (2022)
10. Ruppert, T., Abonyi, J.: Software sensor for activity-time monitoring and fault detection in production lines. Sensors **18**, 2346 (2018)
11. Vogel-Heuser, B., Fischer, J., Hess, D., Neumann, E., Würr, M.: Managing variability and reuse of extra-functional control software in CPPS. In: Proceedings of the 2021 Design, Automation & Test in Europe Conference & Exhibition (DATE), pp. 755–760. Grenoble, France (2021)
12. Chien, C.F., Dauzère-Pérès, S., Huh, W.T., Jang, Y.J., Morrison, J.R.: Artificial intelligence in manufacturing and logistics systems: algorithms, applications, and case studies. Int. J. Prod. Res. **58**, 2730–2731 (2020)
13. Kudelina, K., Vaimann, T., Asad, B., Rassõlkin, A., Kallaste, A., Demidova, G.L.: Trends and challenges in intelligent condition monitoring of electrical machines using machine learning. Appl. Sci. **11**, 2761 (2021)
14. Baskar, K., Venkatesan, G.P., Sangeetha, S., Preethi, P.: Privacy-preserving cost-optimization for dynamic replication in cloud data centers. In: 2021 International Conference on Advance Computing and Innovative Technologies in Engineering (ICACITE), pp. 927–932. IEEE (2021)

15. Liu, R., Yang, B., Hauptmann, A.: Simultaneous bearing fault recognition and remaining useful life prediction using joint-loss convolutional neural network. IEEE Trans. Ind. Inform. **16**, 87–96 (2020)
16. Abdullah, A.S., Selvakumar, S., Manoj, A., Bhubesh, K.R.: Medical steel fault prediction using deep learning techniques. In: Proceedings of the 2021 Federated Conference on Computer Science and Information Systems (FedCSIS), pp. 187–194. Warsaw, Poland (2021)
17. Mythily, D., Renila, R.H., Keerthana, T., Hamaravathi, S., Preethi, P.: Iot based fisherman border alert and weather alert security system. Int. J. Eng. Res. Technol. (IJERT) (2020)
18. Ruan, H., Dorneanu, B., Arellano-Garcia, H., Xiao, P., Zhang, L.: Deep learning-based fault prediction in wireless sensor network embedded cyber-physical systems for industrial processes. IEEE Access **10**, 10867–10879 (2022)
19. Maschler, B., Vietz, H., Jazdi, N., Weyrich, M.: Continual learning of fault prediction for turbofan engines using deep learning with elastic weight consolidation. In: Proceedings of the 2020 25th IEEE International Conference on Emerging Technologies and Factory Automation (ETFA), vol. 1, pp. 959–966. Vienna, Austria (2020)
20. Li, Y.: A fault prediction and cause identification approach in complex industrial processes based on deep learning. Comput. Intell. Neurosci.. Intell. Neurosci. **2021**, 6681496 (2021)
21. Yang, S., Yang, S., Fang, Z., Yu, X., Rui, L., Ma, Y.: Fault prediction for software system in industrial internet: a deep learning algorithm via effective dimension reduction. Commun. Comput. Inf. Sci. **1053**, 258–267 (2019)
22. Barcelos, A., Cardoso, A.J.: Current-based bearing fault diagnosis using deep learning algorithms. Energies **14**, 2509 (2021)
23. Iqbal, R., Maniak, T., Doctor, F., Karyotis, C.: Fault detection and isolation in industrial processes using deep learning approaches. IEEE Trans. Ind. Inform. **15**, 3077–3084 (2019)
24. Li, X., Jia, X., Wang, Y., Yang, S., Zhao, H., Lee, J.: Industrial remaining useful life prediction by partial observation using deep learning with supervised attention. R. Soc. Open Sci. **7**, 200674 (2020)
25. Mansouri, T., Vadera, S.: A deep explainable model for fault prediction using IoT sensors. R. Soc. Open Sci. **9**, 211002 (2022)
26. Zhang, X., Xie, W., Zhen, J., Zong, X., Jiao, T., Zhang, D.: Research on fault prediction and diagnosis of superconducting system based on deep learning. In: Proceedings of the 2021 IEEE 4th International Conference on Automation, Electronics and Electrical Engineering (AUTEEE), pp. 238–241. Shenyang, China (2021)

T. Saravanan from Namakkal, native of Tamilnadu, He did his B.E Computer Science in Selvam College of Technology, M.E Network Engg. from Anna University of Technology, Coimbatore and Ph.D Information Communication from Anna University, Chennai. He had published several papers in SCI, Scopus. He authored several book chapters and several edited book as well. He is an active Member of ISTE, IAENG, IACSIT.

S. Maheswaran has completed his B.E (EIE) from Bharathiyar University in the year 2002, M.E (Applied Electronics) from Anna University in the year 2004 and Ph.D in the field of Embedded Systems from Anna University in the year 2016. He has about 20 years of teaching experience at various levels and presently working as an Associate Professor in the Electronics and Communication Engineering Department, Kongu Engineering College, Perundurai. He has published two Patent, Several papers at International Journals. He presented papers in 15 International and 6 National conferences. He is a reviewer for 5 international journals and conferences. His-area of research includes Embedded Systems and Automation. He is member of IETE, The Institution of Engineers (India) and ISTE. He is the recipient of many award includes Young Scientists' conclave Best hall presentation award 2016 (IISF-2016)—under the theme of "Innovative Agriculture Practices and Livestokes Management" organized by Ministry of Science & Technology, Ministry of Earth Sciences, Vijnana Bharati (VIBHA), CSIR, Science & Technology and Earth Sciences, Government of India.

Saigurudatta Pamulaparthyvenkata is a seasoned Data Engineer with extensive experience in designing and implementing data solutions across diverse industries, including healthcare, finance, and technology. Based in Bryan, Texas, he holds a Master's degree in Computer Science from the University of Illinois, Springfield. Saigurudatta specializes in Big Data technologies, including Hadoop, Apache Spark, Kafka, and Snowflake, with proficiency in programming languages like Python, Scala, and SQL. His expertise extends to cloud platforms like AWS, machine learning, and data-driven decision-making. Throughout his career, Saigurudatta has contributed to building scalable, efficient, and secure data pipelines that enhance operational performance and drive actionable insights. He has consistently demonstrated the ability to tackle complex challenges, such as reducing data latency, optimizing storage costs, and improving data accuracy. A problem-solver at heart, he combines technical acumen with innovative thinking to deliver impactful solutions. Passionate about leveraging technology to address real-world problems, he is committed to continuous learning and contributing to advancements in data engineering and AI.

P. Preethi Associate Professor & Research Coordinator in the Department of Computer Science and Engineering, Kongunadu College of Engineering and Technology, Trichy, Tami Nadu, India. She received B.Tech., degree from Roever Engineering College, Perambalur in 2012. She was awarded with M.E., from Srinivasan Engineering College, Perambalur in 2014. She has 9 years of teaching experience, 2 years of research experience and awarded Ph.D., from Anna University, Chennai in the year of 2021. Her area of interest lies in Cloud Computing, Network Security, Machine Learning and published 28 (6 papers in SCI and 11 papers in SCOPUS) papers in international journals and in national and international conferences in that area. Published 6 books in reputed publications. Recognized and awarded Best faculty Award by Novel Research Academy (registered under MSME, UA No.: PY03D0003488), Puducherry at 2022. Actively aiding as a reviewer (Measurement: Sensors—Journals |Elsevier, Computers materials & continua—Journals|Tech Science Press etc.), Guest Editor (Elsevier-Special Issues) and editor (IJFREE) in various journals. She has obtained funding projects from various funding agencies like CSIR, TNSCST and DRDO. She is an active member of diverse professional bodies like ISTE, IEEE, CSE etc. Also acting as a Co-Prinicipal Investigator for SERB funded project.

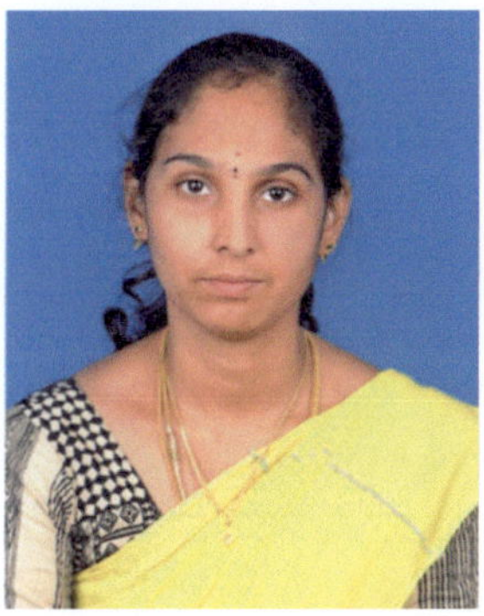

N. Indhumathi has completed her B.E Electronics and Communicatio Engineering from Bannari Amman Institute of Technology, M.E Embedded System Technologies from Anna University of Technology, Coimbatore She had published several papers in Scopus. She authored several book chapters as well. She has actively mentored various Hackathons. She is an active Member of IETE.

Interpretable and Extendible AI Models in Manufacturing for Industrial Processes

P. Jayadharshini, S. Santhiya, M. Parvathi, J. Charanya, J. Rakshitaa, and K. Nithika

Abstract Two especially important challenges that are quite often overlooked are interpretability and extendibility. Increasingly, AI is deeply woven into the fabric of modern manufacturing-automation efficiency and optimization of processes like predictive maintenance and quality control. Yet especially for industrial applications of AI, the challenges of interpretability and extendibility need to be sufficiently appreciated. For instance, in manufacturing, the primacy of transparent operations, occupational safety, and regulatory compliance would make the development of interpretable models of AI rather critical. Given the risk of a lack of interpretability, and specifically for complex AI models like deep learning, agents who make the decisions themselves will not be able to understand or justify their predictions based on the AI. This chapter attempts to show growing demands on manufacturers to produce an interpretable AI system by describing how there is increased ability to foster trust and collaboration between AI systems and human operators as computation improves decision-making processes. Extendibility: It describes the ability of AI models to adapt in the changing and evolving conditions of manufacturing. Production processes continually change themselves in terms of new product lines, changes in market demands, and modifications in operating conditions. The extendible AI

P. Jayadharshini (✉) · S. Santhiya · J. Charanya
Assistant Professor, Department of Artificial Intelligence, Kongu Engineering College, Perundurai 638060, Tamil Nadu, India
e-mail: jayadharshini.ai@kongu.edu

S. Santhiya
e-mail: santhiya.cse@kongu.edu

M. Parvathi
Assistant Professor, Department of Artificial Intelligence and Data Science, Nandha Engineering College, Perundurai, India

J. Rakshitaa · K. Nithika
Student, Department of Artificial Intelligence, Kongu Engineering College, Perundurai 638060, Tamil Nadu, India
e-mail: rakshitaaj.21aim@kongu.edu

K. Nithika
e-mail: nithikak.21aim@kongu.edu

L. Krishnasamy et al. (eds.), *Distributed Deep Learning and Explainable AI (XAI) in Industry 4.0*, Information Systems Engineering and Management 55,
https://doi.org/10.1007/978-3-031-94637-0_12

models adapt the new tasks or datasets with minimal retraining and thus reduce the cost and time to update or modify AI systems. The techniques that will be discussed in detail are transfer learning, modular architectures, and continuous learning in an insight of how AI models might be at a good level of performance yet flexible enough for new challenges and environments. This chapter discusses representative techniques in constructing interpretable and extensible AI models, including decision trees, rule-based systems, LIME (Local Interpretable Model-Agnostic Explanations) and SHAP (SHapley Additive exPlanations), case studies referring to the automotive, electronics, and food manufacturing industries, showing how the model is used in actual applications, from predictive maintenance to quality control, for example. Thirdly, it addresses issues like the integration of AI systems with legacy manufacturing infrastructure, the balancing of interpretability with performance, and the scale of AI models for real-time high-volume data processing. The book concludes with future research directions, envisioned to enhance the role of AI further in transforming manufacturing processes, this time focusing on transparency, adaptability, and efficiency.

Keywords Explainable AI · Extendable AI · Manufacturing · Industrial operations · Predictive maintenance · Transfer learning · LIME · SHAP · Continuous learning · Artificial intelligence in industry · Explainable AI · Industry 4.0

1 Introduction

It becomes a springboard of innovation for the manufacturing sector, which can transform old processes into intelligent systems that can indeed make decisions and operate automatically with minimal human intervention. Predictive maintenance models, quality control models-the list goes on, across a wide spectrum of models of AI-became ubiquitous. However, as against this industrial setting, it presents even stricter demands for this: even high-performance performance but, more importantly, also in terms of interpretability and extendibility.

Explainability—The outputs from the model need to explain why the model arrives at such outputs. It may quite be important in a manufacturing environment where the role of human oversight and decision making is crucial. Extendibility—AI models need to self adapt to new tasks or datasets. This will enable the companies to reuse or apply the model with much fewer efforts than they would have had to make if they were going to rebuild systems, largely different.

The development of interpretable and extendible models has become highly relevant, considering the fact that AI systems are increasingly assuming critical roles in manufacturing industries. This chapter should center on the most relevant issues, approaches, and applications toward the real-world deployment of interpretable and extendible AI models, which represent their critical relevance to enhance operational efficiencies and innovations within the manufacturing domain.

2 The Need for Interpretable AI in Manufacturing

This is the demand for interpretable AI in manufacturing: so-called because many of the decisions taken by AI-driven systems impact production quality, safety, and efficiency [1]. Such AI models are able to detect patterns or anomalies not quite visible or intuitive to human operators. However, when such systems come up with decisions without any rationale for doing so, there is a deep challenge involved. It undermines trust in AI systems-this "black-box" nature of so many of today's models, deep learning in particular. This can also be a very costly oversight if decisions are taken on the basis of predictions that one does not fully understand.

As depicted in Fig. 1, AI plays a critical role in modern manufacturing by improving productivity, reducing costs, and enhancing product quality. AI technologies, such as machine learning and computer vision, are driving automation, enabling faster and more precise production processes. Predictive maintenance powered by AI helps identify potential equipment failures before they occur, reducing unplanned downtime. AI also facilitates optimized supply chains by predicting demand and adjusting production schedules accordingly. Additionally, AI-powered quality control systems enhance defect detection, ensuring higher product standards. Overall, AI is revolutionizing manufacturing by making operations more efficient, cost-effective, and adaptive to changing market demands.

As an example, consider a predictive maintenance application where an AI model predicts a likely failure of a certain piece of machinery [2]. Unless it can explain why this conclusion is drawn, maintenance teams may either ignore the prediction or overcompensate by doing unnecessary maintenance, or fail to take appropriate preventive action in a timely manner. This makes up for inefficiencies and potential financial losses that stem from this lack of interpretability.

Interpretability is critical when output needs to comply with safety and regulatory standards [3]. Most manufacturing processes operate in highly regulated environments where decisions made need to be transparent and justifiable. When an AI model suggests changes to a production process, plant managers and engineers need to understand the basis of recommendations so that they comply with the appropriate standards and do not introduce unforeseen risks.

Interpretable models also stimulate collaboration between AI systems and human operators [4]. In the context of a human–machine collaborative environment, an engineer would use interpretable AI to make decisions on production processes, quality control, and equipment maintenance. In conclusion, interpretable AI bridges the gap between human knowledge and machine intelligence, thus leading to better decision-making and superior operational outcomes.

Fig. 1 Significance of AI in manufacturing

3 Key Techniques for Building Interpretable AI Models

Several approaches have been developed to ensure that AI models remain interpretable while maintaining high performance. Techniques cover both classical machine learning approaches and more advanced methodologies tailored specifically to deep learning models.

3.1 Decision Trees

Decision trees are amongst the most interpretable machine learning models. Given a space of inputs, they recursively partition such space and assign a label to each partition [5]. A decision tree is said to divide a complex decision into a series of decisions at different levels-the decision tree will consist of several nodes corresponding to different features of the input data such that, at each node, the model considers a single feature of the input data, and makes a decision based on that feature.

The hierarchical nature of decision trees makes it easy to understand and follow [6]. For instance, in a quality control application, a decision tree may first check on the temperature of a production machine, next verify the speed at which products are moving along the assembly line, and finally check for visual defects. Each decision is clearly outlined in the tree, providing human operators with a straightforward explanation of how the model arrived at its conclusion.

Decision trees become more difficult to interpret as they grow deeper [7]. Big and deep trees can be hard to understand, and small variations in the input data may have very strong effects on the structure of the tree. Pruning techniques can be useful in reducing the complexity of the tree without loss of accuracy.

3.2 Rule-Based Models

Rule-based models use a set of if–then rules to make decisions [8]. These rules are generated from the training data and can be easily understood by human operators. In the context of manufacturing, rule-based models are particularly useful for tasks such as quality control, where specific conditions (e.g., temperature, pressure, visual characteristics) can be associated with specific outcomes (e.g., pass/fail, defect/no defect).

One advantage of rule-based models is their transparency. Each decision is explicitly linked to a set of conditions, making it easy to trace the reasoning behind a particular prediction. Additionally, rule-based models are often used in expert systems, where domain-specific knowledge is encoded into the rules. However, rule-based models can struggle with scalability and may require frequent updates as manufacturing processes evolve. They may not do very well in complex tasks, where the relationships between variables are not easily captured by simple rules.

3.3 LIME (Local Interpretable Model-Agnostic Explanations)

LIME is a technique for explaining complex models in an agnostic manner, approximating near a given prediction of that model [9]. It perturbs the input data and observes how the predictions change. Based on these perturbations, LIME builds a simpler, more interpretable model that mirrors the behavior of the complex model within the vicinity of the input data point.

In a manufacturing scenario, LIME can be used to explain predictions made by complex models, like neural networks [10]. For instance, it may happen that a neural network says that a machine is likely to fail; LIME can help explain the decision by highlighting the key features that are responsible for this prediction, such as temperature, and vibration. This information can then be used by maintenance teams to take preventive action based on a clear understanding of the underlying factors. Although LIME is good for locally explaining the predictions of a model, it may not offer a global view of how the model behaves. Also, the quality of the explanation obtained essentially depends on how well the locally approximated model matches up with the behavior of the complex model.

3.4 SHAP (SHapley Additive ExPlanations)

Another quite powerful interpretation method of complex machine learning models is SHAP [11]. It gives each feature in the input data a SHAP value, meaning how much that feature contributed to the prediction. SHAP values owe their foundation to cooperative game theory and provide consistent and fair ways to "credit" a prediction to input features.

SHAP can explain predictions of models used for tasks such as defect detection, predictive maintenance, or process optimization in a production setup. For example, if there's a model that predicts a product is defective, the SHAP values will display where exactly those particular features-for example, material quality and machine settings-are actually contributing the most to such a prediction.

One of the major benefits that SHAP possesses is that it gives consistent feature importance, such that people can compare contributions of different features for multiple predictions. Nevertheless, SHAP is computationally expensive, especially for big models or datasets.

4 Extendibility of AI Models in Manufacturing

Extendibility is the ability to adapt AI models into new tasks, datasets, or operational environments with no rebuild from scratch. In a fast-paced manufacturing environment, where it might have a variety of changeable processes and requirements, extendible AI models can be highly advantageous.

4.1 Transfer Learning

Transfer learning is a very interesting technique by which one fine-tuned model for some given task can be used to adapt to a new but related task with minimal retraining [12, 13]. This will especially help in manufacturing where sparsity in data for specific tasks may exist, but on similar processes or machines, valuable information can be extracted.

For example, a model that is trained to predict the failures of any equipment in one type of machine can be transferred to a case of application for failure prediction in a different type of machine by reusing the learned representations and fine-tuning it on a smaller dataset. Transfer learning saves time and resources in terms of labeled data, and with high performance, reduces the need for large labeled data.

4.2 *Modular AI Architectures*

Modular AI architecture should be easily extendible, splitting different components of the model into independent modules [14]. Every module may be updated or replaced or fine-tuned without affecting the overall structure of the model. This modularity is particularly valuable in manufacturing environments where different production lines may require different levels of customization.

For example, in a multi-stage production process, quality control can be managed by one module at the initial stages while another module takes care of process optimization for final stages. In the event of changes happening in the early stage, then only the relevant module needs to be updated, and that would be with minimal disturbance on the entire system.

4.3 *Continuous Learning*

In dynamic environments of manufacturing, a constant need exists to learn so that AI models remain relevant and accurate [15]. Another term associated with continuous learning is online learning, in which models are updated incrementally when new data arrives, without requiring retraining over the entire dataset. Continuous learning is very valuable in applications, such as predictive maintenance, where there is a steady stream of real-time new data regarding performance and failure patterns of the machines. By absorbing new information continually, AI models can adapt to changes in operational conditions and thus make more accurate predictions.

5 Case Study: Predictive Maintenance in Industrial Equipment

In a typical predictive maintenance setup, sensors are installed on the machinery to collect real-time data on parameters such as temperature, vibration, pressure, and rotation speed [16]. Then, this sensor data feeds into AI models that analyze it to determine whether it indicates potential equipment failure. Predictive maintenance systems rely heavily on the interpretability of AI models, however.

Table 1 Predictive maintenance techniques and benefits

Technique	Application	Benefit
SHAP	Explaining failure predictions	Reduced downtime by 30%
Transfer learning	Adapting models to new equipment	Faster deployment with less data
Continuous learning	Updating models with new data	Maintained accuracy over time
Sensor monitoring	Collecting real-time equipment data	Improved fault detection efficiency
Optimized scheduling	Prioritizing high-risk machines	Reduced unnecessary maintenance

As shown in Table 1, predictive maintenance techniques offer key benefits to manufacturing. SHAP helps identify failure factors, reducing downtime. Transfer learning adapts models to new equipment quickly, minimizing retraining. Continuous learning keeps models accurate as conditions change. Sensor monitoring provides real-time data for early failure detection, while optimized scheduling focuses maintenance on high-risk machines, improving efficiency. Together, these techniques enhance productivity and reduce maintenance costs.

5.1 Using SHAP for Predictive Maintenance

For instance, in the large automotive manufacturer's predictive maintenance system, SHAP values were applied to explain a machine learning model that predicted robotic arm failures along the production line [17]. SHAP values allow engineers to drill down to the most important factors in each of their predictions. For instance, when the model predicted an exceptionally high risk of failure of a robotic arm, SHAP revealed that unusually high vibration levels and deviations in temperature were the primary reasons for such inference.

This level of interpretability facilitated a point where the maintenance crew could intervene before the robotic arm failed, thus ensuring that unplanned downtime was reduced by 30%. The corporation also benefited in optimizing the maintenance schedule due to the insights gained from SHAP, focusing on the machines that were most at risk rather than making un-necessary maintenance on all machines.

5.2 Transfer Learning for Maintenance of New Equipment

Again, another example of extendibility in predictive maintenance comes from the electronics manufacturing sector [18]. A very high-precision circuit board manufacturing company applied a deep learning model for forecasting failure of their soldering machines. When they manufactured the next generation with slight changes in specification, those machines were adapted using transfer learning onto the existing model. Instead of training the model from scratch, they fine-tuned the model using a small dataset gathered from the new machines. This not only saved time in training but also reduced the amount of labeled data needed, so the company could now implement the predictive maintenance system on the new machines within weeks rather than months.

5.3 Continuous Learning in Dynamic Production Environments

In manufacturing systems, especially in food and beverage production, the environment is significantly variable [19]. For instance, there may be fluctuations in raw materials availability and seasonality changes, along with demand volatility, demanding that AI models continuously learn from changing circumstances. It follows that continuous learning will play an important role in keeping predictive maintenance systems accurate.

For instance, a soft drink company utilized a continuous learning model to observe and maintain a stock of bottling machines. First, it was trained on previous data; however, since there are more kinds of bottles released and the company reconfigures the production line to meet the seasonal changes, the model is also updated incrementally with the new data so that the model will continue to be valid for forecasting the failure machine even as the operation conditions change.

Thus, while it achieved cuts in unplanned downtime of 25%, improved the accuracy of its predictions, and adjusted maintenance schedules in real time based on what it learned continuously, it also learned to adapt to new data without requiring full retraining, thus reducing the costs of its operations and minimizing the disrupting of production.

6 Ethical and Regulatory Implications of AI in Manufacturing

Adding development and use of AI raises several ethical and regulatory aspects pertinent to proper and responsible deployment. Such deployment is likely to make some decisions that depend on various critical factors about safety, quality control, and efficiency in operations-areas with dire implications if wrong. Such ethical considerations include accountability: who is accountable when an AI system incorrectly or harmfully decides? For instance, decision making in manufacturing needs to explain the results so that human operators can understand and take responsibility for decisions made by AI. This is particularly crucial in highly safety–critical sectors such as aerospace, automotive, and pharmaceuticals, where AI-based decisions can have an impact on human life. Another ethical concern is derived from biased AI models. In the first place, AI systems process data to predict and make decisions regarding issues such as discriminatory hiring practices or even existing inequalities in society. These biases can then proceed to be perpetuated by the model and lead it onto the path of unfair, even discriminatory results.

Even an algorithm with the right intentions may still produce more or less percentages of false defects in products formulated from certain materials or by a particular vendor due to incomplete or partial data. With an ethical stance in favor of manufacturing comes AI models that need to not only be fair but also unbiased. The above mentioned biases can be overcome by using techniques like fairness aware machine learning and auditing the AI models at regular times.

Beyond ethics, regulatory bodies worldwide are now starting to establish guidelines and best practices in AI deployment in industrial environments [20]. An example of this is the plan known as the AI Act in the EU. This is a risk-based approach aimed at regulating AI. Formal rules will be implemented on high-risk applications within AI systems like industrial automation and control machinery. It will ensure safety, transparency, and accountability of AI through the regulations put in place. Manufacturer compliance with changing regulations will demand more interpretable AI models since they will be based on an explanation of how decisions are taken and evidence of conformance to safety and fairness standards. The second part outlines how producers can create AI that should live up to ethical standards comparable to those of legal and regulatory frames. Explainable and bias-free accountable AI is pivotal for responding to calls for both ethical and regulatory markets without sacrificing innovations and competitiveness.

7 AI in Smart Manufacturing: Integrating IoT and Digital Twins

It is rapidly transforming the face of manufacturing and thus has come to be known as "smart manufacturing." [21]. Many sorts of sensors, actuators, and other forms of connected machinery embedded in IoT devices at every step of the manufacturing process will generate high-value real-time data. With such constant inflow of data, AI models can then make dynamic, informed decisions in optimizing production lines in real time. Actually, though, the real power comes when AI is combined with digital twins: virtual representations of physical assets, processes, or systems.

Digital twins offer producers the possibility to set up a simulated environment, allowing AI to predict and optimize and control different elements of production [22]. The automotive assembly line in a large-scale plant might produce an example where the digital twin reproduces the impact of different configurations and settings of machines, parts, and operating conditions on overall assembly efficiency. Then AI models can test different optimization strategies, predict equipment failures, and recommend process adjustments-all without disrupting physical production.

This type of simulation offers a controlled virtual environment for the manufacturer to test optimizations provided by AI without causing any disruption to real in-life production.

In smart manufacturing, AI models should be interpretable and extendible. The process is simulated well with the digital twin of the manufacturing process when in place while volumes of data are collected by the IoT devices. Thereby, there exists the need for human operators to understand the decisions taken by the AI.

Such a model can notify in real-time for an increased speed of a conveyor belt to maximize the flow in production, but then, and this is crucial, the operators must be able to understand why the speed needs to be increased, on what data the AI bases its decision, and what risks are linked with that action. Such insights from interpretive AI models give the operator the information that they need in order to trust or act on recommendations by AI.

Extendibility in smart manufacturing also reflects the fact that production processes keep changing. New products get designed, product designs change, and operation conditions change from time to time [23, 24]. The extensible AI models allow manufacturers to achieve conformity with the changes without having to start up new models from scratch.

Other techniques through which AI models can be updated include transfer learning and continuous learning. With the updating of such models with new data coming from IoT sensors or digital twins, these models can remain precise and effective despite the changing manufacturing environment.

For instance, if there is a new machine installed on the production line, AI learns very rapidly in its adaptation to the data that will be generated from the new machine so that predictions may be made for maintenance needs and optimum performance from the very first day. AI, IoT, and digital twins will thus come together in this ultra-responsive, intelligent manufacturing environment [25]. For example, while AI-driven predictive maintenance through IoT is adequately supplemented with digital twin simulations of how a failure in one particular machine might propagate down the production line, manufacturers can make real-time decisions to avoid costly downtime and ensure ongoing efficiency. The smarter the industries and, more specifically for smart manufacturing, the need for explainable and extensible AI models becomes important for increasing operational efficiency, flexibility, and innovation.

8 Challenges and Future Directions

Many problems remain unsolved despite the fact that partial progress has been made toward the ambitious goal of building interpretable and extendible AI. With manufacturing environments becoming extremely complex and with exploding volumes of machine-generated data, the need for increasingly scalable and efficient models of AI could not be greater. Another very living area of research is ensuring that AI models are interpretable without performance damage.

8.1 Balancing Interpretability and Performance

The first hurdle that has to be crossed in developing an interpretable AI model is finding the right balance between interpretability and performance [26]. Deep neural networks are rather powerful AI models that happen to be rather unintuitive. Among these is LIME and SHAP, which can provide explanations that seem logical but do not necessarily guarantee a full understanding of how the model conducts its business.

This challenge is currently addressed by new techniques for improving the interpretability of complex models. An example of such areas of research explores hybrid models that combine the advantages of simpler models, such as interpretability through decision trees, with the power of deep learning for prediction. Hybrid models have the promise of providing more profound explanations with an improvement in accuracy loss.

8.2 Data Heterogeneity and Scalability

In many manufacturing environments, the data used to train AI models is highly heterogeneous. Differences in machines, production lines, and sensor configurations can generate vast numbers of different types of data, making it challenging to build models that generalize well across different environments. This way, careful data preprocessing, feature engineering, and model adaptation techniques are of utmost importance for making AI models extendible [27, 28]. Highly voluminous data generated by manufacturing systems heavily place critical scalability considerations on the AI models. With the rising complexity of manufacturing environments, it consequently increases the need for real-time analytics, in which AI models can process vast high-dimensional data and streaming data effectively. Researchers and practitioners continue to develop and improve the scalability of AI models that meet these challenges.

8.3 Integration with Legacy Systems

Most manufacturing sites have legacy systems in place that have never been designed for use with AI technologies [29]. Interfacing AI models with them is complex, especially if legacy systems do not have the capability of computational power or other infrastructures to support real-time AI applications. Thus, extendibility and compatibility of AI models with existing systems are very important factors for widespread adoption in this manufacturing sector.

To overcome this challenge, companies are increasingly looking to cloud-based AI solutions that allow AI models to be deployed and managed centrally, but with minimal need for infrastructure upgrades on the factory floor [30]. Edge computing is another emerging solution where AI models can be deployed nearer the data sources, reduce latency, and improve the performance of real-time applications.

8.4 Future Research Directions

Looking forward, there are several promising areas of research that could further enhance the interpretability and extendibility of AI models in manufacturing:

- **Explainability for Deep Learning**: While techniques such as LIME and SHAP have improved the interpretability of deep learning models, there is still much work to be done in understanding how deep neural networks make decisions. New research into model transparency and feature attribution could lead to the development of more interpretable deep learning models.
- **Self-Learning AI Systems**: Self-learning AI systems that may update themselves when new data become available may constitute the other future research area. This would, in itself, stretch the potential of extendibility of AI models while simultaneously reducing man-in-the-loop processes associated with retraining and fine-tuning of the models.
- **Federated Learning**: It is the latest approach through which AI models can be trained on decentralized sources of data without ever sharing the raw data. For example, manufacturing federated learning might allow companies to train their AI models across various production lines or even factories without jeopardizing the security and privacy over the data.
- **Ethical and Regulatory Considerations**: As AI assumes more critical functionalities in manufacturing, the need for transparency, fairness, and accountability of AI decision-making will escalate. The current agenda should look into developing ethical frameworks and regulatory guidelines for applications of AI in industrial environments, especially in high-stakes applications where safety and reliability are of utmost concern.

As shown in Fig. 2, the future of AI in manufacturing focuses on efficiency, sustainability, and innovation. Key trends include autonomous manufacturing, where AI enables self-regulating production, and predictive analytics for proactive maintenance. Collaborative robots (cobots) are enhancing human–machine teamwork, while digital twins allow real-time monitoring and optimization. Sustainable AI solutions are also driving energy efficiency and waste reduction. Together, these advancements are transforming manufacturing into a more intelligent, adaptable, and sustainable industry.

Fig. 2 Future trends of AI in manufacturing

9 Conclusion

Conclusion in the integration of AI in manufacturing processes, it must be interpretable and extendible. With the increasing critical roles in things such as predictions of failure of equipment, optimization of production lines, as well as determining the quality of products, knowing how and being able to modify the models becomes essential for AI systems. Techniques that significantly contribute to improving AI model interpretability are decision tree and rule-based models, LIME, and SHAP. Techniques including transfer learning, modular architectures, and continuous learning add the extendibility capability of AI models.

Challenges persist from trading off interpretability against good performance and scaling the complexity of AI models to handle large and heterogeneous data up to the integration of AI systems with other existing manufacturing infrastructure. Work here will persist in reflecting continuing advances toward more transparent, scalable, and adaptive AI systems that will be able to better match the highly complex demands of modern manufacturing environments.

But with this challenge and embracing the opportunities on offer from interpretable and extendible AI, the capabilities will bring new heights of productivity, efficiency, and innovation and will produce significant progress toward the future of the industry.

References

1. Trakadas, P.: An artificial intelligence-based collaboration approach in industrial IoT manufacturing: key concepts, architectural extensions and potential applications. Sensors **20**(19), 5480 (2020)
2. Bin Akhtar, Z.: Artificial intelligence (AI) within manufacturing: an investigative exploration for opportunities, challenges, future directions. Metaverse **5**(2), 2731–2731 (2024)
3. Deloitteeditor: Artificial intelligence goes mainstream. WSJ. https://deloitte.wsj.com/cio/artificial-intelligence-goes-mainstream-1438142473
4. Krauß, J., Hülsmann, T., Leyendecker, L., Schmitt, R.H.: Application areas, use cases, and data sets for machine learning and artificial intelligence in production. In: Liewald, M., Verl, A., Bauernhansl, T., et al. (eds.) Production at the Leading Edge of Technology, pp. 504–513. Lecture Notes in Production Engineering. Springer International Publishing, Cham (2023). https://doi.org/10.1007/978-3-031-18318-8_51
5. What Does Collaborative Robot Mean? Available online: https://blog.robotiq.com/what-does-collaborative-robot-mean. Accessed on 9 May 2017

6. Monostori, L., Kádár, B., Bauernhansl, T., et al.: Cyber-physical systems in manufacturing. CIRP Ann. **65**(2), 621–641 (2016). https://doi.org/10.1016/j.cirp.2016.06.005
7. Wuest, T., Weimer, D., Irgens, C., et al.: Machine learning in manufacturing: advantages, challenges, and applications. Prod. Manuf. Res. **4**(1), 23–45 (2016). https://doi.org/10.1080/21693277.2016.1192517
8. Lu, S.C.Y.: Machine learning approaches to knowledge synthesis and integration tasks for advanced engineering automation. Comput. Ind.. Ind. **15**(1), 105–120 (1990). https://doi.org/10.1016/0166-3615(90)90088-7
9. Jourdan, N., Longard, L., Biegel, T., et al.: Machine Learning for Intelligent Maintenance and Quality Control: A Review of Existing Datasets and Corresponding Use Cases. Hannover Publishing (2021. https://doi.org/10.15488/11280
10. Sariel, S., Yildiz, P., Karapinar, S., et al.: Robust task execution through experience-based guidance for cognitive robots. In: Proceedings of the 2015 International Conference on Advanced Robotics (ICAR), pp. 663–668 (2015)
11. Krauß, J., Dorißen, J., Mende, H., et al.: Machine learning and artificial intelligence in production: application areas and publicly available data sets. In: Wulfsberg, J.P., Hintze, W., Behrens, B.A. (eds.) Production at the Leading Edge of Technology, pp. 493–501. Springer, Berlin, Heidelberg (2019)
12. Panayotov, V., Chen, G., Povey, D., et al.: Librispeech: an ASR corpus based on public domain audio books. In: Proceedings of the 2015 IEEE International Conference on Acoustics, Speech and Signal Processing (ICASSP), pp. 5206–5210 (2015)
13. OpenAI: GPT-4 Technical Report. arXiv. 2023 arXiv:2303.08774
14. Arinez, J.F., Chang, Q., Gao, R.X., et al.: Artificial intelligence in advanced manufacturing: current status and future outlook. J. Manuf. Sci. Eng. **142**(11) (2020). https://doi.org/10.1115/1.4047855
15. Doltsinis, S., Krestenitis, M., Doulgeri, Z.: A machine learning framework for real-time identification of successful snap-fit assemblies. IEEE Trans. Autom. Sci. Eng.Autom. Sci. Eng. **17**(1), 513–523 (2020). https://doi.org/10.1109/tase.2019.2932834
16. Mozaffar, M., Liao, S., Xie, X., et al.: Mechanistic artificial intelligence (mechanistic-AI) for modeling, design, and control of advanced manufacturing processes: current state and perspectives. J. Mater. Process. Technol. **302**, 117485 (2022). https://doi.org/10.1016/j.jmatprotec.2021.117485
17. Wang, L.: From intelligence science to intelligent manufacturing. Engineering **5**(4), 615–618 (2019). https://doi.org/10.1016/j.eng.2019.04.011
18. Chui, L., Kamalnath, V., McCarthy, B.: An executive's guide to AI, McKinsey. Available online: https://www.mckinsey.com/business-functions/mckinsey-analytics/our-insights/an-executives-guide-to-ai. Accessed on 5 Jan 2024
19. Cardon, D., Cointet, J.P., Mazières, A.: Neurons spike back: the invention of inductive machines and the artificial intelligence controversy. Reseaux **5**(211), 173–220 (2018)
20. LeCun, Y., Bengio, Y., Hinton, G.: Deep learning. Nature **521**(7553), 436–444 (2015). https://doi.org/10.1038/nature14539
21. Lee, J., Davari, H., Singh, J., et al.: Industrial Artificial Intelligence for industry 4.0-based manufacturing systems. Manuf. Lett. **18**, 20–23 (2018). https://doi.org/10.1016/j.mfglet.2018.09.002
22. Li, B., Hou, B., Yu, W., et al.: Applications of artificial intelligence in intelligent manufacturing: a review. Front. Inf. Technol. Electron. Eng. **18**(1), 86–96 (2017). https://doi.org/10.1631/fitee.1601885
23. Ravichandar, H.C., Dani, A.: Human intention inference and motion modeling using approximate EM with online learning. In: Proceedings of the 2015 IEEE/RSJ International Conference on Intelligent Robots and Systems (IROS), pp. 1819–1824 (2015)
24. Zhang, J., Liu, H., Chang, Q., et al.: Recurrent neural network for motion trajectory prediction in human-robot collaborative assembly. CIRP Ann. **69**(1), 9–12 (2020). https://doi.org/10.1016/j.cirp.2020.04.077

25. Wang, W., Li, R., Chen, Y., et al.: Facilitating human-robot collaborative tasks by teaching-learning-collaboration from human demonstrations. IEEE Trans. Autom. Sci. Eng.Autom. Sci. Eng. **16**(2), 640–653 (2019). https://doi.org/10.1109/tase.2018.2840345
26. Zhao, R., Yan, R., Chen, Z., et al.: Deep learning and its applications to machine health monitoring. Mech. Syst. Signal Process. **115**, 213–237 (2019). https://doi.org/10.1016/j.ymssp.2018.05.050
27. Akhtar, Z.B.: The design approach of an artificial intelligent (AI) medical system based on electronic health records (EHR) and priority segmentations. J. Eng. **2024**(4) (2024). https://doi.org/10.1049/tje2.12381
28. Akhtar, Z.B.: Securing operating systems (OS): a comprehensive approach to security with best practices and techniques. Int. J. Adv. Netw. Monit. Controls **9**(1), 100–111 (2024). https://doi.org/10.2478/ijanmc-2024-0010
29. Akhtar, Z.B., Gupta, A.D.: Integrative approaches for advancing organoid engineering: from mechanobiology to personalized therapeutics. J. Appl. Artif. Intell. **5**(1), 1–27 (2024). https://doi.org/10.48185/jaai.v5i1.974
30. Akhtar, Z.B.: Advancements within molecular engineering for regenerative medicine and biomedical applications and investigation analysis towards a computing retrospective. J. Electron. Electromed. Eng. Med. Inf. **6**(1) (2024). https://doi.org/10.35882/jeeemi.v6i1.351

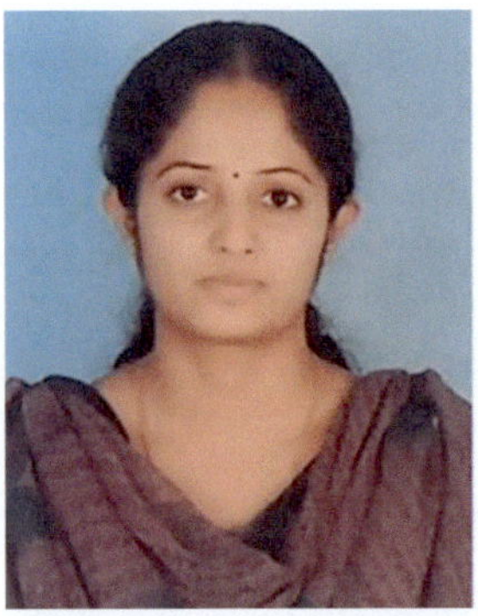

P. Jayadharshini

S. Santhiya

M. Parvathi

J. Charanya

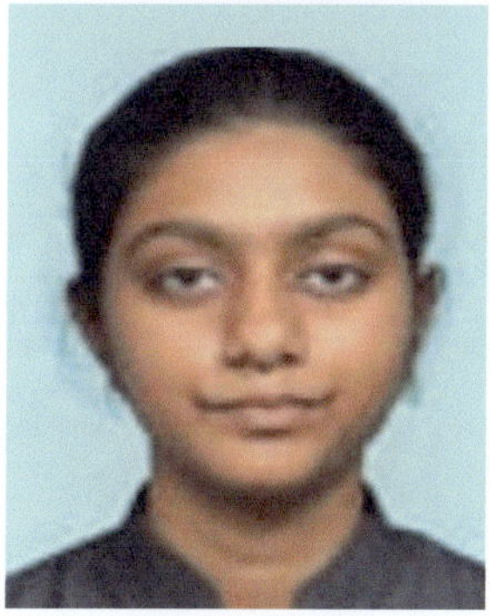

J. Rakshitaa

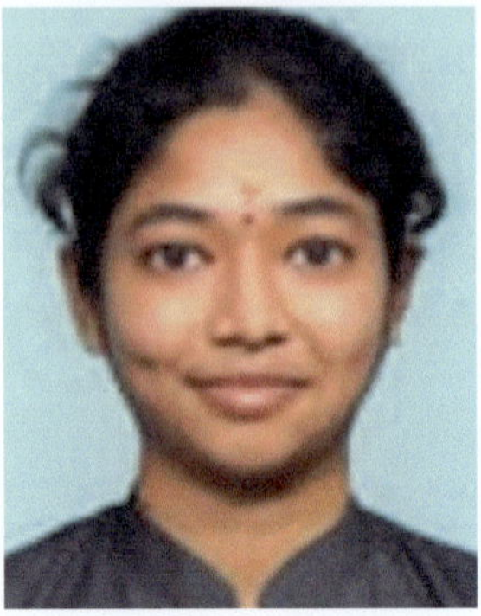

K. Nithika

Cost Analysis of Large Language Models for Different Applications of Industry 4.0: Chatbots and Conversational AI in Manufacturing

Sai Kalyana Pranitha Buddiga and Pushkar Mehendale

Abstract Large language models have attracted so much attention, owing to such amazing capabilities in understanding and generating natural language. As noted above, large costs are associated with the deployment and usage of LLMs, varying according to the target application. This paper achieves a more profound analysis of the diverse costs associated with LLM use in a number of applications from conversational AI to content generation and beyond. The understanding of these diverse costs will allow organizations and professionals to make appropriate decisions on whether to adopt and fine-tune LLMs for their use cases. This paper brings about a general cost landscape of deploying LLMs by offering a detailed case study analysis together with cost analysis. Industry 4.0 technologies are gaining speed into all industrial sectors, and LLMs are rapidly getting into many industrial processes, especially in manufacturing processes. The most valuable applications that such LLMs can imply are conversational AI, chatbots, and virtual assistants applied for production lines, supply chain management, and scheduling of maintenance. Such use of AI-based systems surely delivers efficiency, operations cost savings, and decision-making. However, it charges a lot in terms of costs, such as computational power and energy consumption. Moreover, the deployment of LLMs has further made the concern over data privacy issues. This paper discusses the cost of employing LLMs within smart manufacturing contexts and points out conversational AI applications.

Keywords Large language models · Natural language processing · Cost analysis · Conversational AI

S. K. P. Buddiga
Boston, MA, USA
e-mail: pranitha.bsk3@gmail.com

P. Mehendale (✉)
San Francisco, CA, USA
e-mail: pushkar.mehendale@yahoo.com

L. Krishnasamy et al. (eds.), *Distributed Deep Learning and Explainable AI (XAI) in Industry 4.0*, Information Systems Engineering and Management 55,
https://doi.org/10.1007/978-3-031-94637-0_13

1 Introduction

Large Language Models have introduced an entirely revolutionary impact in the field of natural language processing. They are core components of modern applications, powering everything from conversational AI to content generation and improving information retrieval. Still, their use comes at a very high cost. Costs include running on huge computational resources, energy consumption, data privacy, and ethics. It is important that these costs be recognized and that the aim of exploiting the potential of LLMs is sufficiently weighed against the optimization of resource use in order to allow sustainable and effective deployment.

1.1 Overview of LLMs and Their Growing Importance

Large Language Models have brought a revolution in natural language processing. This has enabled the capability of understanding and generating human-like text with exceptional ability [9]. Models include GPT from OpenAI, Google's Gemini, Anthropic's Claude, and Mixtral's Mixtral, which are designed with their huge amounts of text datasets, thereby enabling them to do pretty complex tasks such as summarizing texts, translations, and creative writing with outstanding accuracy [6, 8].

Large-scale LLM development had incredibly rapid growth with the rapid adaptation across all industries. They power applications like virtual assistants, recommendation systems, and content generation platforms. While organizations continue to leverage LLMs as tools for innovation and improving user experiences, it is also important to understand the cost of using such models. It should be balanced against the benefits and the improvements LLMs make to given domains.

A significant aspect when using LLMs is the computational resources required to train and run them. They are typically trained with vast datasets over powerful hardware, which is utterly expensive [12]. Additionally, their expense to maintain and update them will also sometimes be quite high due to regular fine-tuning they would have to undergo to maintain maximum precision and performance.

Another important cost factor is the human labor cost. So much human effort goes into developing and training further on LLMs and then maintaining them after these. It encompasses the cost of data labeling, cost of model development, and the cost for quality assurance. That is why the benefits of the application of LLMs need to outweigh the costs involved.

1.2 *Motivation for Understanding Costs and Optimizing Resources*

On the other hand, LLMs are very compute-intensive and thus costly in terms of infrastructure and maintenance; they require tremendous computational resources to train and make predictions. Privacy and ethical considerations with respect to the training data may need some serious data governance measures. Cost–benefit analysis: the return on investment against the cost of infrastructure, energy, data privacy compliance, and maintenance. In bottom line terms, then, LLM deployment at scale requires resource optimization and cost-cutting; unpacking the true costs and motivations behind LLMs will help organizations make informed decisions about LLM adoption, optimization, and responsible use; AI adoption will then be intentional and impactful [3, 10].

2 Understanding LLM Costs

2.1 *Overview of Cost Factors and Considerations*

There are various expenses that come with deploying LLMs and which organization needs to consider:

(1) ***Computational Resources***: The training and inference phases for LLMs are computationally intensive and depend on very powerful GPUs or TPUs, which can be very expensive to get hold of and maintain.
(2) ***Energy Consumption***: The running of LLMs largely involves huge and massive energy consumption resulting in high electricity expenses besides environmental impacts.
(3) ***Data Acquisition and Storage***: The acquisition of such a large textual data set required for the training of the LLMs has its cost; not to mention the cost of organizing the data quality and relevance.
(4) ***Data Privacy and Security***: Protection of sensitive information and data privacy regulations require a significant investment in security technologies and practices.
(5) ***Ethical Considerations***: Also, the implementation of LLMs also involves the costs and expenses related to the mitigation of biases, fairness, and transparency.

2.2 Discussion of Cost Metrics and Evaluation Methods

The deployment of LLM would require wide-scale cost evaluation. Main metrics and methods include:

(1) ***Total Cost of Ownership (TCO)***: direct and indirect costs associated with LLMs' lifecycle, in terms of the cost of acquisition, operation, and maintenance.
(2) ***Return on Investment (ROI)***: Computing the ROI would enable organizations to understand how much in financial advantage one would derive from deployment of LLMs relative to its cost.
(3) ***Energy Efficiency Metrics***: Quantities such as inferences per unit of energy consumed and training cycle per unit of energy consumed can be used to measure the environmental and cost implications of LLM operations.
(4) ***Scalability and Flexibility***: As the scalability of LLMs and their flexibility to handle different workload and application scenarios aligns with long-term cost management.
(5) ***Compliance and Security Costs***: The cost of data privacy, security, and compliance with the number of regulations must be understood part of the cost landscape.

3 NLP Applications Costs

The cost of tokens per 1000 tokens for AWS and Azure hotels models is shown in Table 1.

Table 1 Token cost of AWS and Azure hosted models as of August 2024

Models	Input token cost per 1 K tokens	Output token cost per 1 K tokens
AWS Claude 2	$0.008	$0.024
AWS Claude 3 Haiku	$0.00025	$0.00125
AWS Claude 3 Sonnet	$0.003	$0.015
AWS Claude Instant	$0.0008	$0.0024
AWS Llama 3 70B	$0.00265	$0.0035
AWS Llama 3 8B	$0.0004	$0.0006
AWS Mistral 7B	$0.00015	$0.0002
AWS Mixtral 8*7B	$0.00045	$0.0007
AWS Titan Express	$0.0002	$0.006
AWS Titan Lite	$0.00015	$0.0002
Azure GPT-3.5-Turbo-0125	$0.0005	$0.0015
Azure GPT-4-Turbo	$0.01	$0.03
Azure GPT-4o	$0.005	$0.015

3.1 Chatbots and Conversational AI

Such conversational AI systems and chatbots rely on LLMs to simulate the human-like interaction and offer tailored assistance to the user. However, deploying and maintaining such a system has definite cost and considerations. Firstly, such development process of the conversational AI models requires significant computing resources for training and fine-tuning it, which is infrastructure-cost intensive. Similarly, as with the passage of time, maintenance and updating them to ensure relevance and effectiveness triggers large degrees of operational expenses continuously. Moreover, data privacy and security concerns are essential in conversational AI applications since most of these systems deal with sensitive information concerning the users. The "Implementation of robust data protection measures and compliance frameworks." This incurs additional cost in the deployment process [2]. Despite the challenges related to chatbots and conversational AI, they are revolutionizing customer service, making processes easier, and enhancing user experiences.

3.2 Language Translation and Localization

LLMs play a significantly important role in language translation and localization applications by rendering highly accurate and contextually suitable translations across various languages. However, high-quality translation comes at a considerable computation and data overhead, thus implying a cost factor. Training multilingual LLMs generates handling of extremely large volumes of text data across multiple languages with powerful computing hardware and scalable infrastructure requirements. The training of models on specific language pairs or dialects further incurs an additional cost factor. Simultaneously, linguistic and cultural adequacy in translation calls for access to diverse and representative datasets; hence, data acquisition costs [13]. However, the translation of languages and localization with LLMs offer business and other organisations the chance of international communication and cross-cultural understanding and expansion of markets [1].

3.3 Text Generation and Content Creation

LLMs are very important tools for producing texts and content, enabling the creation of high-quality written material with minimal human interaction. Far from being negligible, though, are the costs involved. Inputting computing resources and large datasets needed to train an LLM are pricey up front. Otherwise, perfecting the content to be high in quality while achieving the style and tone desired may demand a fine-tuning in iterations, thus costs. Besides, having ethics such as preventing the creation of harmful or biased content, this requires constant inspection and alteration, thus increasing the working costs. Among such unachievable challenges, the leverage of LLM for content creation comes with increased efficiency, scalability, and developing personalized and engaging content at scale [14, 15].

4 Case Study: Chatbot Cost Analysis

There are two primary pricing models for the use of an LLM—Pay by Token and the Hosting Own Model.

This is to say that most LLM services charge for the processing of tokens-the words and sometimes symbols in input and output.

Other organizations host their models themselves, where they pay for infrastructure and potentially a license fee for the LLM itself. Hosting your model gives you control over data privacy and operational flexibility, but requires significant investments in infrastructure and ongoing maintenance. Key reconciliation is that of OpenAI's tokens-based pricing with that of most GPU services, which charges based on compute time. On face value, OpenAI's pricing using token-based pricing is more expensive but likely more cost-effective and resource-effective than one deploying the LLM locally for extended periods based on utilization.

4.1 Considerations for Commercial Models

There are a few key considerations when discussing commercial models of LLMs. First and foremost, the concern is data privacy. There is an immediate risk that sensitive corporate information could be inadvertently included in prompts by employees that eventually could lead to a security breach or misuse of confidential information. Organizations must, therefore, enforce stringent data governance policies and training programs to ensure employees are made aware of the implications of dealing with sensitive data in these models.

The functionality is the other function. Proprietary models often come with sophisticated features such as function calling and JSON mode, greatly improving the usability in certain applications, although perhaps by being primarily commercial providers, they have fairly limited support for log probability or logprobs APIs, which may form a limitation in tasks required for complete probabilistic outputs. Hence, the scope in terms of feature set of the commercial model, which is in sync with the specific needs of an application, forms the basis for an effective decision.

Another important one is API cost. Commercial models are easy to integrate, convenient, and scalable, but the cost of calling an API escalates very high at scale, especially for applications with high usage frequencies. At a scale, cumulated API requests become a significantly expensive affair for such applications. It is critical that organizations estimate the usage patterns and long-term cost implications of dependency on a commercial API.

Commercial models also vary in fine-tuning ability. A provider can deny permission for the fine-tuning of their model, which could be an adverse point for applications that need customised responses or domain knowledge. The commercial models are not appropriate in such applications where open-source models serve better with high fine-tuning.

Finally, the use cases are unique edge usage challenges for all commercial models. Many proprietary models rely strongly on constant connectivity to function at their best, thus making them unsuitable for this kind of scenario where the devices work in offline environments. For applications that need powerful capabilities of edge computing, the organization may have to shift to alternative solutions that have more flexibility regarding deployment and connectivity.

4.2 Considerations for Open-Source Models

Open-source models are far removed from all the preceding discussions concerning the need for an organization to consider before actually deployment. Data lineage and copyright issues make it less likely that open-source model builders will face legal challenges relative to their commercial counterparts. Their use, however, to generate revenue can invite legal scrutiny, especially when trained on copyrighted data. In other words, organizations need to understand what are the legal effects of using open-source models for business and how a particular country's copyright law might conform to it.

Another important characteristic is functionality. Hosting open-source models also gives control over the functionality, including log probability outputs as well as other direct responses. Such control would be useful in particular applications where scrutiny and customized outputs are required. The tools for calling the functions as well as for performing constrained sampling on the open source are limited to a few; this may require extra development work to make some functionalities realizable.

Real costs associated with large open-source models lie in the engineering. Such infrastructure requires heaps of investment to optimize the performance and keep models running. Such APIs from commercial providers might be pricey, but their engineering cost can surpass that of self-hosted open-source models if not controlled. Some of these costs can be mitigated through model hosting services that support the required open-source models but are still appropriately planned and resource allocated.

In theory, fine-tuning open-source models has the advantage of being able to calibrate those models toward specific use cases. However, in reality, this procedure is complicated, and therefore not an easy process. Access to high-quality training data and computational resources apart from technical capabilities is also required to fine-tune. Organizations will have to be prepared to spend on the infrastructure and skills that they require to fine-tune open source models that bring them to a suitable solution for their needs.

Commercial and open-source models have benefits which are unique to each as well as challenges. Understanding these factors puts organizations in a better place to decide the implementation strategy for LLMs that best fit their operations and strategic objectives.

5 Case Study: Cost Analysis of Large Language Models (LLMS) in Industry 4.0: Chatbots and Conversational AI in Manufacturing

An LLM is absorbed rapidly into industry sectors starting with the manufacturing processes; it gradually spreads to other industries. LLM significantly enriches and adds great value to applications in conversational AI, chatbots, and virtual assistants integrated into production lines, supply chain management, and maintenance scheduling for maximum efficiency at reduced costs of operations. Such AI-facilitated systems may, therefore, enhance the capabilities in decision-making. Of course, there are also considerable costs associated with deploying LLMs in forms of computational expenses, energy consumption, and data privacy. This case study discusses the costs of integrating LLMs into a smart manufacturing setup and their role in conversational AI applications.

In the modern manufacturing environment, communication between operators, machines, and systems is crucial in production processes. Chatbots and conversational AI have assumed a critical role in affecting the auto-communication in the manufacturing environment through real-time responses across sites regarding troubleshooting, scheduling for maintenance, and inventory management.

A large automotive manufacturing group wanted to apply LLMs to manage its predictive maintenance system and assist warehouse people with live updates and questions using an AI-led chatbot. The company has a few specific areas where they want to be improved:

1. Reduced downtime through improving predictive maintenance.
2. The warehouse staff are now equipped with instant, accurate information about parts and stock availability.
3. Factory operators were also supported by automating manual checks and reporting.

5.1 Application of LLMs in Manufacturing

5.1.1 Conversational AI for Predictive Maintenance

- Challenge: This plant experiences a lot of unplanned machine break down since it fails to predict effectively on the failure of most equipment. The current system works relying on manual monitoring and operator feedback that is ineffective and reactive.

- LLM Solution: An AI-based chatbot was installed by the company that would act as a linking and connecting bridge between operators and the central system of maintenance. It would collect data from sensors attached to various machines and automatically analyze performance patterns so as to predict probable failures.
  - Example: For instance, an operator might ask, "When will the next maintenance be required on Machine X?" The chatbot will respond through analysis of historical data and sensor readings that is indicative when maintenance will be required, and it will then suggest ordering certain parts in anticipation of failure.
  - Cost impact: Very high upfront costs for training the LLM and interfacing with factory systems, but the predictive maintenance application saved 25% of downtime, thereby the cost saving offset the up-front investment heavily.

5.1.2 Chatbot for Warehouse Management

- Challenge: Warehouse personnel could only manually look into various systems to know the status of parts and stock levels, which would very often delay the time of retrieval.
- LLM Solution: The chatbot interface was added to the Warehouse Management System so that the workers could instantly check their stock availability, current status of orders, and upcoming deliveries.
  - Example: A warehouse manager asks, "What is the level of stock of Component Y?" The chatbot quickly finds the data and tells the manager whether the reorder needs to be executed or whether the shipment is behind schedule.
  - Cost Impact: The automation reduced manual checks by 30%. This reduces the labor costs while improving overall inventory management productivity. Ongoing infrastructure and model fine-tuning costs were associated with keeping and continually updating the model.

5.2 Cost Analysis of LLM Deployment in Manufacturing

5.2.1 Token-Based Costs

- One significant aspect of their LLM service is how it adopted a pay-per-token model, which would allow scalable propagation with very minor large infrastructure investments. In this model, a charge is enforced upon the input as well as output tokens involved in the processing that is carried out by the chatbot during the time of interaction.

Table 2 Overall accuracy with thresholds (best to worst)

Detectors	Human data (%)	ChatGPT data (%)	Difference (human – ChatGPT) (%)
Copy_Leaks	98.70	96.50	+ 2.20
GPT2_Detector	97.90	94.75	+ 3.15
Check_AI	97.75	93.80	+ 3.95
GLTR	85.60	93.00	– 7.40
GPT_Kit	96.50	72.50	+ 24.00
Originality_AI	90.20	68.00	+ 22.20
AI_Classifier	92.30	65.40	+ 26.90
GPT_Zero	57.80	42.00	+ 15.80

- Cost Example: As can be seen from Table 2, in terms of tokens per 1000, costs ranged from $0.002 to $0.015 depending upon the level of utilization of the chatbot and also the choice of LLM, that is, AWS Claude 2, AWS Mistral 7B, Azure GPT-3.5 Turbo.
- Benefit: With token-based pricing, the company could upscale or downscale depending upon its need for conversations, thereby saving itself from huge up-front costs which were deployed at the cost of hosting their own LLM infrastructure.

5.2.2 Infrastructure and Maintenance

- Hosting Costs: Initially, the company had considered self-hosting an LLM for their chatbot but soon realized that infrastructure, especially what would require GPUs and servers, was going to be too expensive. Thus, it had chosen cloud hosting, which requires lesser upfront investment but is rather more in monthly subscription fees.
- Energy Consumption: The running of the LLMs proved to be very energy-intensive, especially when running the models during training and tuning. Because the company was using a hosted model, though, the costs on energy were absorbed by the cloud service provider. However, these still translated to a 6% increase in IT charges, considering data storage and cloud services.

5.2.3 Human Resources and Data Privacy Costs

- The LLM needed to be updated from time to time so that the chatbot stays updated and accurate. Human resources were required for further model fine-tuning and various data privacy laws, such as GDPR. Sensitive data in the prompts was also subjected to comprehensive management to prevent security breaches, which again added to the compliance cost.

5.3 Results and Impact on Efficiency and Cost

5.3.1 Reduction in Downtime

- It reduced downtime from a predictive maintenance chatbot, since it managed to predict more accurate and probable equipment failures and helped in scheduling proactive maintenance. The company was able to bring down the maintenance cost by 20% because of fewer unplanned repairs and less equipment downtime.

5.3.2 Improved Warehouse Operations

- Introduction of the chatbot warehousing system increased the accuracy of warehouse inventory and reduced labor hours spent in the check process. The company saved $50,000 annually in labor but spent $10,000 annually in maintaining the LLM system.

5.3.3 ROI and Cost Savings

- Even though the running of the LLM system and cloud services involved perpetual expenses, the ROI was positive. The firm saved approximately $250,000 per year due to the resultant improved efficiency and reduced downtime associated with operations while it incurred approximately $75,000 per year in expenses for the use of tokens, cloud services, and personnel for model fine-tuning.

Deployment and maintenance of LLM through means of chatbots or conversational AI greatly affect production efficiency and costs. Very high investment costs weigh down with the cost of deploying and maintaining the LLM, such as infrastructure cost, token-based pricing, and human labor, but the benefit that comes from being operational will outweigh them. This case study evidentially shows that careful cost analysis and optimization lead to the good implementation of Industry 4.0 and long-term productivity gains and downtimes.

5.4 Summary of Text Detectors Works with LLM-Generation

We have discussed an overview of our experience with LLM-generated text detectors on all important issues: intuitiveness for the user, documentation clearness and completeness, extensibility, input diversity, quality of report produced, ability to work in multiple LLM-generated languages as well as cost, and its analysis is presented in Tables 2, 3 and 4 along with graphical representations as Figs. 1, 2 and 3.

Table 3 Overall accuracy based on weighted average (best to worst)

Detectors	Human data (%)	ChatGPT data (%)	Difference (human – ChatGPT)
Copy_Leaks	98.30	93.90	+ 4.40
GPT2_Detector	97.60	92.50	+ 5.10
Check_AI	97.45	92.85	+ 4.60
GLTR	93.90	68.75	+ 25.15
GPT_Kit	94.50	65.80	+ 28.70
Originality_AI	87.90	64.60	+ 23.30
AI_Classifier	66.20	64.25	+ 1.95
GPT_Zero	71.50	54.25	+ 17.25

Table 4 False positives tabulation

Detectors	False positives
GPT_Kit	0
Copy_Leaks	1
GPT2_Detector	3
Check_AI	3
AI_Classifier	7
Originality_AI	8
GLTR	18
GPT_Zero	50

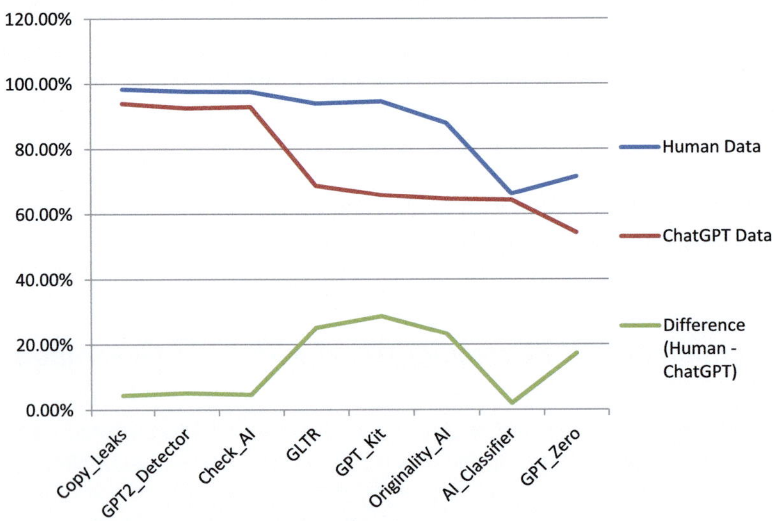

Fig. 1 Graphical representation of overall accuracy with thresholds

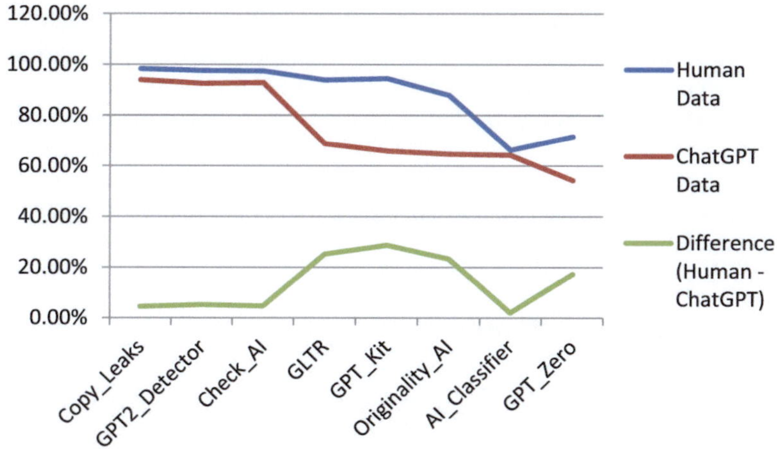

Fig. 2 Graphical representation of overall accuracy with weights

Fig. 3 False positive computation

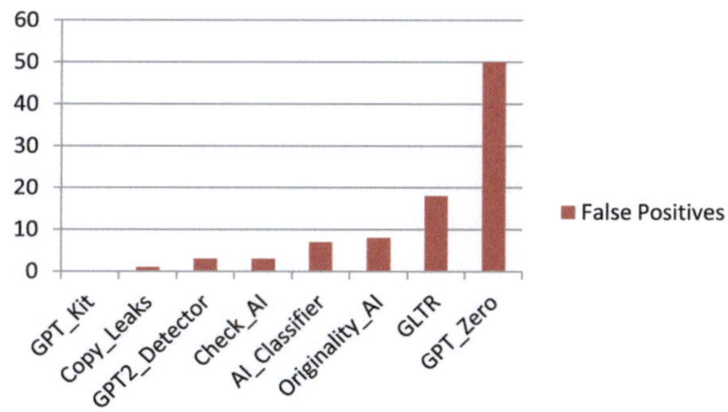

6 Applications of LLM in Industrial 4.0

6.1 Analysed Industrial Data

ChatGPT may be used to analyze large amounts of industrial information in order to identify inefficiency in manufacturing processes. This might help manufacturers improve quality assurance, reduce waste, and streamline their operations. ChatGPT may be trained to seek for patterns and predict when repair might be required using information from sensors from manufacturing machinery. This might save producers money on repairs and downtime. The capacity to customize financial counseling to suit the needs of very specific client profiles has been a blessing for the banking industry.

The reason for this is that finding individuals that might guarantee exceptional profits often requires a significant amount of work. Bankers may evaluate internal client data on financial instruments and habit purchases using ChatGPT. With ChatGPT, acquiring programming skills has become much simpler. Aside from providing precise syntactical suggestions, this artificial intelligence for conversations makes recommendations for better methods of coding utilizing widely used structures of information and techniques. It can create code in nearly all of the languages now available on the marketplace in response to human prompts. System architecture information may be created, edited, and consumed via ChatGPT.

6.2 Analyse Possible Danger

ChatGPT could be used to analyze supplier information and identify potential risks or possibilities. This might help businesses make informed supplier selections and save supply chain costs. ChatGPT might be used by producers to provide automatic customer support. This might include handling inquiries, providing assistance with technology, and answering frequently asked queries. Producers and users need to work together to put security precautions in place and promote transparency in order to lessen risks. OpenAI needs anonymized information as well as strong safety safeguards in order to help protect end-user privacy and preserve trust in AI systems. Overuse of ChatGPT may lead to a reliance on content produced by AI, which compromises human thinking and innovation. Adoption of artificial intelligence technologies such as ChatGPT might have a significant effect on the economy. It may lead to massive corporations centralizing their AI capabilities or, in certain industries, the replacement of staff members. The consistency and relevance of ChatGPT's replies will be improved by improving its capacity to take in and remember background from lengthy text passages and back-and-forth talks, a task that OpenAI scientists are now working on.

6.3 Automate Various Tasks

ChatGPT might streamline processes like as gathering information, IoT device surveillance, and prospective duties like recruiting, onboarding, instruction, and achievement tracking for companies. Companies may utilize this to save time and money while improving their client relationships. ChatGPT is employed for a range of language-related tasks, such as assistance with customers and instruction in the language. Given its ability to understand and generate text that looks human, it is a valuable tool in a variety of fields, like business, education, and amusement. With ChatGPT, users may create original material for a variety of apps, improve interaction, and spend less time on duties. On many topics, it may provide replies that are human-like. It can generate precise and pertinent responses and has exceptional comprehension of language. Its utility and safety are further enhanced by fine-tuning, which integrates input from people, making it an invaluable instrument with a wide range of uses.

6.4 Enhancing the Efficiency

Companies can conserve both time and cash while improving advertising and promotional effectiveness using ChatGPT. It could help businesses achieve their marketing goals and grow their customer base. By using ChatGPT in the creation of applications, companies may produce excellent software that is customized to meet their unique needs while also increasing efficiency and efficacy. By using ChatGPT's sophisticated data analytics capacity, companies may improve their understanding of their inventory levels as well as supply chain oversight to save expenses and boost productivity. It can provide accurate reports on delivery schedules and supplier efficiency, providing companies with the data they want to enhance their warehouse setup and ensure on-time delivery of their products. As more user input is processed, ChatGPT may alter and enhance its responses to better meet the needs of its users. Because of this, chatbots developed using ChatGPT have the ability to enhance customer service over time by developing their effectiveness and efficiency. The model may help students review and reinforce their understanding of the material by offering quizzes, practice inquiries, or memorization. It may help students create effective study plans and provide test-taking suggestions to help them do better on exams.

6.5 Better Planning for Industrial Operations

ChatGPT may provide maintenance prediction schedules and on-the-ground quality assurance discrepancy identification for industrial processes. It may assist traders on the exchange room with its readily accessible and simplified technical material and instructions. Natural language user inputs, including spoken directions, may be produced via ChatGPT as actionable actions for industrial robots and machines. ChatGPT may be used to address client queries in a timely and appropriate manner, reducing wait times and improving customer satisfaction. By providing passengers with knowledge of their journey schedule, anticipated expected arrival and departure instances, and other vital details, it may also serve as a virtual traveling companions. ChatGPT performs better than any AI-based tool that was before accessible. As a result, it's critical to treat fear and anxiety properly and use them as possibilities. There has previously been evidence of ChatGPT's usefulness in the train industry. The advantages can benefit companies and society worldwide. The time has arrived to provide the environment and terrain that such advanced AI requires. Unlike an actual human accountable, this gadget is continually available at all times of the day. Additionally, a chatbot can simply and swiftly react to a variety of questions. This availability guarantees that procedures and operations go as intended while averting disputes between significant parties, which benefits the organization and improves productivity.

7 Future Scope

ChatGPT will soon be able to help with the creation of exact contractual agreements. This chatbot can already write legal documents on par with a few of the greatest human lawyers. In the end, the benefits of AI are evident, yet we have only just started to discover their greatest potential. We may expect these developments to significantly impact the creation of jobs and the community, especially as we go into the fourth industrial revolution and beyond. By accepting and responsibly using these innovations, we can unleash unprecedented levels of imagination, effectiveness, and production that might potentially aid in the resolution of some of the world's most pressing issues and contribute to the construction of a brighter future for everyone. In future generations, AI models for language might access immediate information or conduct live research, enabling ChatGPT to deliver more precise and up to date replies. The future of ChatGPT is quite promising since it is expected that future advancements and innovations will allow it to surpass its present constraints and grow into an even more versatile, potent, and useful tool for many applications. By keeping up its investment in R&D, the AI community can fully use language models and drive the next wave of creativity in NLP and beyond. ChatGPT is built on top of NLP and ML. Based on historical data and trends, particularly industrial and technical trends, it forecasts upcoming changes. Future advancements may be predicted using methods like text categorization, clustering, and time series analysis. Using machine learning techniques, it will interpret and generate natural language as if it were written by a human. It used billions of data points for training in order to create a neural network that can identify patterns and correlations, make predictions about the future, and respond. As new developments in context understanding, rational thinking, bias decrease, immediate data availability, and adaptability are made feasible by continual development and research, ChatGPT's future seems bright. There are several benefits and applications for the powerful ChatGPT language paradigm. Because it can produce text that looks and reads like actual people, can generate text based on certain topics or styles, and may improve the learning of languages and processing, it will be a helpful tool for businesses, content providers, and academics. Because of its versatility, customizability options, and ease of use, it will prove to be a valuable resource for anybody looking to construct personalized language models. In the next days, ChatGPT will assist producers by providing up-to-date details on the state of machinery and processes inside a production facility. This makes it possible for operators to identify issues with machinery or processes as soon as they appear and address them before they become worse. There is little question that ChatGPT will be used in production in the future because to its boundless possibilities.

8 Conclusion

This paper has conducted an in-depth examination of the multiple costs of deploying large language models across a wide range of applications. While LLMs have critically significant advantages in enhancing natural language processing tasks, their deployment entails great costs regarding computational resources, energy consumption, data privacy, and certain ethical considerations. The cost analysis is spread over various applications such as conversational AI, language translation, and content creation. There is an urgent need for a holistic cost–benefit evaluation to use LLMs in a sustainable and effective manner.

Open-source and commercial models have different implications and drawbacks. For instance, the open-source model of Meta Llama has a higher initial cost and high maintenance requirement but is more flexible and inexpensive in the long term. Another commercial model, as in OpenAI's ChatGPT, would be easier to set up and scale and maintained by support but may perhaps be a costlier operation due to a pay-per-token usage structure.

While such cost factors are understood, empowering an organization to make informed decisions towards the adoption of LLMs and optimum resource usage with ethically responsible deployment, this understanding would keep the promised technological benefits in line with financial sustainability and responsible AI practices.

9 Future Directions

However, with multiple costs of LLM deployment, the need to explore directions for improving access to and affordability of such models is necessary. Several of these include ways of optimizing hardware and algorithms to curb energy consumption in LLMs. Other techniques used are model compression, model quantization, as well as the development of more efficient training methods to reduce operational costs and environmental footprint. The federated learning approaches can also distribute the computational load across multiple devices thereby minimizing the need for centralized high-performance infrastructure [4]. This may diminish cost and energy consumption in training and inference processes.

Deep learning algorithms for data optimization, such as pruning, augmentation, and synthetic data generation, reduce the amount of training data without impacting model performance. It will save costs on storage and processing and increase efficiency overall. Encouraging collaborations or shared infrastructure among organizations can help share the costs and resources in deploying LLM. Shared model repositories and cooperative training initiatives are growing in order to bring the power of LLMs within reach of far more users, down to modestly sized organizations with tighter budgets [5].

Clear policies and regulations on data privacy, security, and ethical use of AI would reduce the compliance cost and risks. Governments and regulatory authorities can be supportive by formulating guidelines and frameworks for responsible and cost-effective deployment of LLM. By investing in automated tools and platforms for fine-tuning and model maintenance, investments in human labor and expertise can be reduced for these activities. Automated systems can help in the minimization of customization of models for specific use cases thus aiding in initial cost minimization as well as cost minimization over time.

The AI community will work towards significantly decreasing deployment costs of LLMs, thus making such powerful tools accessible and sustainable. This will enhance the adoption of LLMs across various industries and encourage innovation and responsible AI development.

References

1. Wu, Y., Schuster, M., Chen, Z., Le, Q.V., Norouzi, M., Macherey, W., Krikun, M., et al.: Google's Neural Machine Translation System: Bridging the Gap between Human and Machine Translation. arXiv arXiv:1609.08144 (2016)
2. Xu, A., Liu, Z., Guo, Y., Sinha, V., Akkiraju, R.: A new chatbot for customer service on social media. In: Proceedings of the 2017 CHI Conference on Human Factors in Computing Systems, pp. 2849–2856 (2017)
3. Strubell, E., Ganesh, A., McCallum, A.: Energy and Policy Considerations for Deep Learning in NLP. arXiv arXiv:1906.02243 (2019)
4. Sanh, V., Debut, L., Chaumond, J., Wolf, T.: DistilBERT, a Distilled Version of BERT: Smaller, Faster, Cheaper and Lighter. arXiv arXiv:1910.01108 (2019)
5. Jiao, X., Yin, Y., Shang, L., Jiang, X., Chen, X., Li, L., Wang, F., Liu, Q.: TinyBERT: Distilling BERT for Natural Language Understanding. arXiv arXiv:1909.10351 (2019)
6. Brown, T.B., Mann, B., Ryder, N., Subbiah, M., Kaplan, J., Dhariwal, P., Neelakantan, A., et al.: Language Models Are Few-Shot Learners. arXiv arXiv:2005.14165 (2020)
7. Clark, K., Luong, M.-T., Manning, C.D., Le, Q.V.: ELECTRA: pre-training text encoders as discriminators rather than generators. In: International Conference on Learning Representations (2020)
8. OpenAI. 2020: GPT-3: Language Models Are Few-Shot Learners. arXiv arXiv:2005.14165
9. Rogers, A., Kovaleva, O., Rumshisky, A.: A primer in BERTology: what we know about how BERT works. Trans. Assoc. Comput. Linguist. **8**, 842–866 (2020)
10. Sharir, O., Peleg, B., Shoham, Y., Shashua, A.: The Cost of Training NLP Models: A Concise Overview. arXiv arXiv:2004.08900 (2020)
11. Bender, E.M., Gebru, T., McMillan-Major, A., Shmitchell, S.: On the dangers of stochastic parrots: can language models be too big? In: Proceedings of the 2021 ACM Conference on Fairness, Accountability, and Transparency, pp. 610–623 (2021)
12. Bommasani, R., Hudson, D.A., Adeli, E., Altman, R., Arora, S., von Arx, S., Bernstein, M.S., et al.: On the Opportunities and Risks of Foundation Models. arXiv arXiv:2108.07258 (2021)
13. Dodge, J., Ilharco, G., Schwartz, R., Farhadi, A., Hajishirzi, H., Smith, N.A.: Documenting Large Webtext Corpora: A Case Study on the Colossal Clean Crawled Corpus. arXiv arXiv: 2104.08769 (2021)
14. Sandhya, N., Saraswathi, R.V., Preethi, P., Chowdary, K.A., Rishitha, M., Vaishnavi, V.S.: Smart attendance system using speech recognition. In: 2022 4th International Conference on Smart Systems and Inventive Technology (ICSSIT), pp. 144–149. IEEE (2022)

15. Baskar, K., Venkatesan, G.P., Sangeetha, S., Preethi, P.: Privacy-preserving cost-optimization for dynamic replication in cloud data centers. In: 2021 International Conference on Advance Computing and Innovative Technologies in Engineering (ICACITE), pp. 927–932. IEEE (2021)

Sai Kalyana Pranitha Buddiga is an accomplished Data Science and AI professional with extensive expertise in developing innovative and scalable analytical solutions. With over a decade of experience across industries, including financial services and technology, she has driven advancements in machine learning, natural language processing, and compliance analytics. Pranitha holds a Master of Science in Business Analytics and Project Management from the University of Connecticut School of Business. A Senior Member of IEEE and a BCS Fellow, her contributions extend beyond her professional roles to include co-authored publications, and thought leadership in artificial intelligence and data science.

Pushkar Mehendale is a seasoned Machine Learning Engineer with over six years of experience in designing and implementing advanced ML solutions in production environments. With a proven track record of driving significant business impact through strategic AI applications, Pushkar has worked with leading organizations, including Drift/Salesloft, Sown To Grow, and StubHub. He holds a Master's degree in Computer Science from the University of Illinois at Chicago. Passionate about innovation, Pushkar specializes in scalable ML systems, recommendation engines, and infrastructure optimization.

Mitigating Bias in AI Recruitment Through Explainable AI for Fair and Inclusive Hiring Practices

P. Jayadharshini, P. Karunakaran, S. Santhiya, A. S. Renugadevi, G. Dhanush, and E. Pavithra

Abstract AI-based recruitment software transforms the way recruiting is done, including screening of resumes and assessment of applicants. However, this has brought a lot of exciting changes along with it, and ethical concerns related to how biases are generated in these systems are emerging increasingly. Explaining AI (XAI) makes this process transparent and accountable while keeping track of how AI decides. The case study highlights the dilemma that AI-orientated hiring faces, the generation of bias in algorithms, and how XAI can be used as a means of mitigating some of the biases. We analyze the potential of XAI in making hiring practices fairer and more inclusive through examples from real-world businesses, the legal context in which this will be applied, and technical solutions.

Keywords Artificial Intelligence (AI) · Explainable AI (XAI) · Bias · Transparency · Hiring faces

---

P. Jayadharshini (✉) · S. Santhiya · A. S. Renugadevi
Assistant Professor, Department of Artificial Intelligence, Kongu Engineering College, Perundurai, Erode 638060, Tamil Nadu, India
e-mail: jayadharshini.ai@kongu.edu

S. Santhiya
e-mail: santhiya.cse@kongu.edu

A. S. Renugadevi
e-mail: renugadevi.ece@kongu.edu

P. Karunakaran
Professor, Department of Artificial Intelligence and Data Science, Nandha Engineering College, Perundurai, Erode, Tamil Nadu, India

G. Dhanush · E. Pavithra
Student, Department of Artificial Intelligence, Kongu Engineering College, Perundurai, Erode 638060, Tamil Nadu, India
e-mail: dhanushg.21aim@kongu.edu

E. Pavithra
e-mail: pavithrae.21aim@kongu.edu

L. Krishnasamy et al. (eds.), *Distributed Deep Learning and Explainable AI (XAI) in Industry 4.0*, Information Systems Engineering and Management 55,
https://doi.org/10.1007/978-3-031-94637-0_14

1 Introduction

Hiring through AI has undergone much greater development over the last decade. It has changed the traditional way of recruitment. From resume screening and ranking candidates to scheduling interviews with the aid of AI tools, companies are now in the adoption of automation of the hiring process. Companies such as HireVue, Pymetrics, and LinkedIn have installed AI-driven systems in their processes of hiring to make the processes faster, cheaper, and more efficient. As these systems become more widespread, worries about their possible fairness and ethics creep in.

Explainable AI [1] is the ability of an AI system to provide an easily understandable explanation of their decisions. It has evolved as an integral component of AI in hiring. Using XAI results [2] in hiring managers and the HR team understanding how AI arrives at its decisions, leading to much-desired trust and accountability. It becomes really critical when AI is used for candidate selection, and unexplained rejections lead to accusations of unfair practices.

Bias in AI [3], particularly in hiring software, is a major issue. Generally, the AI learns from past trends and data. These may carry forward existing biases of the recruitment process. To illustrate, suppose that past choices of hiring were biased toward some demographics. Then this AI system would do more or less the same thing as those previous processes and would likely yield outcomes not precisely just. Therefore, such biases have to be tackled while ensuring that AI hiring tools support diversity and inclusivity.

2 Ethical Considerations in AI-Driven Hiring

Hiring and employment functions controlled by AI shall work in ethical boundaries that seem to be just, answerable, and transparent. Ethical AI requires accountability for human decisions, which means increased sensitivity is necessary. The principles of fairness relate to treating all candidates based on their qualifications and abilities rather than their racial, gender, or socioeconomic backgrounds when hiring.

Another very important ethical issue is accountability. It is pretty tricky to establish liability when AI systems take up all the responsibility in making selections; which mistakes and biased results fall under whose accountability? Employers tend to rely on AI tools [4] almost exclusively without knowing how they determine these selections, since that could lead to very unfavorable situations. It calls for accountability on the part of AI developers as well as employers in this regard: who takes responsibility for decisions made through AI.

Why trust such AI-based hiring processes? Transparency-that is why. The candidate needs to understand how decisions are made and why they are not selected. This can lead to skepticism and legal challenges, including an assertion that candidates were discriminated against.

Left unchecked, bias in hiring systems contributed by AI will have more important social and legal consequences. For example, as recently as in 2018, Amazon had to scrap its AI-assisted hiring tool after it had systematically discriminated against female candidates. In this regard, AI systems [5] are seen to perpetuate unintended bias-even when the intent is best. If left unchecked, lawsuits and reputational damage as well as loss of trust are sure to follow.

Discrimination risks arise both in the type of data used to build AI models and the algorithms themselves. In other words, if historical hiring data reflect discriminatory practices, an AI system will replicate those patterns. For example, if an AI system is trained on resumes from a male-dominated industry, it may learn to favor males. Some algorithmic designs may also carry biases [6], such as the weighted importance of certain keywords or traits that reduce the impact for certain groups.

3 Bias in AI Hiring Systems

The modern recruitment systems of AI have become very popular among firms because they wish to automate and make the selection process seamless. However, they carry potential bias and, if formulated and tracked poorly, may either perpetuate the biases already existing in them or inflate them. Sources of bias in AI include biased training data, imperfect algorithms, and lack of transparency. Bias mitigation should remain fairly important measures toward having fair and equitable AI-driven hiring systems to be seen for its sources and forms.

3.1 Sources of Bias in AI Hiring Systems

Biases in AI systems comes at various stages of the AI lifecycle, including the following:

- The bias in training data sets: AI models are trained on historical data. Thus, if the historical data is biased in some way, then artificial intelligence acquires those very biases. Consider an example: Let us assume that a certain company has hired all along based on traditional and conventional selection biased towards men or those coming from prestigious universities. In such a scenario, this artificial intelligence network continues to favor the same, which eventually translates into continued discrimination against women or otherwise students from lesser-known schools. This is known as historical bias, the most common form of bias that is found within AI hiring systems. Barocas and Selbst [19] argue that biases often arise from the inherent nature of big data practices.
- Selection Bias: This occurs when the data to be used for training the AI system lacks generality of the population. If the dataset for the training of the AI is dominantly applicants who hail from a certain geography or industry, then the AI

system could have hard times in properly gauging candidates from other backgrounds. This kind of bias can lead to unjust results for applicants who do not align with the major profiles in the training set.

- Algorithmic bias: Even assuming data was completely unbiased, algorithms themselves could introduce bias. Some algorithms may favor or disadvantage certain groups because of attributes correlated with race, gender, and other protected characteristics. An example would be algorithms that weight experience or educational qualification very highly; the algorithms would, in practice, further disadvantage minority candidates simply because minorities are disadvantaged by systemic barriers to opportunities.
- Label Bias: This occurs when the AI is trained using biased output labels or labeling. If historical hiring leader decisions have been skewed toward males, then an AI that predicts who may turn out to be a good leader based on such history will predict males as good leaders and females as bad leaders.
- Even when unobservables such as gender or race are not available in the database, AI algorithms are likely to find other indirect sources-known as proxies-by which to infer them. Decisions will therefore remain biased even if a system has been programmed to avoid them, since an applicant's name, address, or college can be proxies for gender, or socio-economic status.

3.2 Types of Bias in AI Hiring Systems

There are a number of biases found in AI hiring systems, each with its own implications for issues of fairness and equity. Experiments on ad settings by Datta et al. [24] reveal biases that are comparable to those in hiring systems.

- Gender Bias: It is the most prevalent type of bias in AI hiring systems. Studies have suggested that AI favors male candidates and is more reinforced to favor males, especially in traditional male-dominated fields like technology and engineering. This is pretty unfair and disadvantages female applicants.
- Perhaps the most egregious example of gender bias in AI hiring has been witnessed within a leading tech firm, where an AI system consistently demerits resumes containing references to women's colleges or female-oriented activities. The reason for this bias was that the AI system was trained on 10 years' worth of resumes the company had received, when during those 10 years the majority of the males working there were hired.
- Racial and ethnic biases: AI hiring systems might also possess racial and ethnic biases. It appears when the training data reflects the trend of previous discrimination. For example, when the AI is trained with data from a company that has fewer percentages of minorities, then it's likely to hire more applicants who are white as opposed to multicolored applicants. In the worst possible forms, racial bias in AI recruitment disproportionately rejects candidates with those names that are perceived to be associated with certain racial or ethnic groups.

- Age Bias: Age bias in AI hiring occurs when the recruiting process does less justice to older workers. For example, an AI system may penalize a candidate who has spent many years at work and attended college many decades ago because of the assumption that the candidates are less flexible and technologically empowered than those younger recruits. In reality, such candidates have more experience and skills, and they can be very important to any organization, but the AI using age proxies can unwittingly screen them out.
- Disability Bias: The AI hiring systems also bring about other biases, such as those discriminating against people with disabilities. Such a bias may occur because the system doesn't consider the accessibility needs; therefore, it will disadvantage people with impairments when trying to be evaluated through speech patterns or the speed of typing. Virtual interviews and relying on assessment through the machine significantly disadvantage people with disabilities who require certain accommodations. Machine bias in decision systems, as shown by Angwin et al. [16], highlights the potential for discriminatory outcomes even in non-hiring contexts
- Socioeconomic Bias: Socioeconomic bias involves the kinds of AI systems where the applicant from the upper socio-economic background is confirmed due to focusing on specific qualifications, experiences, or educational institutions provided for the affluent. For instance, those applicants who have graduated from elite universities or have worked at prestigious companies get high marks while other applicants who are from less privileged backgrounds get a penalty even though they have the skills and potential to fill the position. Table 1 summarizes the key types of bias, their descriptions, and real-world examples.

Table 1 Common biases in AI hiring systems

Type of bias	Description	Example
Gender bias	AI Favors one gender over another due to training on biased historical data	Resumes with "women's college" devalued by Amazon's hiring tool
Racial bias	AI reflects discriminatory practices in the training data, disadvantaging certain racial groups	AI rejects names associated with specific ethnic groups
Age bias	Older candidates penalized due to assumptions about flexibility or tech adaptability	AI undervalues resumes with long work gaps or older education years
Disability bias	AI disadvantages candidates with disabilities due to reliance on inaccessible evaluation metrics	Virtual interviews penalizing slower speech or typing speeds
Socioeconomic bias	Applicants from less affluent backgrounds disadvantaged due to algorithmic preferences	Higher marks for candidates from elite universities or prestigious companies

3.3 Impact of Bias on the Hiring Process

It can indirectly influence the prospects and companies, but the impact would be huge in all cases, with some including:

- Reduced Diversity: Bias-crafting systems will ultimately undo the practice of diversity in business. Underrepresented groups are systematically excluded or disadvantaged candidates. This means that an underrepresentation of underrepresented groups may undermine the firm's innovation and competitiveness in the global market. A lack of diversity will also affect the reputation of the firm, therefore less appealing to top talent.
- Legal and Ethical Implications If the biased AI hiring system companies do not comply with the antecedent anti-discrimination laws, then they are at a risk of facing several court procedures. Most countries of the world have placed accountability on employers to avoid any processes of discrimination in the recruitment process against applicants on various characteristics such as race, gender, age, or disability. The inability to correct bias in AI systems for hiring should also lead companies to litigations, fines, and unwarranted damage to their brands.
- Negative Candidate Experience Candidates who may have been maltreated by an AI hiring system will develop a negative attitude toward the company. It can lead to a lack of trust and damage the reputation of the company in the talent market. Moreover, if candidates perceive that the hiring process is unfair, they are less likely to apply for future job openings or refer friends to the company. Chouldechova [17] discusses how prediction tools with disparate impact can exacerbate existing inequalities.

3.4 Mitigating Bias in AI Hiring Systems

- Firms should be proactive at all stages of AI development and deployment and reduce the risks of bias. Some of the key strategies that will be applied for mitigation of bias are as follows:
- Bias Audits and Fairness Testing: Regularly test and audit the decisions of AI systems for bias with fairness metrics such as demographic parity or equal opportunity. Testing for fairness could help identify regions where an AI system might be over favoring or over penalizing a given group or sets.
- Debiasing Training Data: This should ensure that training data is sufficiently representative of the diverse candidate pool the company would be interested in attracting to work for them. This may require collecting more data from underrepresented groups, elimination of biased features in the dataset, or even synthetic data generation techniques to balance the dataset.
- Fairness-aware Algorithms: Using fairness-aware ML algorithms encompasses both constraints designed to enforce equity. Those algorithms can thus help

prevent AI systems from taking unfairly discriminatory actions in favor of or against candidates on protected characteristics.

- Explainable AI: The application of XAI is supposed to be implemented by an HR professional. It may make the AI system more transparent and more explainable while making decisions. It can enlighten one about which features or factors contributed to a certain decision and to what extent those factors led to biases that need correction.
- Human Oversight: Engage human oversight in the hiring decision-making process through the use of AI by allowing recruiters to review and override AI-decision made in cases such as biased or selective information that the AI may use to base its decisions on.

3.5 The Path Forward

The critical challenge ahead will indeed be to address bias as artificial intelligence hiring systems begin to appear in greater numbers so that these systems are fair, ethical, and comply with legal standards. Companies that take proactive measures to mitigate bias will fare better in their quest to build diverse, inclusive, and innovative teams, with a lower risk to their legal and reputational well-being. Development of such fairness-aware AI systems alongside regular monitoring and human oversight may lead to a hiring process that is as efficient as it is fair.

4 Role of Explainable AI (XAI)

Explainable AI, or XAI for short, plays a crucial role in reducing bias in hiring software. These are not like the so-called 'black box' models since XAI systems provide clear and interpretable explanations for their decision-making processes. For instance, to target and correct these biases, it is important to understand how the AI system evaluates candidates, and this would best be explained by HR professionals and job applicants.

XAI can help with bias because making the decision-making process more transparent helps [7]. For example, hiring managers can see how the AI got to its decision and are thus in an ideal position to spot biases in the system. For instance, if the XAI system reveals over-weighting certain keywords or traits in favor of one group more than another, these biases can be corrected.

The benefits of XAI are far greater than only in terms of bias mitigation. XAI develops trust among employers, candidates, and AI systems by providing explanations on the latter's decisions. The results of the hiring process will also be more acceptable to candidates as they know why they were selected or rejected. This transparency can also allow for audits of employer's performance and ensure that it is truly aligned with ethical standards. Rudin [22] emphasizes the importance of interpretable models over black-box approaches for high-stakes decisions.

In addition to bias reduction, there is fairness in it [8], as all candidates are treated equally and evaluated under the same standards. This is particularly important in high-stake cases such as hiring since bias might have long-term effects not only for an individual but also for organizations at large.

5 Legal and Regulatory Framework

Legal and regulatory frameworks govern AI in hiring to prevent discrimination of candidates. For instance, under the General Data Protection Regulation in Europe, there would be provisions that mandate transparency and accountability of decisions from algorithms. Candidates under the same regulation have a right to know how such algorithms arrived at the final decision and can seek human involvement if such decisions result from an unfair or biased decision. Wachter et al. [25] argue that the GDPR's explanation requirements for automated decisions are insufficient for meaningful transparency.

In the United States, employment laws are based on the Federal Equal Employment Opportunity Commission that enforces federal laws prohibiting any form of discrimination in hiring. This means, therefore, that biased results arising from the AI systems could be deemed to be violative of such laws and open to legal challenges. For example, when an AI system rejects disproportionately persons from a given racial or ethnic background, the employer may be held liable for discrimination.

A new legal issue for the contemporary era revolves around the legal liability of biased AI. Employers bear the responsibilities of making sure the systems they implement communicate the norms and laws about antiracism, antisexism, and all other forms of anti-discrimination. When the systems developed result in biased outputs, employers may be compelled to answer lawsuits from applicants who thought they were unfairly rejected. With the increasing use of AI in hiring, the legal fronts are bound to shift with subsequent new rules that regulate the fair and sensible use of AI.

Several potential regulatory approaches to mitigating bias in AI have been proposed. First, one might require that AI systems undergo regular auditing to ensure that they are not producing biased results. Another potential approach would be to require developers of AI systems to include fairness constraints in models, so that a model could only be created that would tend to produce equitable results from the outset.

6 Bias Mitigation Strategies in Hiring Software

The bias of AI hiring software is something that has to be dealt with on various dimensions, both in the data and the algorithms implemented in such systems. The most effective way to reduce the bias is in the form of training data that should depict themselves to be diverse and representative of the whole population. This can be achieved through data balancing, which involves adding underrepresented groups into the dataset, and data anonymization, which removes identifiable information that might lead to outcomes biased towards a particular group. Zemel et al. [20] propose learning fair representations to address discrimination in predictive models.

The fairness-aware machine learning algorithms can be used at the algorithmic level to remove biases. These algorithms are designed to treat all candidates equally regardless of their background, so at this level, fairness constraints can be introduced in the model to avoid such discriminatory actions taken by the AI system. The mitigation strategies are summarized in Fig. 1, highlighting the interplay between fairness-aware algorithms and data preparation techniques.

Equally required is the post-deployment bias mitigation. AI needs to be constantly monitored for it to work fairly as well as correct the biases in real-time. Feedback loops, that feed new data into the system all the time and updates its performance accordingly, and model audits, where experts check the system performance for biases and mends the issues accordingly, are a couple of the means by which this can be done.

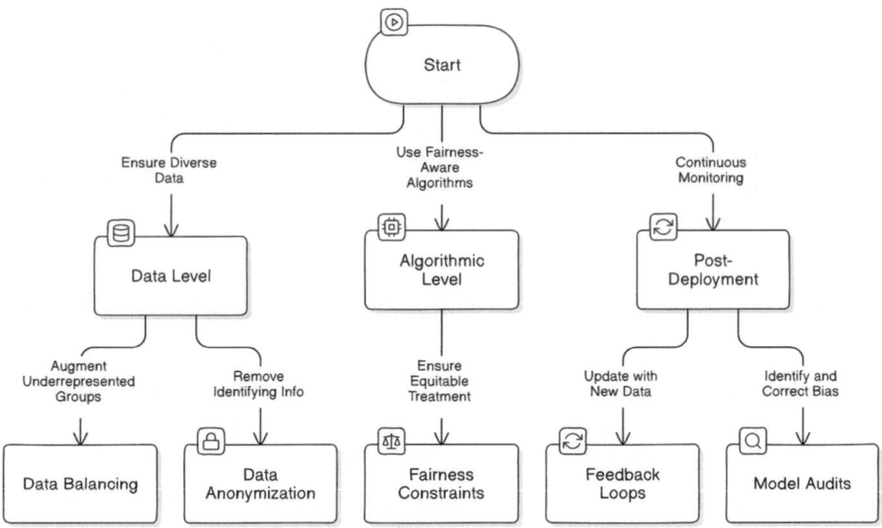

Fig. 1 Mitigating bias in AI hiring software

7 Best Practices for Developing Ethical AI in Hiring

Ethical AI in hiring systems is in many stages from data gathering up to model training and its deployment and monitoring [9]. Best practices below ensure that the AI systems foster fairness, reduce bias, and keep up with the ethical guidelines.

7.1 Diverse and Representative Training Data

The quality and diversity of the training data is reflected straight into the fairness of AI models: biased training data leads an AI model to mirror those biases and lead to discriminatory outcomes. Improving on this:

- Data coming from Diverse Sources: Training data is collected from a wide variety of sources that ensure the diversity in race, gender, age, socioeconomic background, education, and experience are reflected.
- Data Augmentation for Underrepresented Groups: In cases where a particular group is underrepresented in the dataset, some techniques would be to apply data augmentation in an artificial sense, which would otherwise be creating balance in the data. This could be oversampling the minority class or generating synthetic data that represents the underrepresented demographics.
- Remove Any Identifiable Personal Characteristics Data should be anonymised in the sense that names, gender and any other form of address not be given to the model as identification features so that a decision is not taken based on these attributes. But look into skills, qualifications and appropriate experience.
- Historical Bias Detection: Also, before exposing any historical hiring data, it's also necessary to detect and remove potential biases. For example, historical trends existing against the recruitment of women in leadership could be unconsciously retained by the AI unless the same bias is corrected in the data set. Implement bias detection tools that would flag the same as early as during data preparation.

7.2 Fairness-Aware Machine Learning Algorithms

The choice of algorithms used to develop AI models in hiring is critical and one of the key ways to mitigate bias. Algorithms exist that are designed to foster fairness by not playing favoritism with one group over another.

- Use fairness metrics: Introduce fairness constraints in the AI model that evaluate candidates appropriately without any bias. Some of the common fairness metrics include demographic parity (equal selection rates across groups), equalized odds (equal opportunity across groups), and fairness through unawareness (sensitive attribute exclusion from the decision-making process). Kleinberg et al. [21] explore the trade-offs involved in creating fair risk scores for decision-making.

- Algorithmic transparency: Whether the stakeholders are technical or non-technical, transparent algorithms allow them to understand how their decisions come about. For instance, people can use LIME (Local Interpretable Model-Agnostic Explanations) or SHAP (SHapley Additive exPlanations), which explain an individual's predictions and tell which features impacted the outcome the most.
- Counterfactual Fairness: Another avenue to ensuring fairness is through the use of counterfactual fairness. An AI is said to be counterfactually fair if the recommendations issued would not change if the candidate's demographic attributes were also to change, say, their gender or race. Use of this principle will ensure that it uses only job-related criteria as a basis for making such decisions.

7.3 Continuous Monitoring and Auditing

Indeed, even after deploying an AI system [10], its performance must be monitored periodically so that the fairness of the functioning can be ensured.

- Regular Bias Audits: The system should be audited from time to time to detect emerging biases that crop in with new data entering the system. Audits should have two interlinked components: technical evaluation to assess the performance of the algorithm and human oversight to review individual cases where bias may have occurred. Auditing can also include testing decisions yielded by the system with predefined fairness metrics.
- Feedback Loops: Put in feedback mechanisms that enable users-users may include HR professionals, hiring managers, and applicants to mark potential biases or unfair decisions. Use this feedback in redesign so that changes are made to keep the AI aligned with ethics standards over time. Algorithm audits, as Raghavan [18] suggests, are vital for uncovering systemic biases in AI tools.
- Adaptive learning models: Use AI models that can self-improve and adapt over time by learning from the users, which may even be in the form of observation of the mistakes through incorrect or biased judgments.

7.4 Human Oversight and Collaboration

AI should not replace human judgment but rather assist and enhance the decision-making process. Human involvement ensures that AI-driven hiring decisions are ethical and aligned with the company's values.

- **Human-in-the-Loop Systems**: Implement AI systems that allow human recruiters or HR professionals to review and override decisions made by the AI, particularly in cases where the AI's reasoning is unclear or appears biased. This allows for a balance between AI-driven efficiency and human intuition.
- **Cross-Functional Collaboration**: The development of ethical AI systems in hiring requires collaboration between various teams, including data scientists,

ethicists, HR professionals, legal advisors, and diversity experts. By working together, these teams can ensure that the AI system aligns with organizational goals for diversity, equity, and inclusion (DEI) while complying with legal and ethical standards.

- **Transparency to Candidates**: Ensure that candidates are informed about the use of AI in the hiring process and provide them with explanations for decisions where appropriate. If a candidate is rejected by an AI system, they should be given an explanation of why they were not selected and allowed to request a human review.

7.5 Ethical Frameworks and Governance

The creation of guidelines in ethics and governance creates order by ensuring that AI development and deployment correlate with the company's values, as well as those set out in law.

- Ethical AI Principles: Develop guiding ethical principles on the use of AI in hiring decisions. The principles should stress fairness, accountability, transparency, and respect for candidate privacy. A public commitment to ethical AI practices could well be one of the best ways to get candidates and stakeholders on your side.
- AI Governance Committees: Establish governance committees responsible for the design and application of AI systems in the hiring process. Such committees should comprise diverse stakeholders who review the system from time to time and address any ethical issues that may come up.
- Regulators. AI systems should be in the bounds of regulations like GDPR in the EU and EEOC guidelines in the U.S. In general, these laws would request that all the processed decisions are transparent enough and that there must not be discrimination in hiring.
- Impact Assessments: Before the deployment of an AI system, there should be ethical impact assessments carried out in such a way that all possible risks versus potential benefits can be weighed. It is mostly focused on questions that provide whether bias and discrimination are matters of concern, and the system does not have merely a neutral or negative impact on efforts of diversity and inclusion.

7.6 Regular Updates and Retraining

The truth is, AI models are only as good as the data on which they are trained, and hiring landscapes change constantly. For these reasons, it always remains important to update and retrain AI systems in such a way that they may remain relevant and fair.

- Dynamic Model Updates: The AI models must alter the jobs, industry trends, and the candidate pool from time to time in order to avoid making biased, aged decisions.

- Integrate new data: Immediately after obtaining new hiring data, the AI model needs to be retrained to ensure it continues to classify with the same success rate as expected and in a fair manner. The reason is that the new data presence reduces bias since it brings on board all the latest trends and demographics in the workforce.
- Model version control refers to the management of versioning for AI models for monitoring changes and improvements in model production over time. As a result, accountability is created, and organizations are able to compare differences between model versions.

7.7 Training and Awareness

In addition, it is noted that organizations should train their workforce, mainly the human resource professionals, on the ethical implications of using AI in hiring and how these systems should be used responsibly.

- AI Ethics Training for HR Teams: Educating the HR teams on ethical use of AI when interpreting AI-driven recommendations, spotting potential biases [11], and making sure that AI does not contradict diversity and inclusion goals can help make better decisions in hiring.
- Candidate Education: The candidate should be aware of his rights, including the role which AI is going to play in hiring. Transparency on the use of AI builds confidence and gives the candidate a feeling that he or she will be treated justly.

8 Case Studies in Bias Mitigation

Integration of AI in hiring processes revolutionized recruitment since companies may streamline their operations, cut down on the cost, and eventually identify the right candidates. However, if not efficiently managed, AI systems have emerged as significant contributions to augmenting bias and discrimination, such as in the widely reported cases of Amazon and Unilever's hiring systems. Two corporate giants in the two respective sectors offer lessons about the challenges and benefits inherent in AI-driven hiring.

8.1 Case Study 1: Amazon's AI Recruitment Tool: A Cautionary Tale

In 2014, it launched a project to completely automate the hiring process through designing an AI recruitment tool that evaluates resumes and should identify the best talent. It trained the AI system on ten years of resumes submitted to the company, then looked to recreate the decisions that Amazon's hiring team had made over that time. The system would rate candidates from one star to five stars based on the match between the resumes and the company's ideal applicant profiles.

8.1.1 Bias in Training Data

The problem lies in the data itself with which the hiring AI was trained. The system fed upon a resume of previously rejected applicants [12], most of whom were male, because, after all, tech is mostly populated by men. The AI system began penalizing the resumes containing the word "women" or those that refer to all-women's institutions or activity, as participation in women's sports teams. According to definition, male-dominated profiles were the ideal standard, while female-centric resumes are devalued. Amazon's AI recruiting tool demonstrated significant gender bias, as detailed by Dastin [23].

- Historical Bias: This hiring AI fed back into the historical bias that had created overwhelmingly male hires in the first place. Instead of closing that imbalance, the AI simply perpetuated it by downgrading female applicants. The reason for this problem is that the data on which the AI was trained reflected previous human decisions, which have proven biased toward males.
- Algorithmic Bias: While designed as a "gender-neutral" AI, not excluding gender as a feature when it was not necessary, the AI learned that several terms (such as "women's chess club") were associated with negative outcomes. Proxy bias–the AI algorithm reached some gendered conclusions based on other features in the resume, such as extracurricular activities or the type of institution attended.

8.1.2 Lack of Human Oversight

Initially, the system for Amazon was working mostly without human intervention. So, the absence of human steps simply made the issue more severe since when it is trained, the bias was not observed [13]. Probably if human recruiting existed beforehand, they might have seen such patterns and corrected the AI before such actions took place.

8.1.3 Outcomes and Fallout

The company soon scrapped the AI system in 2017 after it realized that it was systematically downgrading female applicants. This is because it recognized that the system by itself was biased, and it cannot provide able candidates without deepening inequalities.

The recent case of Amazon's failed AI hiring tool-the tool used biased data to train hiring models-indicates that even more mistakes can be done with biased AI models. On the other hand, further concentration on transparency, human oversight, and fairness checks on AI recruitment systems is a lesson from this debacle.

8.1.4 Lessons Learned from Amazon's AI Failure

- Training data should be representative, and it is the actual training data to the AI system that should be representative of the workforce that the company is trying to build. Biased AI models result from biased training data.
- Algorithmic Transparency: AI models should be explainable and transparent. The system would have identified the possible gender bias much earlier if only Amazon had understood the process behind such a system.
- Human Oversight is Unavoidable: AI should act as a facilitator for the human's decision and not in place of the human. The recruiter has the right to override the AI decision; else, it goes against the principles of law.

8.2 Case Study 2: Unilever's AI Hiring Process

What can go wrong with AI is wonderfully exhibited in Amazon's case. On the other hand, Unilever's approach to AI in recruitment offers a model on how ethical, bias-resistant AI hiring systems might be developed. Starting 2017, Unilever has used AI in hiring processes, particularly in early stages of recruitment regarding internships and entry-level positions. In the setup of Unilever's AI system, candidates are to be evaluated based on their answers to online games, video interviews, and even personality assessments. The comparison between Amazon's failed attempt and Unilever's successful AI hiring process is illustrated in Fig. 2.

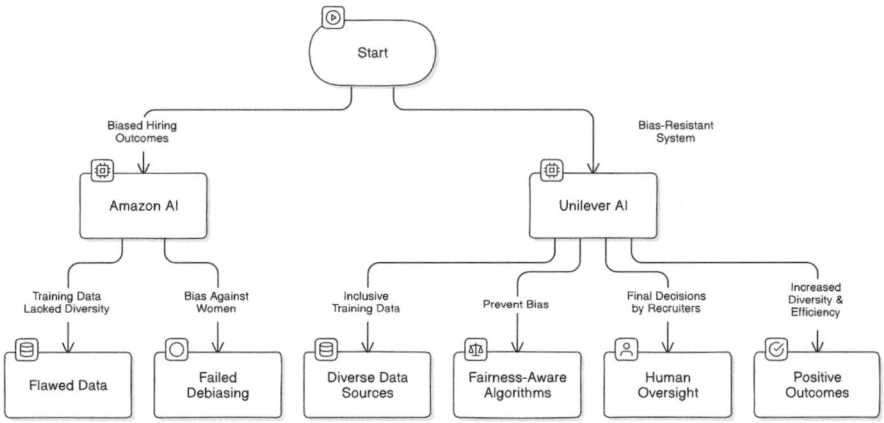

Fig. 2 Amazon versus Unilever AI Hiring Systems

8.2.1 Use of Diverse Data Sources

Unlike Amazon, Unilever made a concerted effort to ensure that the data used to train its AI system was diverse and representative. The company incorporated behavioral and psychometric data from a wide range of candidates, reducing the risk of bias based on race, gender, or educational background. Additionally, Unilever's AI was trained to evaluate attributes such as cognitive ability, emotional intelligence, and problem-solving skills, which are more neutral indicators of a candidate's suitability.

- **Debiasing Training Data**: Unilever's use of psychometric tests and behavioral assessments helps to reduce bias in its AI hiring system. These assessments focus on candidates' abilities rather than demographic characteristics, helping to level the playing field for candidates from different backgrounds.

8.2.2 Fairness-Aware Algorithms

It is designed with fairness constraints not to favor one group above others, and the algorithms are tested against fairness metrics such as equal opportunity and demographic parity ensuring that a particular group benefits or suffers more than others.

Algorithmic Transparency: Unilever also uses tools to explain the decision-making process of the AI system. The insights of how the system makes a judgment are shared with both recruiters and candidates, and the feedback of candidates' performance is communicated from the online games and video interviews.

8.2.3 Human Oversight and Collaboration

Unilever emphasizes that AI is deployed to assist human decision-makers and not to replace them. The company is using AI in ranking and shortlisting candidates, but it is a human recruiter who decides on the final selection based on the recommendations presented by the AI system. This will ensure that human judgment will more significantly influence hiring decisions, and the recruiters will be able to pin-point if any biases or mistakes occur through the AI system.

Human-in-the-Loop: The AI system uses human oversight at major stages in the hiring process. Recruiters can review or adjust the ranking by the AI with more context or factors that may not be captured by the algorithm, such as cultural fit or unique experiences.

8.2.4 Positive Outcomes

The AI-based hiring process initiated by Unilever has resulted in numerous positive outcomes, including more efficient hiring, lower hiring cost, and better candidate diversity [14]. In terms of diversity, the company had raised its workforce level by 16 percent, and it also reduced hiring time and bias by leaps and bounds. Unilever relies on neutral, scientifically validated assessments that help attract a more diverse pool of candidates and ensure hiring decisions are based on merit rather than demographic characteristics.

8.2.5 Lessons Learned from Unilever's AI Success

- **Skills and Competence Emphasis:** The AI used by Unilever considers more cognitive and emotional competencies in assessing the prospects rather than a proxy such as gender or educational background.
- **Regular Bias Audits:** The structure of Unilever is regulated periodically so that no form of bias creeps into the system. Ongoing vigilance makes it easier to detect biases as it develops in time.
- **Human observation also ensures fairness:** with human recruiters in the final act of making a decision, Unilever ensures that the AI complements, rather than replaces, human judgment.

9 Future of XAI in Hiring Software

XAI looks very promising for the future of an AI-driven hiring process; various trends are emerging to enhance its use. One such emerging trend is through reinforcement learning, where AI systems learn and improve with the data they receive. Thus, AI systems, in this approach, continuously improve and adjust their accuracy with time, as they have a reduced bias due to feedback from evaluators.

Another promising development is the collaboration of AI systems with human recruiters. While AI will process countless applications effortlessly, oversight by humans ensures that the final decisions made by the system reflect values and ethical standards of the company. In this manner, companies can have a hiring process that is even more balanced and fair by combining the efficiency of AI with human judgment.

However, there are numerous significant challenges lying ahead. For example, despite the continuous evolution and advancement of the AI landscape, strategies for bias mitigation need to be placed abreast with the advancing technological fronts [15]. Furthermore, standardized guidelines and frameworks pertaining to the role of AI in hiring will be well needed in ensuring that best practices regarding ethics are adopted by different companies in various industries.

In the future, XAI can play a critical role in transforming recruitment processes; they will be made more transparent, accountable, and fair. Through clear explainability of its decision-making, XAI will help companies build trust with candidates and ensure that AI-driven hiring systems promote diversity and inclusivity.

10 Conclusion

AI hiring holds great promise to assist with the efficiency of recruiting and help eliminate much human bias in the hiring process, but increasingly prevalent AI adoption poses some new ethics- and bias-related challenges. XAI addresses these challenges through transparency and accountability in AI decision-making processes.

This end can be achieved through the use of data-level and algorithm-level mitigation strategies in fair and inclusive AI systems. Employer use of XAI can ensure that AI systems used comply with legal requirements as well as ethical standards.

With increasing reliance of firms on AI in hiring decisions, clear caution over such ethical issues is very crucial. However, embracing the best practices of developing ethical AI and cooperation with a diverse team serves organizations well to create hiring systems which will be relatively efficient, fair, and inclusive as well.

References

1. Reddy, G.P., Kumar, Y.V.P.: Explainable AI (XAI): explained. In: IEEE Open Conference of Electrical, Electronic and Information Sciences (eStream), Vilnius, Lithuania, pp. 1–6 (2023). https://doi.org/10.1109/eStream59056.2023.10134984
2. Reddy, G.P., Sinha, S., Park, S.-H.: Generative AI for the maritime environments. In: 15th International Conference on Ubiquitous and Future Networks (ICUFN), Budapest, Hungary, pp. 618–623 (2024). https://doi.org/10.1109/ICUFN61752.2024.10625100
3. Wang, S., He, Z.: A prediction model of vessel trajectory based on generative adversarial network. J. Navig.Navig. **74**(5), 1161–1171 (2021). https://doi.org/10.1017/S0373463321000382
4. Grgić-Hlača, N., Zafar, M.B., Gummadi, K.P., Weller, A.: Beyond distributive fairness in algorithmic decision making: feature selection for procedurally fair learning. In: Proceedings of the AAAI Conference on Artificial Intelligence, vol. 32(1) (2018). https://doi.org/10.1609/aaai.v32i1.11296
5. Mujtaba, D.F., Mahapatra, N.R.: Ethical Considerations in AI-based recruitment. In: IEEE International Symposium on Technology and Society (ISTAS), Medford, MA, USA, pp. 1–7 (2019). https://doi.org/10.1109/ISTAS48451.2019.8937920
6. Binns, R.: Fairness in machine learning: lessons from political philosophy. In: Proceedings of the 2018 Conference on Fairness, Accountability, and Transparency, pp. 149–159 (2018). https://doi.org/10.1145/3287560.3287583
7. Cowgill, B., Dell'Acqua, F., Deng, S., Epstein, Z., Lequien, M.: Biased Programmers? Or Biased Data? A Field Experiment in Operationalizing AI Ethics. arXiv preprint arXiv:2011.07447 (2020)
8. Mehrabi, N., Morstatter, F., Saxena, N., Lerman, K., Galstyan, A.: A survey on bias and fairness in machine learning. ACM Comput. Surv. (CSUR) **54**(6), 1–35 (2021). https://doi.org/10.1145/3457607
9. Raji, I.D., Buolamwini, J.: Actionable auditing: investigating the impact of publicly naming biased performance results of commercial AI products. In: Proceedings of the 2019 AAAI/ACM Conference on AI, Ethics, and Society, pp. 429–435 (2019). https://doi.org/10.1145/3306618.3314244
10. Doshi-Velez, F., Kim, B.: Towards a Rigorous Science of Interpretable Machine Learning. arXiv preprint arXiv:1702.08608 (2017)

11. Holstein, K., Wortman Vaughan, J., Daumé III, H., Dudik, M., Wallach, H.: Improving fairness in machine learning systems: what do industry practitioners need? In: Proceedings of the 2019 CHI Conference on Human Factors in Computing Systems, pp. 1–16 (2019). https://doi.org/10.1145/3290605.3300830
12. Lepri, B., Oliver, N., Letouze, E., Pentland, A., Vinck, P.: Fair, transparent, and accountable algorithmic decision-making processes: the premise, the proposed solutions, and the open challenges. Philos. Technol. **31**(4), 611–627 (2018). https://doi.org/10.1007/s13347-017-0279-x
13. Kim, P.T.: Data-driven discrimination at work. William Mary Law Rev. **58**(3), 857–936 (2017)
14. Binns, R., Veale, M., Van Kleek, M., Shadbolt, N.: 'It's reducing a human being to a percentage': perceptions of justice in algorithmic decisions. In: Proceedings of the 2018 CHI Conference on Human Factors in Computing Systems, pp. 1–14 (2018). https://doi.org/10.1145/3173574.3173951
15. Obermeyer, Z., Powers, B., Vogeli, C., Mullainathan, S.: Dissecting racial bias in an algorithm used to manage the health of populations. Science **366**(6464), 447–453 (2019). https://doi.org/10.1126/science.aax2342
16. Angwin, J., Larson, J., Mattu, S., Kirchner, L.: Machine bias. ProPublica, May 2016 [Online]. Available at: https://www.propublica.org/article/machine-bias-risk-assessments-in-criminal-sentencing
17. Chouldechova, A.: Fair prediction with disparate impact: a study of bias in recidivism prediction instruments. Big Data **5**(2), 153–163 (2017)
18. Raghavan, P.: The algorithm audit: scoring the algorithms that score us. Commun. ACM. ACM **64**(9), 62–71 (2021)
19. Barocas, S., Selbst, A.D.: Big data's disparate impact. Calif. Law Rev. **104**(3), 671–732 (2016)
20. Zemel, R., Wu, Y., Swersky, K., Pitassi, T., Dwork, C.: Learning fair representations. In: Proceedings of the 30th International Conference on Machine Learning (ICML), Atlanta, GA, USA, 2013, pp. 325–333
21. Kleinberg, J., Mullainathan, S., Raghavan, M.: Inherent trade-offs in the fair determination of risk scores. In: Proceedings of the 8th Innovations in Theoretical Computer Science (ITCS), Berkeley, CA, USA, 2017, pp. 43:1–43:23
22. Rudin, C.: Stop explaining black box machine learning models for high-stakes decisions and use interpretable models instead. Nat. Mach. Intell. **1**, 206–215 (2019)
23. Dastin, J.: Amazon scraps secret AI recruiting tool that showed bias against women. Reuters, Oct 2018 [Online]. Available at: https://www.reuters.com/article/us-amazon-com-jobs-automation-insight-idUSKCN1MK08G
24. Datta, A., Tschantz, M.C., Datta, A.: Automated experiments on ad privacy settings: a tale of opacity, choice, and discrimination. In: Proceedings on Privacy Enhancing Technologies, vol. 2015, no. 1, pp. 92–112 (2015)
25. Wachter, S., Mittelstadt, B., Floridi, L.: Why a right to explanation of automated decision-making does not exist in the General Data Protection Regulation. Int. Data Privacy Law **7**(2), 76–99 (2017)

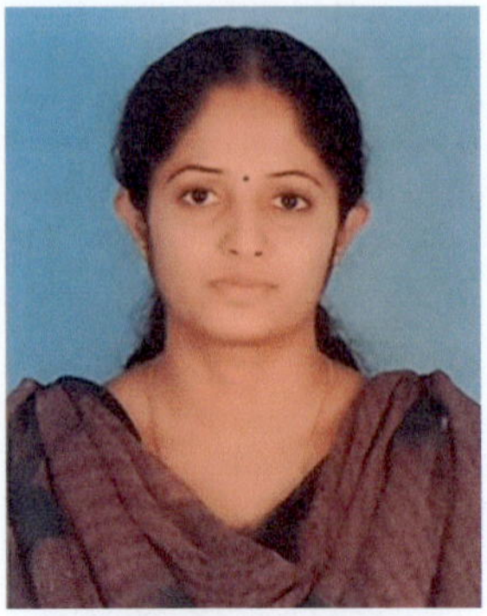

P. Jayadharshini

P. Karunakaran

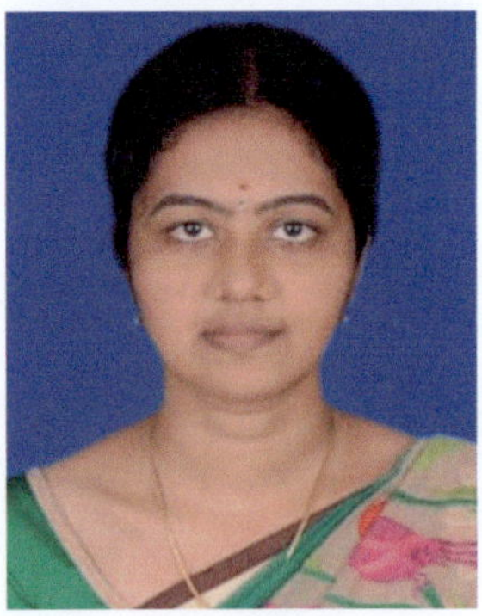

S. Santhiya

A. S. Renugadevi

G. Dhanush

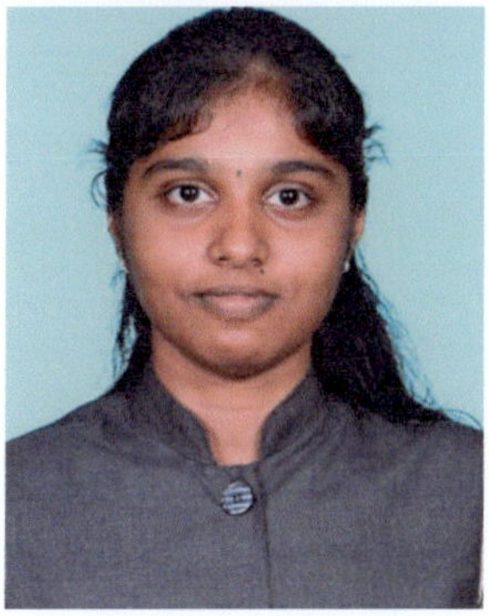

E. Pavithra

From Insights to Action: Interpretable AI as a Catalyst for Manufacturing Innovation

R. Madhumith, S. B. Mahalakshmi, and P. Hemashree

Abstract This chapter examines the crucial role of interpretable AI models in transforming manufacturing processes, the examination of the relationship between interpretability and extendibility within industrial settings is of particular significance. It begins by providing an explanation of how AI has evolved in the manufacturing sector, highlighting the need for interpretable AI in light of traditional manufacturing challenges. The discussion then explores the fundamental principles that underpin interpretable AI, carefully considering the trade-offs between transparency and accuracy, and elucidating various techniques for explainability. Additionally, the article discusses the concept of extendible AI frameworks, emphasizing their adaptability and scalability within industrial domains. It proceeds to showcase a broad range of applications for interpretable AI in manufacturing, including maintenance and quality control, supported by insightful case studies from the automotive, semiconductor, and food industries. Simultaneously, it addresses the existing challenges and envisions future directions, taking into account ethical considerations and regulatory requirements. Ultimately, the article synthesizes the key insights into a comprehensive conclusion, providing recommendations for promoting the adoption of interpretable AI models to enhance manufacturing efficiency, quality, and sustainability.

Keywords Interpretable AI · Extendible AI · Industry 4.0 · Manufacturing 4.0

R. Madhumith · S. B. Mahalakshmi (✉) · P. Hemashree
Department of Artificial Intelligence and Machine Learning, Coimbatore Institute of Technology, Coimbatore, Tamil Nadu, India
e-mail: mahalakshmi@cit.edu.in

P. Hemashree
e-mail: hemashree@cit.edu.in

L. Krishnasamy et al. (eds.), *Distributed Deep Learning and Explainable AI (XAI) in Industry 4.0*, Information Systems Engineering and Management 55,
https://doi.org/10.1007/978-3-031-94637-0_15

1 Introduction

Artificial intelligence (AI) has arisen as a key strength driving the transformation across multiple industries, with manufacturing standing out as a key sector reaping the benefits of its innovations. This section offers an in-depth exploration of AI's role in manufacturing, highlighting the importance of interpretable AI models and the vast opportunities presented by extendible AI in industrial processes. The main objective of this analysis is to provide insights into the critical elements and future potentials of AI in the manufacturing domain [1].

1.1 Background of AI in Manufacturing

The manufacturing sector has transformed significantly due to Artificial Intelligence (AI), revolutionizing product design, production, and management. Machine learning, a key AI technology, enhances efficiency, precision, and productivity. While the concept of intelligent machinery has long existed, recent advancements in computing and data analysis have made AI a powerful force in manufacturing.

Rise of Machine learning

The significant ascent of machine learning can be attributed to its capacity to learn from data and produce predictions autonomously, with minimal human intervention. This allows algorithms to inspect sensor data, identify patterns, and predict potential breakdowns, supporting proactive maintenance. AI's impact on production includes streamlining the supply chain, ensuring quality control with machine vision systems, using AI-driven robotics for complex tasks, and improving processes by analyzing data to identify inefficiencies and suggest remedies.

However, the utilization of AI in manufacturing operations gives rise to challenges that call for thoughtful consideration. These challenges involve guaranteeing data accuracy and reliability, integrating AI systems with existing infrastructure, and addressing concerns related to potential job displacement. Despite these hurdles, the influence of AI on manufacturing cannot be denied, as it instils intelligence into different aspects of production processes. With the ongoing evolution of AI technology, we can foresee even more transformative applications that will influence the future of manufacturing in the years to come.

1.2 Importance of Interpretable AI Models

Transparent and easily understandable AI models are essential in the manufacturing industry to facilitate clarity and insight into the decision-making processes, despite the benefits AI brings to the industry.

- **Explainability and trust**: Understanding how AI models make predictions in safety–critical settings like manufacturing is essential. This knowledge builds confidence and allows human professionals to verify the justification of the model's forecasts [2].
- **Trust and Transparency**: The decision-making process of a sophisticated AI model may pose challenges in comprehending the rationale behind its choices. This opacity in its operations may result in a diminished level of confidence from human users, impeding its acceptance and giving rise to potential safety issues.
- **Debugging and Improvement**: When an AI model yields unexpected results, it's essential to understand the reasons behind them to effectively address the issue. Opaque models significantly hinder this understanding.
- **Explainability to Regulators**: The interpretability of artificial intelligence models is essential in specific industries where regulatory requirements compel producers to provide rationale for the decisions rendered by AI systems. Employing interpretable models can assist in fulfilling these regulatory obligations.

1.3 Overview of Extendible AI for Industrial Processes

The domain of industrial artificial intelligence, known as Extendible AI, focuses on developing adaptable AI systems for complex industrial environments. These systems can integrate new data, tasks, and functions without extensive reprogramming, crucial for operations experiencing ongoing changes.

Key characteristics of Extendible AI

- **Modular Design**: The structure is made up of autonomous components that can be easily integrated, removed, or changed. Each component is designed to handle specific functions within the overall operation [3].
- **Lifelong Learning**: The system has the capability to acquire knowledge from fresh data gathered while in operation, enabling it to alter to the varying conditions and improve its efficiency over time.
- **Explainability and Transparency**: To foster trust and enhance effective association between humans and machines, it is vital that the decision-making processes of the system are comprehensible to human operators [4].

2 The Need for Interpretable AI in Manufacturing Sector

The growth of AI in manufacturing offers great potential for improved efficiency and productivity. However, the lack of explainability in many AI models, specifically those using complex deep learning structures, is a significant obstacle to widespread adoption. This article will explore the need for Interpretable AI in manufacturing and its benefits.

Why Is Black-Box AI a Problem in Manufacturing?

Traditional AI models, also known as "black-box" models, function as opaque systems. While they can produce highly accurate results, understanding the reasoning behind their decisions remains a challenge. This lack of transparency creates several challenges in a manufacturing setting:

- **Limited Trust and Adoption**: If production personnel and management don't understand why an AI system makes decisions, it can lead to lack of trust and approval. This may cause reluctance to adopt the technology or a continued need for human validation, which could negate efficiency gains [5].
- **Debugging and Improvement**: Identifying and fixing issues with a black-box model can be difficult, leading to operational delays and hindering AI model improvement.
- **Safety and Explainability**: Transparency is crucial in safety–critical AI applications to ensure physical safety. Without understanding how AI makes decisions, it is difficult to guarantee its actions are safe and reliable [6].
- **Bias and Fairness**: AI models can be biased by the data used for training, making it difficult to identify and correct unjust outcomes.

Benefits of Interpretable AI in Manufacturing

By adding interpretability to AI models, manufacturers can access a range of benefits. It builds trust and approval among human experts by helping them understand the reasoning behind AI suggestions, fostering cooperation and faster integration into workflows [7].

Interpretable models serve an essential function in improving the debugging and maintenance processes. By providing a clear understanding of the system, these models make it easier to pinpoint errors and biases, leading to quicker troubleshooting and ongoing enhancements in the performance of the AI model.

Interpreting serves an essential function in ensuring safe operations, particularly in high-stake situations. It helps understand factors influencing AI decision-making, promoting transparency, dependability, and security. Interpretable models also help identify and address biases in AI training data, ensuring ethical and fair implementation across manufacturing processes.

2.1 Challenges in Conventional Manufacturing Processes

Conventional manufacturing methods frequently encounter a number of limitations, including:

- **Limited visibility**: Manufacturing procedures can be complex, with many factors affecting the final result. Conventional techniques often lack the ability to constantly monitor these factors, making it difficult to identify obstacles and improve production efficiency.

- **Reactive maintenance**: Traditional maintenance schedules based on time intervals can lead to unexpected machinery issues and downtime
- **Quality control challenges**: Manual QA procedures are prone to subjectivity and errors, while traditional techniques may miss minor flaws that impact product quality.

These limitations can lead to inefficiencies, production delays, and increased costs.

2.2 Role of AI in Addressing Manufacturing Challenges

AI possesses the capability to transform the manufacturing industry by tackling numerous obstacles as mentioned earlier. Several significant uses of AI in manufacturing encompass:

- **Predictive maintenance**: AI algorithms can analyze sensor data from machines to predict malfunctions, allowing for preventive maintenance and reducing downtime and costs [8].
- **Real-time process optimization**: AI can analyze real-time manufacturing data to improve production and increase efficiency, ultimately cutting costs.
- **Improved quality control**: AI-driven visual technology enables the examination of products for imperfections with enhanced precision and uniformity in contrast to traditional manual techniques [9].

Through the utilization of artificial intelligence, manufacturers have the opportunity to gather insightful data on their operations, streamline procedures, and realize substantial enhancements in efficiency, quality, and overall productivity.

2.3 Role of Explainability in AI-Powered Manufacturing Systems

AI integration in manufacturing has improved productivity, quality, and creativity, but concerns about the interpretability of complex AI models, especially deep learning algorithms, have arisen. This study explores the importance of interpretability in AI models in manufacturing, addressing the benefits, challenges, and potential solutions to enhance their clarity.

AI has revolutionized the manufacturing sector by enhancing predictive maintenance, improving quality control, and optimizing processes. Nevertheless, the denseness of AI models, especially deep neural networks, presents substantial obstacles to their integration within the industry. Interpreting AI models is essential for trust, regulation compliance, and human-AI collaboration. This study explores research on AI interpretability in manufacturing, focusing on improving model transparency and its impact. Model interpretability is crucial for successful AI implementation in manufacturing, enhancing trust, collaboration, compliance, and error detection. Future research should aim for accuracy and interpretability balance using post-hoc and intrinsic methods for transparent AI in manufacturing.

3 Principles of Interpretable AI Models

This section delves into the core principles that depict the development and utilization of interpretable AI models. Grasping these principles is crucial for constructing dependable AI systems, particularly in sectors where the capacity to offer justifications holds significant value.

3.1 Transparency Versus Accuracy Trade-Off

Balancing model interpretability and accuracy is a key challenge in explainability. Complex black-box models often prioritize accuracy over interpretability, leading to trust issues and concerns about fairness and bias in decision-making [10]. Interpretable models prioritize explanations over accuracy, making them simpler and easier to understand, though they may sacrifice some accuracy compared to complex black-box models [11]. The balance between transparency and accuracy varies depending on the context. In industries like healthcare or finance, simpler models may be preferred for easier understanding, even if they sacrifice some accuracy. In tasks like image recognition, a higher level of complexity may be acceptable to achieve the best accuracy [12].

3.2 Explainability Techniques in AI Models

Researchers have developed methods to make black-box models more interpretable, bridging the gap with interpretable models. These methods fall into two main categories:

1. **Model-agnostic techniques**: The methods described are universally applicable and not reliant on the model's structure. Model interpretability techniques include feature importance analysis and LIME, which simplify complex models by identifying influential features and providing understandable predictions.
2. **Model-specific techniques**: These methods employ the intrinsic structure of a specific model to produce explanations. For instance, in decision tree models, the explanation can be directly derived from the tree structure, while in rule-based models, the explanations can be presented as a set of rules.

The choice of an interpretability method is influenced by variables like the preferred depth of explanation, the nature of the model in use, and the particular field of application.

3.3 Model Complexity and Interpretability

Model complexity is a multidimensional notion that involves different factors impacting the complexity of a model. Several crucial elements are taken into account.

Initially, the number of features in a model greatly affects its complexity. Generally, a model with more features is considered more intricate because it must understand the relationships among a larger set of variables, making it harder to determine the specific impact of each feature [13].

Additionally, a model's complexity is influenced by its architectural design. Some architectures are inherently more complex than others. For instance, deep learning models with multiple layers and non-linear activation functions are more complex than simpler models like linear regression.

Finally, the complexity of a model can be impacted by both the size and characteristics of the data utilized for training. Models that are trained on large datasets or complex data types, like images or text, often demonstrate higher levels of intricacy.

It is significant to understand the complexity of a model by considering various factors. This understanding helps individuals in research and practical fields make informed decisions about which models are suitable for specific purposes.

The concept of interpretability pertains to the degree to which we are able to grasp the predictions made by a model. A model that is interpretable enables us to comprehend the rationale behind its choices and pinpoint the variables that have a substantial impact on its results. This is especially crucial in a variety of situations, such as:

(i) **Debugging Errors**: Interpretability plays a vital role in pinpointing the root cause of errors and providing guidance for enhancements when a model produces inaccurate predictions.
(ii) **Regulatory Compliance**: In specific, sectors such as healthcare, understanding the rationale behind a model's conclusions may be essential due to regulatory requirements.
(iii) **Building Trust**: Models that are interpretable are crucial in fields where transparency is vital, as they contribute to fostering trust in the decision-making process.

A fundamental trade-off exists between the complexity of a model and its interpretability, where increasing model complexity often comes at the cost of reduced interpretability.

Conversely, simpler models tend to be more comprehensible, albeit at the cost of reduced accuracy.

To illustrate this trade-off, consider the following examples:

(i) **Linear Regression Versus Deep Neural Network**: Linear regression is known for its interpretability through direct insight from coefficients, but may encounter difficulties when addressing complex problems. In contrast, deep neural networks excel at complex tasks despite being seen as opaque due to their intricate internal mechanisms.

(ii) **Decision Trees Versus Support Vector Machines**: Decision trees provide clear rules for predictions but have rigid decision boundaries, while support vector machines have complex boundaries but lack transparency in their decision-making process.

4 Extendible AI Frameworks for Industrial Processes

4.1 Understanding Extendible AI

Extendible AI, also called modular or extensible AI, refers to AI systems designed with flexibility, modularity, and scalability. These systems can be continuously improved, adapted to new tasks, and expanded with additional components as needed. The key principles of extendible AI include modularity, interoperability, scalability, and ongoing learning.

4.1.1 Key Characteristics of Extendible AI

- **Modularity**: The AI system can independently create, modify, or replace parts, making personalization and enhancements easier without affecting the entire system. For instance, a production AI system could have separate modules for different tasks that can be upgraded or replaced as needed [14].
- **Interoperability**: Extendable AI systems are designed to work with various data sources, software applications, and hardware environments, ensuring seamless integration with existing industrial frameworks. For example, they can connect with outdated systems in a manufacturing facility to extract data, analyze it, and provide recommendations without requiring a complete overhaul of the current setup [15].
- **Scalability**: Scalability in AI frameworks allows for handling larger data volumes and more complex tasks as industrial activities grow. This can be achieved through horizontal scaling (adding more computational resources) and vertical scaling (improving current resources). This ensures that the AI remains effective as manufacturing operations become more intricate and data-heavy [16].
- **Continuous Learning**: A scalable AI system must continuously learn from new data and experiences to improve efficiency and remain relevant. Techniques like

reinforcement learning, online learning, and transfer learning are commonly used for this purpose [17].

4.1.2 Benefits of Extendible AI in Industrial Processes

- **Enhanced Flexibility**: Manufacturers have the ability to customize AI solutions to meet their unique requirements due to the modular design of extendible AI. By adjusting functionalities according to evolving needs, they can maintain alignment between the AI system and business objectives [18].
- **Improved Efficiency**: extendable AI can improve industrial processes by integrating with current systems and adapting to new data, leading to increased efficiency and productivity [19].
- **Cost-Effectiveness**: AI frameworks that are extendable minimize the need for extensive overhauls in industrial systems, allowing manufacturers to gradually implement AI solutions and reduce disruptions and upgrade costs [20].
- **Future-Proofing**: Ensuring that extendable AI systems remain effective and relevant in the long run involves the capability to integrate new technologies and adapt to evolving industry standards [21].

4.1.3 Challenges and Considerations

- **Complexity in Implementation**: The creation and implementation of scalable AI systems may present challenges, necessitating a comprehensive grasp of AI technologies as well as industrial procedures [21].
- **Data Integration**: The creation and implementation of scalable AI systems may present challenges, necessitating a comprehensive grasp of AI technologies as well as industrial procedures [22].
- **Security and Privacy**: It is essential to guard sensitive industrial data and adhere to data privacy regulations when deploying extendible AI systems [23].

4.2 Adaptability and Scalability in Industrial Settings

The adaptability and growth potential are fundamental traits of scalable artificial intelligence frameworks, particularly in the field of industrial operations where demands and technological landscapes are in a constant state of flux.

4.2.1 Adaptability

The adaptability of an AI system pertains to its capacity to acclimate to fresh tasks, circumstances, and data inputs. Within industrial environments, this adaptability plays a critical role in upholding efficiency and competitiveness.

- **Dynamic Process Adjustments**: AI systems with extendible capabilities can quickly adapt to manufacturing operations by considering current circumstances and new data inputs. If an issue is detected in a production line, the AI can adjust machine settings or production schedules to address the problem, reducing downtime and waste [24].
- **Customization for Specific Use Cases**: Different industries have specific needs in their manufacturing processes. Tailored AI systems can address these requirements. For example, AI in the automotive sector can focus on predictive maintenance and defect detection, while in pharmaceuticals, it may emphasize quality assurance and regulatory compliance [25].
- **Integration with Legacy Systems**: Many industrial processes depend on outdated systems that are hard to upgrade. AI frameworks have been developed to work alongside these old systems, extracting information and enhancing their capabilities without the need for a complete overhaul. This smooth integration ensures that industrial companies can benefit from AI without significant disruptions to their operations [26].

4.2.2 Scalability

The capacity of an AI system to manage larger volumes of data and more intricate tasks as industrial activities grow is known as scalability. This attribute is essential in guaranteeing the AI system's sustained efficiency amidst escalating demands.

- **Data Handling Capabilities**: As industrial processes become more digital, there is a growing amount of data being generated. Scalable AI systems have been created to accomplish and analyze huge volumes of data efficiently. In a smart factory, AI can process data from multiple sensors at once to improve production efficiency [27].
- **Resource Management**: Scalable AI systems must efficiently allocate computational resources to maintain strong performance as data and task complexity grow. Strategies like distributed and edge computing are commonly used to optimize resource management in extendible AI [28].
- **Future-Proofing**: Developers create flexible AI frameworks that anticipate future needs and technological advancements, allowing for the integration of upcoming algorithms, sensors, and data formats. This forward-looking approach ensures the longevity and effectiveness of the AI system, giving manufacturers a competitive advantage in a constantly changing technological landscape [29].

5 Applications of Interpretable AI Models in Manufacturing

Explainable AI systems are crucial in manufacturing for promoting transparency, building trust, and ensuring regulatory compliance. They help manufacturers understand AI predictions, ensuring reliability. This analysis explores the use of interpretable AI in prognostic maintenance, quality control, defect detection, and improving operational efficiency.

5.1 Predictive Maintenance and Fault Diagnosis

The amalgamation of artificial intelligence in the manufacturing sector is essential for improving predictive maintenance and identifying faults. Its main goals include reducing operational downtime, extending the longevity of equipment, and decreasing maintenance costs.

5.1.1 Importance of Proactive Maintenance

Proactive maintenance leverages AI to forecast equipment failures and plan maintenance ahead of time, unlike reactive maintenance which only fixes equipment after it breaks. Explainable AI models are important for clear and understandable predictions.

5.1.2 Techniques and Technologies

Machine Learning Algorithms: Various machine learning models, including decision trees, random forests, and support vector machines, are used in predictive maintenance. These models analyze historical sensor and equipment data to identify patterns indicating potential failures. Maintenance engineers must understand these models to identify components needing maintenance [14].

Explainable AI (XAI): SHAP and LIME are instrumental in providing valuable insights into model predictions. These techniques assist maintenance teams in comprehending the pivotal factors influencing anticipated equipment failures, thereby improving the dependability of maintenance determinations [30].

5.1.3 Benefits and Case Studies

- **Cost Savings**: By predicting failures in advance, organizations can reduce unforeseen interruptions and extend the lifetime of their machinery. A study on a top car manufacturer found a 20% reduction in maintenance costs and a 15% increase in equipment durability using explainable AI for preemptive maintenance [31].
- **Improved Safety**: Predictive maintenance is essential for enhancing operational safety by preventing catastrophic equipment failures. For instance, a chemical processing plant used an interpretable AI model to monitor critical valves and pressure systems, resulting in early detection of potential failures and improved plant safety [32].

5.2 Quality Assurance and Anomaly Detection

Quality assurance and anomaly detection are essential for maintaining product standards and reducing waste in manufacturing. Interpretable AI models provide transparency to ensure quality and uniformity.

5.2.1 Role of Interpretable AI in Quality Control

Quality assurance involves ongoing assessment of product quality during manufacturing. AI algorithms quickly identify defects, ensuring only items meeting quality criteria move through production. Transparent AI models explain defect identification, providing valuable insights for quality control teams.

5.2.2 Methods and Technologies

- **Computer Vision**: AI-powered computer vision technology uses cameras and algorithms to inspect products for defects. CNNs are commonly used to categorize images and detect flaws. Explainable models help technicians understand why certain items are identified as faulty, revealing the unique characteristics recognized by the AI.
- **Statistical Process Control (SPC)**: SPC methodologies combined with AI are used to oversee manufacturing operations. Algorithms utilized in explainable artificial intelligence, such as Bayesian networks, are essential for clarifying the connections between process parameters and the resulting quality of the final product [33].

5.2.3 Benefits and Case Studies

- **Reduced Waste**: AI-powered quality control systems can significantly reduce waste by detecting defects early in the manufacturing process. In the semiconductor industry, the use of explainable AI led to a 30% decrease in defect occurrences [25].
- **Enhanced Product Quality**: Interpretable AI models are essential for improving product quality by providing insights into defect causes. In the food processing industry, AI-driven quality assurance systems are crucial for maintaining consistent quality and safety compliance [27].

5.3 Production Optimization and Process Efficiency

Efficient production and optimized processes are crucial for maximizing output, cutting costs, and maintaining high quality standards in manufacturing. Easily understandable AI models provide valuable insights for optimization strategies, boosting confidence and enabling continuous improvements.

5.3.1 Importance of Production Optimization

Production optimization involves improving different facets of the production process to increase efficacy and productivity. AI models analyze extensive data to identify bottlenecks and suggest improvements. Interpretable models provide valuable insights, helping manufacturers understand and trust AI-powered optimizations.

5.3.2 Techniques and Technologies

- **Process Simulation and Modeling**: AI models replicate manufacturing processes and forecast results across various situations using methods like digital twins and simulation-based optimization. Explainable AI models help engineers understand how different variables influence process efficiency [34].
- **Optimization Algorithms**: AI models replicate manufacturing processes and forecast results across various situations using methods like digital twins and simulation-based optimization. Explainable AI models help engineers understand how different variables influence process efficiency [29].

5.3.3 Benefits and Case Studies

- **Increased Throughput**: Enhanced production procedures led to a 25% increase in output and reduced cycle times in the electronics manufacturing sector through the use of interpretable AI [21].
- **Cost Reduction**: AI process enhancements can significantly lead to cost savings in the automotive industry, with a 20% decrease in material wastage and a 15% reduction in labor expenses [26].
- **Sustainable Manufacturing**: By implementing an explainable AI system, a textile manufacturing facility reduced energy consumption by 18%, helping the company meet sustainability goals by improving production methods and reducing environmental impact.

6 Case Studies: Interpretable AI Solutions in Manufacturing

6.1 Case Study 1: Predictive Maintenance in Automotive Manufacturing [35]

Introduction

Industry 4.0 and Predictive Maintenance (PdM)

Industry 4.0 brings together cyber-physical systems, IoT, and AI in manufacturing to enhance automation, information exchange, and real-time processing. Predictive Maintenance (PdM) uses data analytics to forecast equipment failures and optimize maintenance, reducing downtime and costs while improving operational efficiency.

Objectives and Scope

This study seeks to deliver an extensive evaluation of Predictive Maintenance (PdM) in relation to Industry 4.0. It will explore the core principles, methodologies, and practical implementations of PdM, alongside a thorough examination of its challenges and possible pathways for future development.

Fundamental Concepts of PdM

Data-Driven Business Strategy

Predictive maintenance is based on the acquisition of data from sensors embedded in industrial equipment, employing machine learning algorithms and statistical models to evaluate the Remaining Useful Life (RUL) of assets and to detect possible failure points. Key principles to consider include:

- **Condition-Based Maintenance (CBM)**: Maintenance procedures are carried out based on the live condition of the equipment, identified through sensor data and machine learning models.
- **Prognostics and Health Management (PHM)**: A comprehensive strategy involving the stages of observation, analysis, and action is utilized to oversee the health and upkeep of industrial systems.

Key Technologies

- **Internet of Things (IoT) and Industrial IoT (IIoT)**: IoT devices gather and send information from industrial equipment, enabling immediate monitoring and analysis.
- **Cyber-Physical Systems (CPS)**: The amalgamation of tangible operations with digital simulations in order to develop systems that can adapt and configure themselves autonomously.
- **Artificial Intelligence (AI) and Machine Learning (ML)**: Algorithms utilize both historic and real-time data in order to forecast potential failures and enhance the efficiency of maintenance timetables.

Methodologies and Applications of PdM

Predictive Models and Algorithms

- **Supervised Learning**: Employs labelled historical data for the purpose of training algorithms that forecast potential equipment malfunctions by analyzing present circumstances.
- **Unsupervised Learning**: Recognizes trends and irregularities within data sets lacking predefined results, beneficial for uncovering unfamiliar malfunctioning mechanisms.
- **Hybrid Approaches**: Combines machine learning techniques and domain knowledge to boost forecasting precision.

Applications in Manufacturing

1. **Predictive Maintenance and Fault Detection**

    (a) Real-time monitoring of machinery in order to anticipate malfunctions and plan for maintenance tasks
    (b) **Case Study Example:** An automotive manufacturer implemented PdM to monitor the health of assembly line robots, reducing unplanned downtime by 30%.

2. **Quality Control and Defect Detection**

    (a) The assessment of sensor data in order to pinpoint abnormalities and imperfections in merchandise.
    (b) **Case Study Example**: A semiconductor manufacturer used ML models to identify defects in chips during production, improving yield rates by 15%.

3. **Production Optimization and Process Efficiency**
   (a) Enhancing manufacturing operations through the anticipation of machinery malfunctions and arranging maintenance tasks for off-peak periods.
   (b) **Case Study Example**: A chemical plant used PdM to monitor pumps and valves, optimizing maintenance schedules and reducing operational costs by 20%.

Challenges and Future Directions

Current Challenges

1. **Data Validation and Aggregation**
   (a) Warranting the accuracy, completeness, and coherence of data retrieved from several sources.
   (b) The consolidation of information from outdated legacy systems and contemporary IoT devices into a cohesive platform.
2. **Model Accuracy and Interpretability**
   (a) Achieving a delicate equilibrium between the complexity and correctness of predictive models and their comprehensibility for maintenance engineers.
   (b) Addressing the black-box nature of certain ML models in order to establish trust and enhance their adoption.
3. **Scalability and Real-Time Processing**
   (a) Adapting predictive maintenance systems to achieve substantial amounts of data stemming from expansive industrial activities.
   (b) Guaranteeing the capability to process data and make decisions in real-time.

Future Trends and Research Directions

1. **Advanced AI and ML Techniques**
   (a) Enhancing algorithms to improve prediction accuracy and enhance anomaly detection capabilities.
   (b) Investigating reinforcement learning and transfer learning methods for the development of adaptive and context-aware maintenance strategies.
2. **Edge Computing and Distributed Architectures**
   (a) Utilizing edge computing enables the local processing of data on devices, thereby decreasing latency and minimizing bandwidth demands.
   (b) The deployment of distributed frameworks to enhance the scalability and robustness of predictive maintenance solutions.
3. **Ethical Considerations and Regulatory Frameworks**
   (a) Ethical deliberations surrounding data privacy, security, and the influence of artificial intelligence on employment are being discussed.

(b) Establishing regulatory structures to guarantee the secure and equitable implementation of predictive maintenance technologies.

Conclusion

Predictive maintenance in Industry 4.0 can revolutionize manufacturing by reducing equipment downtime, improving maintenance planning, and boosting production efficiency. Advances in AI, machine learning, IoT, and edge computing are expected to tackle current challenges and lead to more advanced solutions. Future research should focus on improving model accuracy, interpretability, and scalability, while considering ethical and regulatory issues. Tiago Zonta and colleagues' literature review provides a comprehensive overview of the current and future landscape of predictive maintenance in Industry 4.0.

6.2 Case Study 2: Quality Control in Semiconductor Fabrication

Comprehensive Case Study: Enhancing Semiconductor Manufacturing with Explainable AI [36]

Introduction

The semiconductor manufacturing sector demands precision and consistency for its complex processes. Quality management is crucial but challenging due to the high-dimensional and nonlinear nature of the data. Traditional statistical methods struggle to identify and manage process quality factors. This case study explores how Hitachi ABB used explainable AI to enhance the process quality of their semiconductor production, focusing on high-power transistors.

Background

Hitachi ABB's Challenge

Hitachi ABB, a prominent manufacturer of high-power semiconductors, encountered notable reductions in yield during the production of their transistor chips. The intricate nature of the manufacturing process, which involves numerous interconnected measurements for each individual product, posed challenges in identifying the root causes of quality fluctuations through traditional approaches.

Objective

The main goal was to create a strong, data-driven decision-making model utilizing explainable AI in order to pinpoint and address the causes of quality fluctuations, ultimately enhancing the total output and minimizing waste.

Methodology

Explainable AI and SHAP

The project's foundation relied on SHAP values, a technique from explainable AI that reveals the importance of individual features in machine learning outcomes, promoting transparency and understanding of key factors. The methodology was implemented in two key steps:

1. **Prioritizing Processes for Quality Improvement**
    (a) **Data Collection**: Data on production history was collected, including numerous measurements for each individual product.
    (b) **SHAP Analysis**: SHAP values were calculated in an effort to assess the significance of different manufacture parameters, aiding in the identification of the processes that had the greatest impact on yield.
2. **Selecting and Implementing Improvement Actions**
    (a) **Action Selection**: Selected were specific improvement actions aimed at addressing the key drivers of quality variation, as identified through the SHAP analysis.
    (b) **Field Experiment**: The efficacy of the chosen measures was evaluated in a controlled, authentic setting.

Implementation

Step 1: SHAP Analysis and Process Prioritization

- The analysis of historical data for transistor chip production used SHAP values after a rigorous data preprocessing procedure to ensure accuracy and consistency.
- The SHAP analysis elucidated the influence of individual production parameters on yield fluctuation, enabling the team to identify key processes for focused enhancements.

Step 2: Field Experiment and Action Validation

- Improvement measures were chosen following the SHAP analysis, which involved making modifications to certain manufacturing parameters that were deemed crucial for enhancing yield.
- An experimental study was carried out to execute these measures and assess their influence on crop production.

Results

Impact on Yield

The field trial led to a notable decrease in yield loss by 21.7% in comparison to the mean yield within the sample. Motivated by these findings, Hitachi ABB opted to incorporate the decision-making model into their comprehensive quality management framework.

Post-Experimental Rollout

The model was further implemented on a separate transistor chip product line, resulting in a significant decrease in yield loss, with a notable 51.3% enhancement.

Operational Integration

The decision model, integrated into Hitachi ABB's quality management framework, remains instrumental in driving process enhancements. Leveraging explainable AI has enhanced comprehension of production dynamics, leading to the implementation of more accurate and efficient quality management tactics.

Conclusion

Significance and Future Directions

Hitachi ABB's initiative shows the impact of explainable AI in manufacturing. By providing clear explanations of production variables, explainable AI improves decision-making and enhancements. This example demonstrates immediate benefits in increased yield and long-term potential for quality improvement and operational effectiveness.

Challenges and Considerations

Despite achievements, there are still obstacles to overcome, including consistent data quality control, merging AI models with current systems, and regular model enhancements and verification. Future studies should prioritize tackling these hurdles and exploring the wider uses of explainable AI in diverse manufacturing settings.

6.3 Case Study 3: AI in Food Industry [37]

Introduction

The article "Application of Artificial Intelligence in the Food Industry—a Guideline" offers a detailed examination of the amalgamation of AI technologies in the food industry to optimize different functions like quality assurance, safety measures, and operational effectiveness. Nidhi Rajesh Mavani and colleagues delve into the various uses of AI, analyzing their benefits, constraints, and methodologies. This overview acts as a valuable resource for choosing suitable AI techniques to enhance food industry activities.

AI in the Food Industry

AI has been incorporated into the food sector to tackle the growing need for food caused by the expanding global population. AI systems are employed for various purposes such as assessing food quality, implementing control mechanisms, classifying items, and forecasting. The utilization of AI has demonstrated its advantages in numerous aspects.

1. **Food Sorting and Classification**

Advanced technologies like Artificial Neural Networks (ANN) and Fuzzy Logic (FL) are utilized in the sorting and classification of food products, guaranteeing that only top-notch items are delivered to consumers.

2. **Prediction and Control Tools**

Machine Learning algorithms play a vital role in forecasting quality of food and its shelf life, thereby aiding in supply chain management and minimizing food wastage.

3. **Food Quality Control and Safety**

Intelligent packaging systems powered by AI play a critical role in improving food safety by persistently observing the quality of food products throughout their storage and transportation processes, thereby guaranteeing their suitability for consumption.

Knowledge-Based Expert Systems in Food Industry

Insight-based expert systems represent one of the initial triumphs in the application of AI within the food industry. These systems leverage information from diverse origins to address intricate challenges, mimicking the Policy-making processes of expert members. Comprising a knowledge base and an implication engine, these systems employ IF–THEN rules to tackle problems.

Applications include

- **White Wine making**: Expert systems oversee the fermentation procedure, offering intelligent management and information retrieval.
- **Nutritional Value Calculation**: Users can access web-based applications that utilize expert systems to determine the nutritional content of various foods.
- **Food Safety Management**: Expert systems play a crucial role in process design, safety management, quality control, and risk assessment within the food industry.

Fuzzy Logic in Food Industry

Fuzzy Logic (FL) is utilized to effectively address problems involving imprecise, uncertain, and ambiguous data. Its applications span across food modelling, control, classification, and addressing various food-related concerns. The process of FL modelling encompasses fuzzification, inference, and defuzzification, rendering it a treasured tool for decision-making within the food industry.

Integration of AI with Sensors

The integration of artificial intelligence with sensory technologies such as the electronic nose (E-nose), electronic tongue (E-tongue), near infrared spectroscopy

(NIRS), and computer vision systems (CVS) has significantly advanced the food industry. This integration facilitates the collection of data and allows for AI-based assessment and categorization of food samples based on quality.

Summary

The article emphasizes the important impact of AI on enhancing different operations in the food sector. AI tools like ANN, FL, expert systems, and ML have enhanced tasks including the categorization, classification, forecasting, quality assessment, and safety evaluation of food. Moreover, the blending of AI with advanced sensors has equipped industry stakeholders with effective instruments to guarantee food quality and safety, ultimately serving the interests of both producers and consumers.

7 Challenges and Future Directions

Incorporating interpretable AI models in manufacturing has great potential, but also comes with obstacles and opportunities for further exploration. Overcoming these challenges is essential to fully benefit from AI technology while maintaining ethical standards. This section discusses current adoption hurdles, future trends, research opportunities, and the ethical and regulatory frameworks guiding AI development and application.

7.1 Current Challenges in Implementing Interpretable AI Models

Implementing explainable AI models in manufacturing faces various technical and organizational challenges that must be addressed to improve their effectiveness.

7.1.1 Technical Challenges

- **Complexity of Manufacturing Processes**: Manufacturing processes frequently consist of intricate, interconnected systems with a multitude of variables. Developing artificial intelligence models that can accurately represent and assess these complexities present a considerable challenge. The guarantee of accuracy and interpretability in these models necessitates the utilization of advanced methodologies and considerable computational power [38].
- **Data Quality and Availability**: It is essential to have high-quality data to train effective AI models. In manufacturing, quality data may be noisy, incomplete, or inconsistent, hindering the development of accurate models. The ongoing challenge is to maintain data integrity and fill in gaps.

- **Balancing Accuracy and Interpretability**: Balancing model precision and explainability is a common challenge in machine learning. Deep neural networks provide superior accuracy; however, they lack interpretability, whereas more straightforward models trade off accuracy in favor of greater interpretability. Finding the right balance is a major hurdle in model development [39].

7.1.2 Organizational Challenges

- **Integration with Existing Systems**: The incorporation of explainable AI models with current manufacturing systems and processes may present challenges. This task requires more than just technical integration; it also entails ensuring that AI projects are in line with business objectives and operational procedures [40].
- **Skill Gaps and Training**: Creating and deploying explainable AI models requires expertise in AI, machine learning, and industry-specific knowledge of manufacturing. Many companies face the challenge of finding and training individuals with these skills [41].
- **Resistance to Change**: The hesitation of an organization to adopt new technologies can deter the effective implementation of AI models that are designed to be user-friendly. Implementing robust change management strategies is essential for overcoming this resistance and cultivating a culture that promotes innovation.

7.2 Future Trends and Research Directions

The trajectory of interpretable AI in manufacturing is being influenced by emerging trends and ongoing research focused on tackling current challenges and uncovering new opportunities.

7.2.1 Advanced Interpretability Techniques

- **Hybrid Models**: Integrating simple and complex models can enhance prediction accuracy and clarity. Combining rule-based systems with neural networks provides clear explanations for forecasts while maintaining high precision.
- **Visualization Tools**: Enhanced visualization methods and strategies have the potential to simplify intricate model results. Utilizing tools that offer dynamic visual representations of model forecasts and decision-making procedures can improve comprehensibility.

7.2.2 Enhanced Data Management

- **Data Fusion and Integration**: Integrating data from sources like IoT devices, production logs, and quality assurance systems improves model accuracy and clarity, making data synthesis a key research focus.
- **Real-Time Data Processing**: The utilization of real-time data processing and analysis facilitates prompt insights and decision-making. Progress in edge computing and real-time analytics is poised to bolster the implementation of interpretable AI models within ever-changing manufacturing settings.

7.2.3 Ethical and Social Implications

- **Fairness and Bias Mitigation**: It is crucial to guarantee fairness and impartiality in AI models. Ongoing research is being conducted to develop techniques for identifying and addressing bias in AI models, which is essential for ethical AI implementation.
- **Transparency and Accountability**: Research focused on creating frameworks that improve transparency and accountability in artificial intelligence decision-making processes is crucial. This includes establishing protocols for model documentation and implementing audit procedures.

7.3 Ethical Considerations and Regulatory Frameworks

It is essential for interpretable AI in manufacturing to be implemented in accordance with ethical standards and in compliance with regulatory structures in order to guarantee its responsible and equitable utilization.

7.3.1 Ethical Principles

- **Transparency**: AI systems ought to be transparent, offering lucid and understandable explanations for their decisions. This fosters trust and empowers stakeholders to grasp the rationale behind AI-powered interventions.
- **Fairness**: AI models must be developed to ensure fair treatment of all individuals and groups, avoiding biases that could lead to discrimination. Continuous monitoring and adjustments are essential to maintain fairness in AI systems.
- **Accountability**: It is crucial to have clear accountability measures in place to manage the impact of AI decisions. This includes defining the responsibilities for developing, implementing, and overseeing AI systems.

7.3.2 Regulatory Frameworks

- **General Data Protection Regulation (GDPR)**: Adherence to GDPR is crucial for AI technologies in the EU, as it covers regulations on automated decision-making and profiling, requiring transparency and explanations for individuals affected by AI determinations.
- **Ethics Guidelines for Trustworthy AI:** The European Commission guidelines outline key aspects for creating AI systems that meet legal, ethical, and robustness standards, including transparency, fairness, and accountability.
- **Industry-Specific Regulations**: Certain sectors must follow specific regulations that impact the use of AI systems. For instance, the automotive industry must comply with safety and quality standards set by regulatory authorities, which affect the creation and implementation of AI models.

8 Conclusion

8.1 Summary of Key Findings

The exploration of interpretable AI models within the manufacturing sector reveals a significant trade-off between accuracy and explainability. While sophisticated models, such as deep neural networks, provide exceptional accuracy, they often lack the transparency necessary to foster trust and ensure compliance. Conversely, more straightforward models can deliver clearer insights but may compromise on accuracy. This dichotomy underscores the need for a balanced strategy that aligns with the specific requirements of the manufacturing industry.

Case studies illustrate the advantages of interpretable AI across diverse sectors, including automotive, semiconductor, and food manufacturing. In the automotive field, the application of interpretable AI for predictive maintenance has markedly decreased instances of unexpected downtime and associated costs. Furthermore, adaptable AI frameworks are essential for maintaining the scalability and flexibility of AI systems in an industrial environment that is rapidly changing.

The implementation of these models encounters several obstacles, including problems related to data quality, system integration, and a lack of qualified personnel. Additionally, as AI becomes increasingly embedded in industrial processes, the importance of ethical considerations and adherence to regulatory standards continues to grow.

8.2 Final Remarks and Recommendations

Implementing interpretable AI models in manufacturing can significantly enhance efficiency, quality, and sustainability. Their transparent decision-making fosters trust between human operators and AI, crucial for effective integration. This leads to reduced downtime, improved quality control, and streamlined production processes.

To fully leverage interpretable AI, manufacturers should invest in balanced AI models that combine precision and clarity, while also developing adaptable frameworks to meet changing industrial needs. Addressing data integrity, system interoperability, and workforce skills is essential for effective implementation. Continuous research and development will be essential for addressing challenges and fully harnessing the capabilities of these technologies, resulting in substantial progress in innovation, operational efficiency, and sustainability within the manufacturing industry.

References

1. Meesublak, K., Klinsukont, T.: A cyber-physical system approach for predictive maintenance. In: 2020 IEEE International Conference on Smart Internet of Things (SMARTIOT), pp. 337–341. IEEE (2020)
2. Amodei, D., Olah, C., Steinhardt, J., Christiano, P., Schulman, J., Mané, D.: Concrete problems in AI safety (2016). arXiv preprint arXiv:1606.06565
3. Chai, J., Zeng, H., Li, A., Ngai, E.W.: Deep learning in computer vision: a critical review of emerging techniques and application scenarios. Mach. Learn. Appl. **6**, 100134 (2021)
4. Rudin, C.: Stop explaining black box machine learning models for high stakes decisions and use interpretable models instead. Nat. Mach. Intell. **1**(5), 206–215 (2019)
5. Alexander, Z., Chau, D.H., Saldaña, C.: An interrogative survey of explainable AI in manufacturing. IEEE Trans. Ind. Inf. (2024)
6. Ahmad, M.A., Eckert, C., Teredesai, A.: Interpretable machine learning in healthcare. In: Proceedings of the 2018 ACM International Conference on Bioinformatics, Computational Biology, and Health Informatics, pp. 559–560 (2018)
7. Petersen, E., Potdevin, Y., Mohammadi, E., Zidowitz, S., Breyer, S., Nowotka, D., Henn, S., Pechmann, L., Leucker, M., Rostalski, P., Herzog, C.: Responsible and regulatory conform machine learning for medicine: a survey of challenges and solutions. IEEE Access **10**, 58375–58418 (2022)
8. Lee, J., Bagheri, B., Kao, H.A.: A cyber-physical systems architecture for industry 4.0-based manufacturing systems. Manuf. Lett. **3**, 18–23 (2015)
9. Sharma, A., Zhang, Z., Rai, R.: The interpretive model of manufacturing: a theoretical framework and research agenda for machine learning in manufacturing. Int. J. Prod. Res. **59**(16), 4960–4994 (2021)
10. Lipton, Z.C.: The mythos of model interpretability: in machine learning, the concept of interpretability is both important and slippery. Queue **16**(3), 31–57 (2018)
11. Kohoutová, L., Heo, J., Cha, S., Lee, S., Moon, T., Wager, T.D., Woo, C.W.: Toward a unified framework for interpreting machine-learning models in neuroimaging. Nat. Protoc. **15**(4), 1399–1435 (2020)
12. Das, A., Rad, P.: Opportunities and challenges in explainable artificial intelligence (xai): a survey (2020). arXiv preprint arXiv:2006.11371

13. Christoph, M.: Interpretable Machine Learning: A Guide for Making Black Box Models Explainable. Leanpub (2020)
14. Lee, J., Kao, H.A., Yang, S.: Service innovation and smart analytics for industry 4.0 and big data environment. Procedia CIRP **16**, 3–8 (2014)
15. Delsing, J. (ed.): IoT Automation: Arrowhead Framework. CRC Press (2017)
16. Ristoski, P., Paulheim, H.: Semantic Web in data mining and knowledge discovery: a comprehensive survey. J. Web Semantics **36**, 1–22 (2016)
17. Li, H., Ota, K., Dong, M.: Learning IoT in edge: deep learning for the Internet of Things with edge computing. IEEE Net. **32**(1), 96–101 (2018)
18. Brous, P., Janssen, M., Herder, P.: The dual effects of the Internet of Things (IoT): a systematic review of the benefits and risks of IoT adoption by organisations. Int. J. Inf. Manage. **51**, 101952 (2020)
19. Feki, M.A., Kawsar, F., Boussard, M., Trappeniers, L.: The internet of things: the next technological revolution. Computer **46**(2), 24–25 (2013)
20. Schlechtendahl, J., Keinert, M., Kretschmer, F., Lechler, A., Verl, A.: Making existing production systems Industry 4.0-ready: holistic approach to the integration of existing production systems in Industry 4.0 environments. Prod. Eng. **9**(1), 143–148 (2015)
21. Jeschke, S., Brecher, C., Meisen, T., Özdemir, D., Eschert, T.: Industrial Internet of Things and Cyber Manufacturing Systems, pp. 3–19. Springer International Publishing (2017)
22. Zhou, K., Liu, T., Zhou, L.: Industry 4.0: towards future industrial opportunities and challenges. In: 2015 12th International Conference on Fuzzy Systems and Knowledge Discovery (FSKD), pp. 2147–2152. IEEE (2015)
23. Conti, M., Dehghantanha, A., Franke, K., Watson, S.: Internet of Things security and forensics: challenges and opportunities. Futur. Gener. Comput. Syst. **78**, 544–546 (2018)
24. Wuest, T., Weimer, D., Irgens, C., Thoben, K.D.: Machine learning in manufacturing: advantages, challenges, and applications. Prod. Manuf. Res. **4**(1), 23–45 (2016)
25. Qin, J., Liu, Y., Grosvenor, R.: A categorical framework of manufacturing for industry 4.0 and beyond. Procedia CIRP **52**, 173–178 (2016)
26. Kolberg, D., Knobloch, J., Zühlke, D.: Towards a lean automation interface for workstations. Int. J. Prod. Res. **55**(10), 2845–2856 (2017)
27. Verdouw, C.N., Wolfert, J., Beulens, A.J.M., Rialland, A.: Virtualization of food supply chains with the internet of things. J. Food Eng. **176**, 128–136 (2016)
28. Shi, W., Cao, J., Zhang, Q., Li, Y., Xu, L.: Edge computing: vision and challenges. IEEE Internet Things J. **3**(5), 637–646 (2016)
29. Yin, S., Kaynak, O.: Big data for modern industry: challenges and trends [point of view]. Proc. IEEE **103**(2), 143–146 (2015)
30. Ribeiro, M.T., Singh, S., Guestrin, C.: "Why should I trust you?" Explaining the predictions of any classifier. In: Proceedings of the 22nd ACM SIGKDD International Conference on Knowledge Discovery and Data Mining, pp. 1135–1144 (2016)
31. Bousdekis, A., Magoutas, B., Apostolou, D., Mentzas, G.: A proactive decision making framework for condition-based maintenance. Ind. Manag. Data Syst. **115**(7), 1225–1250 (2015)
32. Khan, F., Rathnayaka, S., Ahmed, S.: Methods and models in process safety and risk management: past, present and future. Process. Saf. Environ. Prot. **98**, 116–147 (2015)
33. Montgomery, D.C.: Introduction to Statistical Quality Control. Wiley (2007)
34. Rosen, R., Von Wichert, G., Lo, G., Bettenhausen, K.D.: About the importance of autonomy and digital twins for the future of manufacturing. IFAC-Papersonline **48**(3), 567–572 (2015)
35. Zonta, T., Da Costa, C.A., da Rosa Righi, R., de Lima, M.J., da Trindade, E.S., Li, G.P.: Predictive maintenance in the Industry 4.0: a systematic literature review. Comput. Ind. Eng. **150**, 106889 (2020)
36. Senoner, J., Netland, T., Feuerriegel, S.: Using explainable artificial intelligence to improve process quality: evidence from semiconductor manufacturing. Manage. Sci. **68**(8), 5704–5723 (2022)
37. Mavani, N.R., Ali, J.M., Othman, S., Hussain, M.A., Hashim, H., Rahman, N.A.: Application of artificial intelligence in food industry—a guideline. Food Eng. Rev. **14**(1), 134–175 (2022)

38. Del Zotto, L., Tallini, A., Di Simone, G., Molinari, G., Cedola, L.: Energy enhancement of solid recovered fuel within systems of conventional thermal power generation. Energy Procedia **81**, 319–338 (2015)
39. Doshi-Velez, F., Kim, B.: Towards a rigorous science of interpretable machine learning (2017). arXiv preprint arXiv:1702.08608
40. Westerman, G., Bonnet, D., McAfee, A.: Leading Digital: Turning Technology into Business Transformation. Harvard Business Press (2014)
41. Manyika, J., Chui, M., Bughin, J., Dobbs, R., Bisson, P., Marrs, A.: Disruptive Technologies: Advances that Will Transform Life, Business, and the Global Economy, vol. 180, pp. 17–21. Mckinsey global Institute, San Francisco, CA

R. Madhumith is currently pursuing the Post-Graduate program at Coimbatore Institute of Technology, Coimbatore, where he is specializing in Artificial Intelligence and Machine Learning. His work encompasses a range of cutting-edge technologies, such as Explainable AI, Adversarial Attacks, Responsible AI, and Cybersecurity. He is driven to investigate the applications of Explainable AI in cybersecurity, with the ultimate goal of developing more secure and trustworthy AI-powered systems. He possess a versatile skill set characterized by a collaborative spirit, strong communication abilities, innovative thinking, a positive attitude, and analytical reasoning. He can be reached at www.linkedin.com/in/madhumith-ravikumar.

Dr. S. B. Mahalakshmi is an Assistant Professor and a researcher with an expertise in Machine Learning, Deep Learning, Natural Language Processing. She holds M.Sc. and M.Phil. in Computer Science from Bharathiar University and Ph.D. degree from Periyar University. She has more than 18 years of experience and has published many technical papers in National, International Conferences and Journals. Currently, she is working as an Assistant Professor in the Department of Artificial Intelligence and Machine Learning at Coimbatore Institute of Technology, Coimbatore. She can be reached at https://www.linkedin.com/in/dr-mahalakshmi-s-b-94749152.

P. Hemashree is an Assistant Professor and a researcher with an expertise in Machine Learning, Deep Learning, Cyber Security and Metaheuristic Optimization. She explores the convergence of Deep Learning algorithms and metaheuristic optimization in Cyber Security, aiming to develop more robust and effective intrusion detection and prevention systems. She holds a B.Sc. Degree (Computer Science, Mathematics, Statistics) from Mount Carmel College, Bangalore and an MCA degree from Coimbatore Institute of Technology, Coimbatore. She has presented her research works at several conferences and published research articles in reputable journals. Currently, she is working as an Assistant Professor in the Department of Artificial Intelligence and Machine Learning at Coimbatore Institute of Technology, Coimbatore. She can be reached at www.linkedin.com/in/hemashreepaulraj.

MIX
Papier aus verantwortungsvollen Quellen
Paper from responsible sources
FSC® C105338

If you have any concerns about our products,
you can contact us on
ProductSafety@springernature.com

In case Publisher is established outside the EU,
the EU authorized representative is:
Springer Nature Customer Service Center GmbH
Europaplatz 3, 69115 Heidelberg, Germany

Printed by Libri Plureos GmbH
in Hamburg, Germany